Recent Advances in Animal Nutrition

Recent Advances in Animal Nutrition

Edited by Jasper Mitchell

SYRAWOOD
PUBLISHING HOUSE
New York

Published by Syrawood Publishing House,
750 Third Avenue, 9th Floor,
New York, NY 10017, USA
www.syrawoodpublishinghouse.com

Recent Advances in Animal Nutrition
Edited by Jasper Mitchell

International Standard Book Number: 978-1-64740-376-8 (Hardback)

Cataloging-in-publication Data

Recent advances in animal nutrition / edited by Jasper Mitchell.
p. cm.
Includes bibliographical references and index.
ISBN 978-1-64740-376-8
1. Animal nutrition. 2. Animal culture. I. Mitchell, Jasper.
SF95 .R43 2023
636.085 2--dc23

TABLE OF CONTENTS

PREFACE

Animal nutrition refers to the science of preparing or formulating the feed for animals that provide food (such as meat and milk) or non-food materials (such as wool). It is concerned with fulfilling the nutritional needs of food-producing, companion, or service animals. Proteins, minerals, and vitamins are the major dietary nutrients required by animals. The integrative science of animal nutrition is also involved in different steps by which the animals assimilate feed or food, and use it for their growth, health and performance. Animal nutrition significantly contributes to several economic and environmental aspects. It promotes the safety and dietary quality of animal products for human consumption. The science of animal nutrition also aids in producing nutrient efficient livestock that help reduce greenhouse gas emissions and land use. This book is a valuable compilation of topics, ranging from the basic to the most complex advancements in the study of animal nutrition. With its detailed analyses and data, it will prove immensely beneficial to professionals and students.

After months of intensive research and writing, this book is the end result of all who devoted their time and efforts in the initiation and progress of this book. It will surely be a source of reference in enhancing the required knowledge of the new developments in the area. During the course of developing this book, certain measures such as accuracy, authenticity and research focused analytical studies were given preference in order to produce a comprehensive book in the area of study.

This book would not have been possible without the efforts of the authors and the publisher. I extend my sincere thanks to them. Secondly, I express my gratitude to my family and well-wishers. And most importantly, I thank my students for constantly expressing their willingness and curiosity in enhancing their knowledge in the field, which encourages me to take up further research projects for the advancement of the area.

Editor

PREFACE

Animal nutrition refers to the science of preparing or formulating the feed for animals that provide food (such as meat and milk) or non-food materials (such as wool). It is concerned with fulfilling the nutritional needs of food-producing, companion, or service animals. Proteins, minerals, and vitamins are the major dietary nutrients required by animals. The integrative science of animal nutrition is also involved in different steps by which the animals assimilate feed or food, and use it for their growth, health and performance. It promotes the safety and dietary quality of animal products for human consumption. The science of animal nutrition also aids in producing nutrient efficient livestock that help reduce greenhouse gas emissions and land use. This book is a valuable compilation of topics, ranging from the basic to the most complex advancements in the study of animal nutrition. With its detailed analyses and data, it will prove immensely beneficial to professionals and students.

After months of intensive research and writing, this book is the end result of all who devoted their time and efforts in the initiation and progress of this book. It will surely be a source of reference in enhancing the required knowledge of the new developments in the area. During the course of developing this book, certain measures such as accuracy, authenticity and research focused analytical studies were given preference in order to produce a comprehensive book in the area of study.

This book would not have been possible without the efforts of the authors and the publisher. I extend my sincere thanks to them. Secondly, I express my gratitude to my family and well-wishers. And most importantly, I thank my students for constantly expressing their willingness and curiosity in enhancing their knowledge in the field, which encourages me to take up further research projects for the advancement of the area.

Editor

1

Effect of Phlorotannins from Brown Seaweeds on the In Vitro Digestibility of Pig Feed

Lauren Ford [1], Chloe Curry [1], Mairead Campbell [2,3], Katerina Theodoridou [2,3], Gary Sheldrake [1], Jaimie Dick [2,3,4], Lorenzo Stella [1,5] and Pamela J. Walsh [1,4,6,*]

1 School of Chemistry and Chemical Engineering, Queen's University Belfast, Northern Ireland BT9 5AG, UK; l.ford@imperial.ac.uk (L.F.); chloe011358@hotmail.co.uk (C.C.); g.sheldrake@qub.ac.uk (G.S.); l.stella@qub.ac.uk (L.S.)
2 Institute of Global Food Security, Queen's University Belfast, 19 Chlorine Gardens, Northern Ireland, Belfast BT9 5DL, UK; mcampbell105@qub.ac.uk (M.C.); k.theodoridou@qub.ac.uk (K.T.); j.dick@qub.ac.uk (J.D.)
3 School of Biological Sciences, Queen's University Belfast, 19 Chlorine Gardens, Northern Ireland, Belfast BT9 5DL, UK
4 Queen's Marine Laboratory (QML) Queen's University Belfast, 12-13 The Strand, Northern Ireland, Portaferry BT22 1PF, UK
5 Atomistic Simulation Centre (ASC), School of Mathematics and Physics, Queen's University Belfast, University Road, Belfast BT7 1NN, UK
6 School of Mechanical Engineering, Queen's University Belfast, The Asbhy Building, Stranmillis Road, Northern Ireland, Belfast BT9 5AJ, UK
* Correspondence: pamela.walsh@qub.ac.uk

Simple Summary: In large-scale pig farming, alternative and safe antimicrobials are needed to enhance pig nutrition. Seaweed bioactives, and in particular phlorotannins, have been reported to have antimicrobial properties. However, their effect on the digestibility of pig feed is not well understood. This study investigates the effect of these phenolics on the in vitro dry matter digestibility of seaweed using an in vitro pig digestibility model. The effect of the phenolics when extracted into their purified phlorotannin form, and blended directly into pig feed, was also tested using the same model. The results found that, when added to the pig feed, purified phlorotannins had a more pronounced effect on digestibility than seaweeds containing phenolics. In addition, the results showed that given the seasonal variation within seaweeds, inclusion of whole seaweeds should be based on phenolic concentration as opposed to percentage inclusion of seaweeds.

Abstract: Phlorotannins have been reported to have positive effects on pig health, including improved gut health and digestibility. In this study, we investigate the effect of phenolics found in two brown seaweeds, *Ascophyllum nodosum* and *Fucus serratus*, on in vitro dry matter digestibility of seaweeds and commercial pig feed. Phlorotannin extracts and whole seaweeds were supplemented into pig feed to test their effect on digestibility. Solid-phase extraction was used to purify the phenolics to phlorotannins. The results showed a slight decrease in the digestibility of pig feed that was found to be significant when phlorotannin extracts were added from either seaweed. However, when whole *A. nodosum* was added to the pig feed, the effect on digestibility was less pronounced. Specifically, no significant difference in digestibility was observed at inclusion rates up to 5%, and thereafter results varied. A difference in digestibility was also observed in the same species at the same inclusion rate, collected from different seasons. This suggests that other compounds, e.g., polysaccharides, are having an effect on digestibility when whole seaweeds are supplemented to animal feed. This research has also highlighted the need to base supplementation on phenolic concentration as opposed to a standardised percentage inclusion of seaweeds to ensure that digestibility is not adversely affected.

Keywords: IVDMD; digestibility; brown seaweeds; pig nutrition; phenolics; phlorotannins; gut health

1. Introduction

Due to the increasing human population across the world, the need to supply food products using more innovative and sustainable techniques has become of paramount importance. Edible seaweeds have been used widely in oriental cuisine for human consumption [1] and have shown in recent years that, if farmed or harvested sustainably, they are renewable [2,3]. The seaweed used in animal feeds commercially, however, has been limited to date. Seaweed as an animal feed supplement has been shown to enhance pig growth and help improve the digestibility in animal feeds by providing bioactive phytochemicals [4]. Seaweeds have also been used to feed livestock for thousands of years from the Ancient Greeks to modern-day Icelandic famers [5,6]. In Iceland, sheep, cattle and horses have all been fed processed seaweed, dried using geothermal heat, which reduces processing costs [7]. On the west coast of Scotland *Pelvetia*, a brown seaweed (*Phaeophyceae*) is fed to pigs in order to fatten them up before market [7]. Whilst there are some historical accounts of animal rearing using seaweed as a high proportion of the diet, there are fewer known commercial and scientific trials. A review by Corino et al. [8] summarises studies to date that have reported improvements in digestibility and overall animal health through the supplementation of seaweeds (and/or extracts) to animal feeds in pigs. In addition to their nutritional benefits, in recent years, seaweed compounds, specifically phenolics have gained significant attention for their potential antimicrobial [9–11] and anti-methanogenic [12,13] effects in livestock. Although pigs have the lowest methane emission when compared to other livestock, they still produce an estimated CH_4 emission of 0.8, 2.4 and 8.2 g/head for weaned piglets, fattened pigs and sows, respectively [14].

Seaweed is an advantageous aquatic plant to be incorporated into feeds from a sustainability aspect as it does not compete with land space for food crops and does not require fertiliser in order to grow [15,16]. Seaweed can also be sustainably farmed or harvested from naturally growing sources, allowing natural regeneration [15]. However, seaweed incorporation into animal diets, especially for monogastric animals such as pigs, has had largely varied results. An early study (1979) by Jones et al. [17] reported that the supplementation of brown seaweed into pig diets at high inclusion rates of 10% has a negative effect on weight gain. A study by Michiels et al. [18], however, showed weight gain after 11 days post weaning when compared to the control diet when the pig feed was supplemented with between 2.5 and 10 g/kg of *A. nodosum* (whole feedstock). Inclusion of sun-dried *A. nodosum* in the diet at 20 g/kg for 7 days has also been shown to reduce faecal shedding of *E. coli* 0157:H7 in cattle [19]. Other studies have used seaweed extracts that contain the polysaccharides laminarin and fucoidan. The inclusion of these extracts allowed for a lower lactose content in the feed given to post weaning piglets and was found not to adversely affect the growth rate [20]. McDonnell et al. [21] reported that by supplementing the diet of weaning piglets with laminarin- and fucoidan-containing seaweed species, they also showed an increased daily weight gain. Most digestibility studies, however, do not access the impact of seasonal variation, which is important for commercial application. In addition, there is a lack of understanding of the effect of different seaweed compounds on dry matter digestibility, particularly in the case of phenolics.

Seaweeds are grouped into three phyla according to their pigmentation, known commonly by colour: Chlorophyta (green algae), Phaeophyta (brown algae) and Rhodophyta (red algae) [22]. Each hosts unique phytochemicals (e.g., lipids, phenolics, and alkaloids) and complex carbohydrates, mainly in the form of sulphated (e.g., carrageenan) and non-sulphated (e.g., alginate) polysaccharides. The carbohydrate content in all seaweeds is high. However, brown seaweeds tend to have the highest non-starch polysaccharide content, whereas red and green seaweeds have higher amino acid content [23]. Livestock diets do not typically contain these complex polysaccharides and literature studies to date of their digestibility in both ruminant and monogastric animals are limited.

However, some brown seaweeds, including *Laminaria digitata*, *A. nodosum* and *Fucus vesiculosus*, have also been used commercially in animal feeds [24], suggesting that their dietary properties are comparable to terrestrial-based fed ingredients. One commercial seaweed feed, *OceanFeed Swine®*, which is a blend of different seaweed species, has been reported to have positive results in terms of daily weight gain and feed efficiency in nursery and pig-fattening stages of development [25].

Seaweeds are a largely seasonal plant containing varied amounts of phenolics, fatty acids, protein and minerals throughout the year [26–29]. Brown seaweeds contain phenolic compounds known as phlorotannins, which are unique to this class of seaweed. Although these phenolic compounds are commonly compared to the terrestrial tannins, they are structurally very different [5]. The phlorotannins in seaweeds, however, do have some effects similar to terrestrial plant tannins such as binding to protein [30,31]. Gulzari et al. [32] hypothesised that phlorotannins from seaweed-supplemented soybean meal formed complexes with proteins and fibres in the meal and reduced its overall digestibility in the rumen. However, few studies have looked specifically at the effect of phlorotannins on the digestibility of meal in ruminant or monogastric animals.

The aim of this study was to test the in vitro dry matter digestibility of the brown seaweeds that were rich in phenolics in order to assess the correlations between their digestibility and the phlorotannins levels. Previous studies have shown that phlorotannins could potentially help improve the gut health of pigs [8,10]. However, a review by Huang et al. [33]. highlighted that terrestrial tannins can inhibit digestion when added at high concentrations in livestock diets. Therefore, it is important to understand the effect of phlorotannins on digestibility in order to ensure that their addition does not negatively impact on the weight gain of the animals. In addition, whilst phlorotannins act in a similar way to condensed and hydrolysed tannins from terrestrial plants, they are structurally very different, and little is currently known about their effects in the digestive tract of livestock, in particular pigs. This study investigates the dry matter digestibility of two brown seaweeds, *A. nodosum* and *F. serratus*, that are known to be rich in phenolics and the effect of phenolic content on the digestibility of the whole seaweed biomass. The phenolic content in brown seaweeds is known to vary seasonally [10,28], and therefore seaweed collected from different seasons were used to study this effect. To better understand the effect of phenolics on the digestibility of the seaweed biomass, a further in vitro study was performed with and without polyethylene glycol (PEG). Some methane migration studies have shown that PEG inhibits the activity of phenolics [13,34]. This approach was therefore taken to help better understand the role of phenolics in seaweed biomass digestibility. Phenolic compounds and tannins are known to bind to proteins and fibre, with a higher affinity for the protein fractions [35]. Therefore, purified phlorotannin extracts were isolated from the phenolic fraction and blended with protein-rich commercial pig feed to assess the specific effect of the phlorotannins as opposed to the whole seaweed that contains other macromolecules including chlorophylls, carotenoids and complex mono-/polysaccharides that might affect digestibility.

2. Materials and Methods

2.1. Collection of Seaweed Samples

A. nodosum and *F. serratus*, two intertidal brown seaweeds, were collected by hand at the end of each month from an artificial lagoon behind a breakwater in Bangor, Northern Ireland (54°39′58.6′′ N 5°39′53.4′′ W). Samples were identified by a marine phycologist, collected and stored in plastic sample bags to be transported to the laboratory. Within 3 h of collection, the samples were washed with fresh water to remove particulates and grazing species before being frozen for storage at −20 °C. The whole plant was used for each sample collected. Approximately 200 g wet weight was collected for each species at each collection point. The dry matter was the mass of sample after freeze drying taken away from the mass of the sample before freeze drying.

2.2. Seaweed Processing

Frozen seaweed samples were lyophilised, then ground using a Polymix PX-MFC 90D mechanical grinder fitted with a 2 mm sieve. Samples were weighed before and after lyophilisation to determine the dry matter (DM) of samples. They were then placed in plastic bags and stored in the freezer at −20 °C until used for analysis. Samples for commercial feed study were purified to ensure that the activity measured was due to the phlorotannins only. Samples were purified by solid-phase extraction (SPE) using Amberlyst XAD 7HD. Dried seaweed (approximately 10 g) was stirred using a magnetic flee with acetone/water (7:3 *v*/*v*, 200 mL) for 3 h at room temperature. The seaweed extracts were then centrifuged (5000 rpm for 10 min). The supernatant was placed on the rotary evaporator to remove the acetone. The aqueous mixture was then added to the Amberlyst XAD 7HD beads (10 g) in a solid-phase extraction (SPE) column. The column was first eluted with water (3 × 200 mL) and then with ethanol (3 × 100 mL). The ethanol was then removed by rotary evaporation and then the dried material was stored in the freezer until it could be used for biological analysis.

2.3. Proximate Analysis

2.3.1. Proximate Measurements

The ash content, acid detergent fibre (ADF) and neutral detergent fibre (NDF) were measured using the same methods as reported by Van Soet et al. [36]. Briefly, ground seaweeds were analysed for NDF by boiling in neutral detergent (ND) solution, amylase and sodium sulphite for 1 h. For ADF, the ground seaweed was boiled for 1 h in acid detergent solution. Crude protein (CP) was calculated by measuring the nitrogen content and multiplying it by 5 [37]. The nitrogen content was measured using the Dumas method using Leco Protein/N Analyser (FP-528, Leco Corp., St Joseph, MI, USA) [38]. Gross energy (GE) content was determine by AFBI Hillsborough Analytical Services using an isothermal automated bomb calorimeter (PARR Instrument, Model 6300, Illinois, IL, USA).

2.3.2. In Vitro Dry Matter Digestibility (IVDMD)

In vitro digestion was measured using a technique adapted from Tiwari and Jha [39]. A three-step enzymatic digestion technique was used to determine the apparent total tract digestibility of dry matter (DM). An amount of 1 g of digestible sample was placed in a flask. Then 50 mL of phosphate buffer solution 1 (0.1 mol/L, pH 6.0) was added to the flask followed by 20 mL of hydrochloric acid (HCl) solution (0.2 mol/L). The pH was adjusted to 2.0 by mixing with HCl solution (1.0 mol/L) or sodium hydroxide (NaOH) (1.0 mol/L). A volume of 1 mL of chloramphenicol (34 mg/mL in ethanol) was added to the solution to prevent bacterial growth during hydrolysis. A volume of 2 mL of freshly prepared pepsin solution was added to the solution which was prepared by adding pepsin (0.75 g, Sigma CAT NO 9001-75-6, Dorset, UK) to 30 mL of ultra-pure water. The flask was then closed and incubated in a water bath at 39 °C for 2 h. A volume of 20 mL phosphate buffer solution 2 (0.2 mol/L, pH 6.8) was then added to the samples and 10 mL of NaOH solution (0.6 mol/L). The pH was adjusted to 6.8 with HCl (1 mol/L) or NaOH (1 mol/L) and 6 mL of pancreatin solution was added. Pancreatin solution was made by adding 3 g pancreatin (Sigma CAT NO 8049-47-6) to 90 mL ultra-pure water. Hydrolysis was then continued under the same conditions for a further 4 h.

The IVDMD was calculated as follows:

$$\text{IVDMD} = (\text{DWH} - \text{DWR})/\text{DWH} \times 100$$

where

DWH = dry weight of sample before hydrolysis;

DWR = dry weight of residue.

To counteract the effect of the phlorotannins on digestibility, a solution of PEG (MW 4000) (Fluka Research Chemicals) in water was added in different volumes to achieve the required PEG: phlorotannin (PT) ratio. Phlorotannins were extracted from seaweeds collected in May as this month had the highest concentration of phlorotannins and required less biomass to obtain the concentration required. A stock solution of PEG 4000 was prepared at a concentration of 5 mg/mL PEG in water. The stock solution was sonicated until fully solubilised. The concentration of the phenolic content of seaweeds were calculated using the Folin–Ciocalteu (FC) assay which gives the concentration of phenolics in phloroglucinol equivalents mg/g of seaweed (see Section 2.4). The volume required of PEG stock solution was then calculated for each seaweed digestibility based on weight/weight calculations. Molar equivalents are not possible when using biological matrixes such as these due to the unknown molecular mass of the mixture of compounds present in the extracts.

2.4. Folin–Ciocalteu (FC) Assay

The phenolic content was measured using the FC assay using a method described by Ford et al. [10]. Briefly, phenolic extracts were prepared as per Section 2.2 (seaweed processing). The resultant extract was then diluted ×40 before it was added to the FC reagent at a dilution of 1:1 with deionised water (0.5 mL). After 5 min, 20% aqueous sodium carbonate was added (2.5 mL). The solution was then left to stand for 40 min in the dark before the absorbance was recorded at wavelength of 755 nm on a Jenway 6305 fixed-wavelength spectrophotometer. The absorbances were converted to a concentration in phloroglucinol equivalents using a calibration curve ($R^2 > 0.95$) of phloroglucinol.

2.5. Gel Electrophoresis

Samples were prepared using seaweed polyphenols previously extracted and purified using solid-phase extraction and produced a water-soluble powder. Standard samples were prepared at a concentration of 1 mg/mL in deionised water of the phlorotannins (extracted from seaweeds collected in May) and pepsin enzyme (Sigma) and PEG (2000 and 4000 MW). When mixtures were used, samples were prepared at 1:1 ratio (w/w). After preparation, the samples were centrifuged for 5 min at 5000 RPM to remove particulates. Sample solution (30 μL) was added to LDS buffer (10 μL, Invitrogen) and vortexed. This solution was then added to 10 well gel plates (NuPAGE 4–12% BisTris gels) along with a PageRuler pre-stained protein ladder (Thermo Scientific, Loughborough, UK) which was run in NuPAGE MES SDS running buffer at 200 V for 30 min. A Powerease 300 W powerpack was used with a Xcell sure-lock gel chamber for running the gel plates. After 30 min, the gel plates were removed from the sure-lock chamber and carefully removed from the plastic casing. The gel was then placed in staining solution (Coomassie® brilliant blue; 0.25 g in 125 mL ethanol, 100 mL deionised water and 25 mL acetic acid) for 4 h. After 4 h, the gel was removed and left overnight in de-staining solution 1:1:8 *v*/*v* methanol:acetic acid:water). The gel was then photographed.

2.6. Statistical Analysis

All the ANOVA statistical calculations were performed using IMB SPSS v25 (SPSS Inc, Chicago, IL, USA). All data sets were analysed using one-way ANOVA and post hoc with LSD multiple comparison test, with the exception of Figure 1. For this data set, linear regression analysis was performed with a significant test for regression on the data in excel to test the null hypothesis that phlorotannin content will have no effect on digestibility. A significance level of $p < 0.05$ was used for statistical analysis.

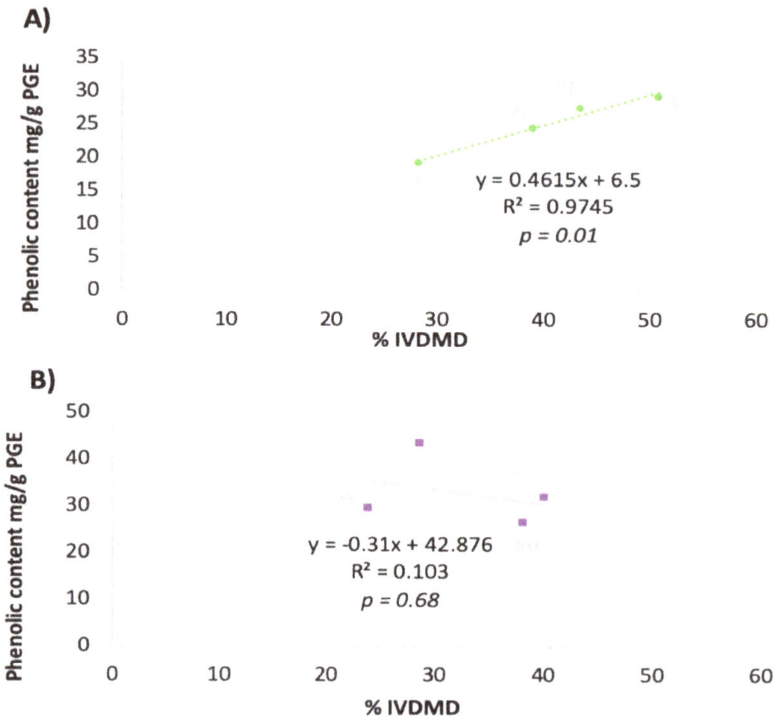

Figure 1. Correlation of phenolic content in phloroglucinol equivalents mg/g (DM) to in vitro digestibility of dry matter (%IVDMD) for *A. nodosum* (**A**) and *F. serratus* (**B**).

3. Results

3.1. Proximate Analysis

The in vitro dry matter digestibility of the brown seaweeds was studied to determine the effect of phenolics in a monogastric model. The proximate analysis investigated included GE, CP, ADF, NDF, DM, ash, IVDMD and phenolic content. The results of this analysis are shown in Table 1. In terms of CP, there was no significant difference found between *A. nodosum* and *F. serratus*, throughout the year, with the exception of *F. serratus* in spring, which showed a significant ($p < 0.001$) increase in CP from 14.5 ± 1.2 to 20.7 ± 1.2%. Overall, the CP was found to be significantly higher ($p < 0.05$) in winter and spring for both species.

The NDF and ADF measurements are given in Table 1. The effect of seasonal variation was apparent in *A. nodosum* in ADF data, with an exponential increase from summer to winter from 28.4 ± 5.8 to 48.9 ± 0.8%, although the increase was only found to be statistically significant between summer and autumn ($p = 0.002$), whereas no seasonal effect was observed in the ADF results for *F. serratus*. No statistically significant difference was found in the NDF results for *F. serratus* between seasons.

When comparing the phenolic content between seaweed species (Table 1), no significant difference was observed between the phenolic content of *A. nodosum* when compared to *F. serratus* in spring and summer ($p = 0.120$), but there was a significant difference ($p < 0.05$) between species in both autumn and winter. In autumn, *F. serratus* showed the greatest increase in phenolic content to 43.4 ± 2.6 ($p < 0.001$) mg/g compared to compared to 29.4 ± 1.4 ($p < 0.001$) for *A. nodosum*. To better understand the relationship between the phenolic content and digestibility, the phenolic content was correlated to the IVDMD, as shown in Figure 1. A linear response was observed for the relationship between phlorotannin content and IVDMD for *A. nodosum* with R^2 value 0.9745 (Figure 1A), which was found to be statistically significant ($p = 0.01$). In *F. serratus*, the relationship between phenolics and IVDMD showed the opposite trend, whereby a negative correlation was observed. This, however, was not statistically significant for IVDMD (Figure 1B).

Table 1. Proximate analysis of *A. nodosum* and *F. serratus*. All parameters are expressed in terms of % of the overall biomass unless otherwise stated.

Species of Seaweed	Season	Gross Energy MJ/kg DM	Dry Matter		Ash		Crude Protein (CP)		Acid Detergent Fibre (ADF)		Neutral Detergent Fibre (NDF)		IVDMD		Phenolic Content FC Assay mg/g (DW)	
			Ave	SD	Ave	SD	Ave	SD	Ave	SD	Ave	SD	Ave	SD	Ave	SD
A. nodosum	Winter (Jan)	13.92	89.23	0.45	30.20	2.45	11.58	3.06	16.90	0.58	48.48	0.82	28.24	6.78	19.20	2.74
	Spring (April)	14.02	88.50	0.05	28.36	0.38	14.50	1.19	15.34	4.68	40.26	3.29	43.32	1.26	27.46	2.95
	Summer (July)	15.08	88.66	0.20	22.61	1.23	7.27	2.22	15.04	1.68	28.39	5.80	50.79	1.47	29.28	1.26
	Autumn (Nov)	14.61	88.38	0.06	23.49	0.39	6.79	1.34	18.79	0.87	41.63	1.32	39.01	2.41	24.53	1.39
F. serratus	Winter (Jan)	14.39	89.70	0.06	33.15	0.39	13.29	1.34	21.94	0.87	42.83	1.32	23.85	2.41	29.42	1.39
	Spring (April)	15.16	90.78	0.60	26.94	3.37	20.73	1.53	24.54	3.51	45.04	7.73	39.99	2.42	31.70	2.96
	Summer (July)	16.55	89.83	0.53	19.34	0.66	7.77	0.69	22.81	1.82	43.73	4.69	38.10	7.93	26.49	1.42
	Autumn (Nov)	15.06	89.50	0.07	26.44	1.28	9.85	1.08	25.71	2.15	44.49	5.21	28.69	2.88	43.40	2.56

3.2. Effect of Phenolics on Digestibility with and without PEG

Purified phlorotannin extracts were complexed with different molecular weights (MW) of PEG to test which would bind the best to the phlorotannins in *A. nodosum* and *F. serratus* (Figure 2). The complexes were tested with the FC assay.

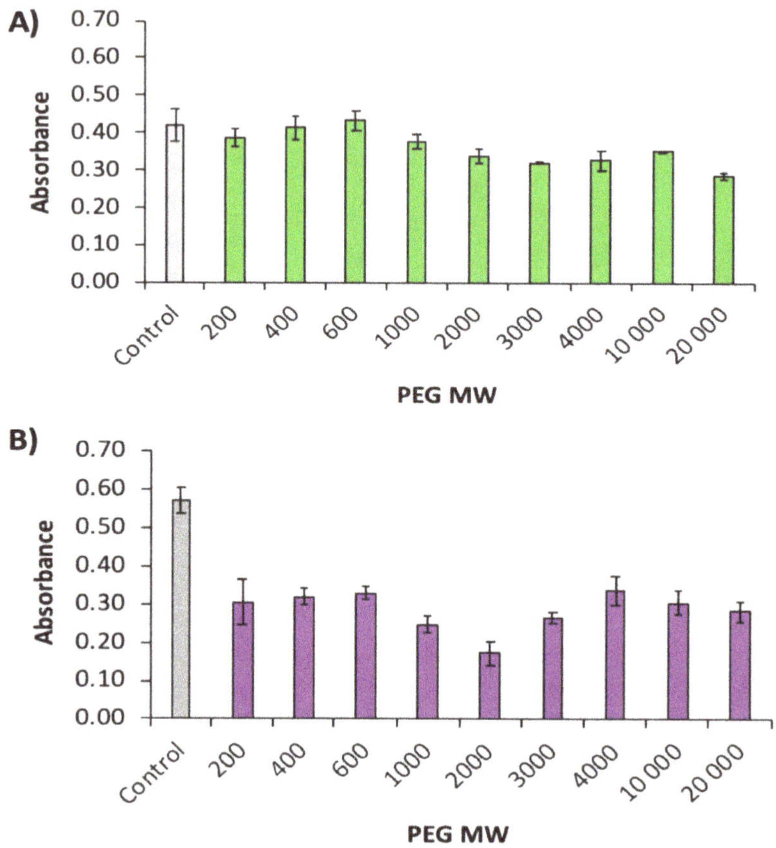

Figure 2. Change in absorbance of the FC assay upon binding to different PEG MW for purified phlorotannins from (**A**) *A. nodosum* and (**B**) *F. serratus* extracted from samples collected in May.

In *A. nodosum*, there was no significant difference between the control and PEG of a MW of 200 to 600. However, for PEG 1000 MW upwards, a statistically significant decrease ($p < 0.005$) in absorbance was observed. However, in *F. serratus*, all molecular weights were found to have a significant decrease in absorbance ($p < 0.001$) with respect to the control.

To verify that PEG was inhibiting the effect of the phlorotannins, gel electrophoresis was used to measure the change in MW of proteins when using pepsin, which was used as an enzyme in the in vitro study. Pepsin was bound to purified phlorotannin and showed a distinct change in the MW when compared to the protein ladder. When PEG was incorporated into the matrix, the spot for the phlorotannin–protein complex was diminished and the pure pepsin spot was recovered (Figure 3). The results suggest that the phlorotannins are binding to the enzymes in the digestive tract but the enzyme was recovered when PEG is added, therefore confirming that PEG is counteracting the effect of the phlorotannins on the proteins.

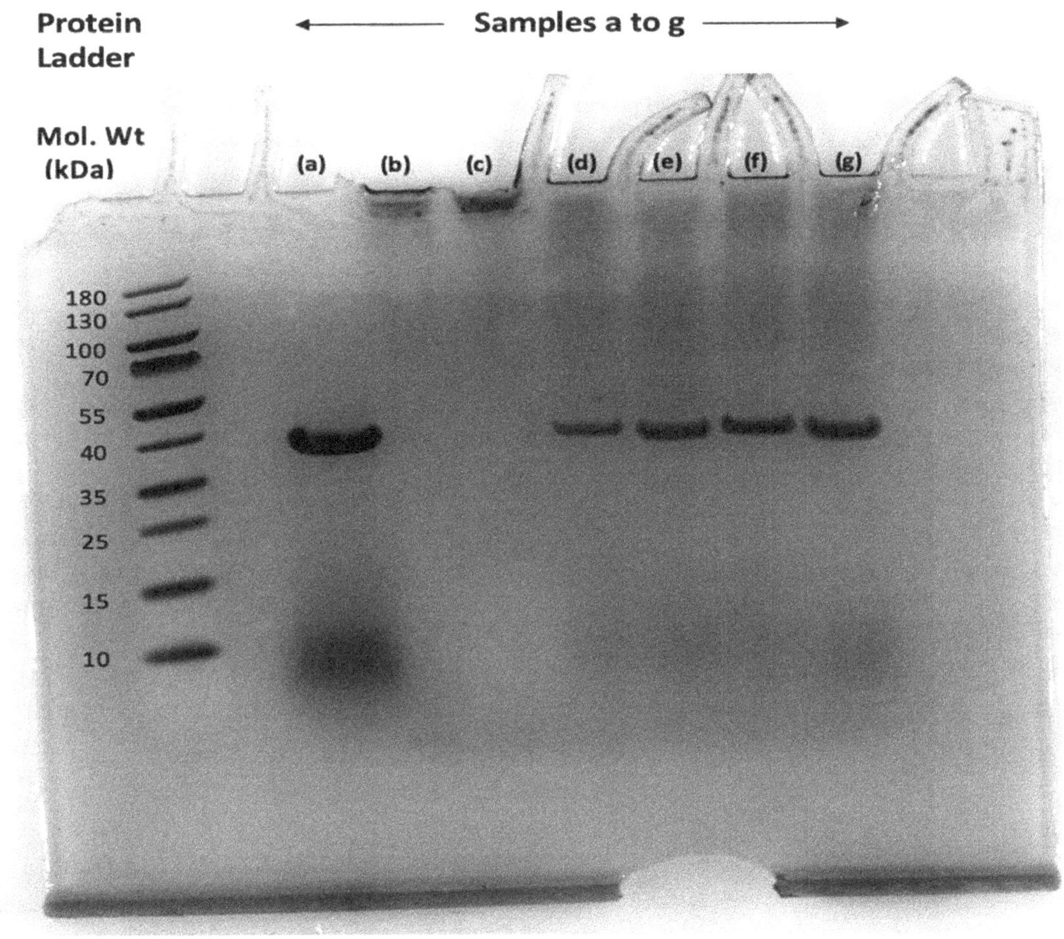

Figure 3. Gel electrophoresis of pepsin with and without binding to the tannins and PEG. From left to right after protein ladder: (**a**) pure pepsin, (**b**) pepsin + Asco phlorotannin, (**c**) pepsin + Fucus phlorotannin, (**d**) pepsin + Asco phlorotannin + PEG 2000MW, (**e**) pepsin + Asco phlorotannin + 4000MW, (**f**) pepsin + Fucus phlorotannin + PEG 2000MW and (**g**) pepsin + Fucus phlorotannin + PEG 4000MW.

Digestibility experiments were set up with and without PEG to measure the effect that the phlorotannins are having on the digestibility of the two seaweed species over the year. Given that gel electrophoresis showed no difference, 4000 MW was selected for the following studies. PEG was added into the seaweed at ratios of 1:1, 1:3 and 1:5 phlorotannin: PEG, the phlorotannin content was calculated using the FC assay and expressed in mg/g PGE.

Figure 4 shows the effect on digestibility when PEG 4000 MW is added to the seaweed samples in the in vitro digestibility study. There was no significant decrease in the IVDMD when PEG was added at any ratio, in any season accept for winter for *A. nodosum*. In autumn, however, there was a statistically significant decrease in IVDMD when PEG was added at a 1:1 ratio ($p = 0.015$) and 1:5 ratio ($p = 0.003$) when compared to the PEG 0 control. In *F. serratus*, a significant decrease was observed in winter at a 1:1 ratio ($p = 0.028$), whereas there was a significant increase in digestibility in summer at all ratios: 1:1 ($p = 0.012$), 1:3 ($p = 0.017$), 1:5 ($p = 0.011$). The phenolic content (Table 1) in *F. serratus* was found to be the lowest ($p < 0.05$) at 26.49 ± 1.42 mg/g in summer compared to the other seasons tested in this species.

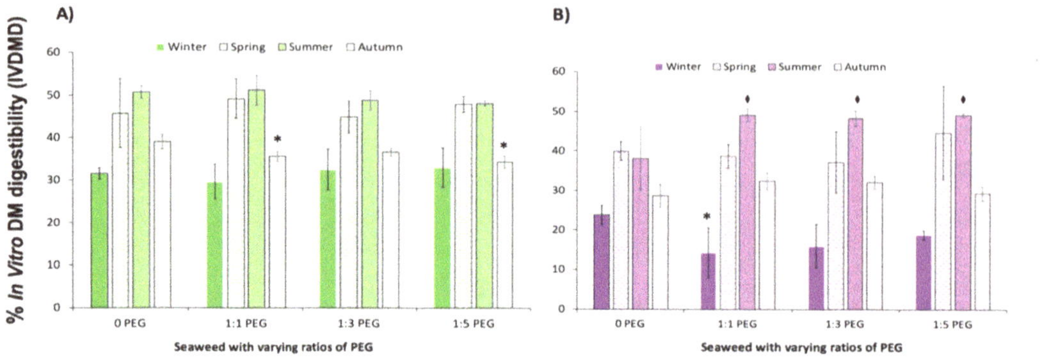

Figure 4. In vitro DM digestibility (IVDMD) profiles of (**A**) *A. nodosum* and (**B**) *F. serratus* with and without PEG. Three different ratios of PEG were used: 1:1, 1:3 and 1:5 phlorotannin:PEG ($p < 0.05$, * = significant decrease; ♦ = significant increase).

Given the mixed results in this study, purified extracts were isolated to study the specific effect of the phlorotannins on the digestibility, which is discussed in the next section.

3.3. Effect of Phlorotannins on Commercial Pig Feed

Purified phlorotannins extracts were mixed at different concentrations with a commercial feed and their effect on digestibility was recorded. The results are shown in Figure 5. A significant decrease ($p < 0.05$) in IVDMD was observed in *A. nodosum* (Figure 5) between the control and all concentrations of phlorotannins added. The same trend was also observed in *F. serratus*. However, when comparing *A. nodosum* to *F. serratus*, there was an observed difference in digestibility between both species when phlorotannins were added. This was found to be statistically significant at concentrations of 0.781, 3.125, 20 and 50 mg/mL. However, at lower concentrations of 0.781 and 3.125 mg/mL, IVDMD was found to be higher in *A. nodosum* than *F. serratus*, whereas the opposite trend was observed at higher concentrations of 20 and 50 mg/mL. When pure phlorotannin extracts were added directly to the pig feed, the results clearly show that phlorotannins found in both *A. nodosum* and *F. serratus* affect the digestibility of the pig feed. At lower concentrations of <3.125 mg/mL, *F. serratus* was less detrimental to the digestibility of a high protein feed than *A. nodosum*, whereas the opposite occurs above 3.125 mg/mL.

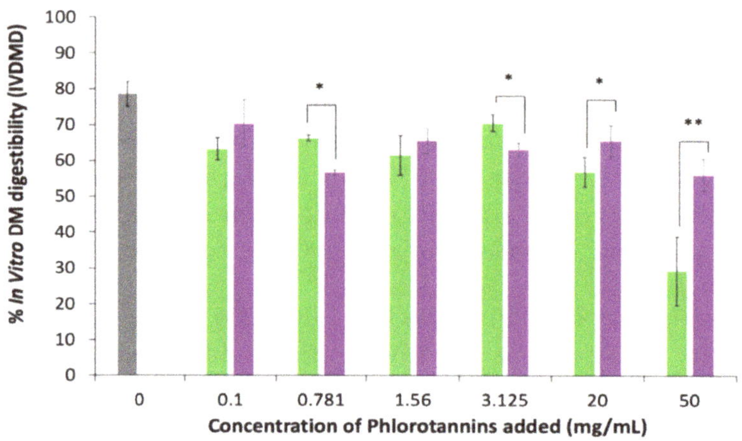

Figure 5. In vitro DM digestibility (IVDMD) of a commercial pig feed when different concentrations of phlorotannins isolated from *A. nodosum* (purple) and *F. serratus* (green) in May were added. Control (commercial pig feed, grey) (* = $p < 0.05$; ** = $p < 0.001$).

3.4. Effect of Seaweeds Added to Commercial Pig Feed

Finally, the same digestibility study was repeated using the whole seaweed plant (without PEG), as opposed to just the phlorotannin extracts, in order to investigate the effects of other compounds found in *A. nodosum* on the digestibility of the pig feed, in addition to the phlorotannins. Our results (Figure 6) show that there was no significant difference in digestibility between samples collected in winter and spring at up to 5% inclusion, which is equivalent to 0.95 and 1.9 mg/g (Table 2) of phlorotannins, respectively, in comparison to the control (pig feed only). At 10% inclusion, a significant drop ($p < 0.01$) in digestibility was observed from 77.8 (±3.5) % to 61.0 (±3.3) % in winter (TPC ≈1.9 mg/g), whereas no significant difference in digestibility was observed in May, which had a significantly higher phenolic content (TPC ≈3.8 mg/g). At 20% inclusion, the opposite trend was observed, whereby the higher concentration of phenolics, ≈7.6 mg/g in May showed a lower digestibility of 51.0 (±2.8) %, compared to winter which was 59.2 (±8.1) %. However, this was not found to be significant.

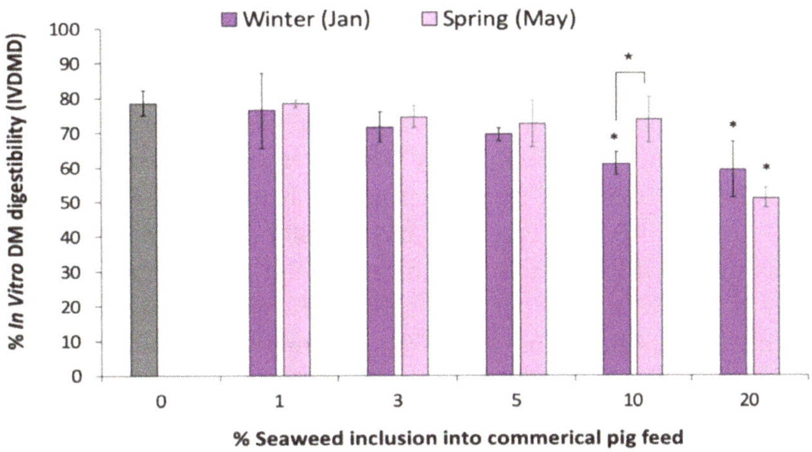

Figure 6. Change in in vitro DM digestibility (IVDMD) of a commercial pig feed when different percentages of *A. nodosum* were added from samples collected in Jan and May. * = $p < 0.05$ between control and % inclusion; ★ = $p < 0.05$ between pairs.

Table 2. Concentration of phlorotannins at each inclusion rate.

% Inclusion		**1**	**3**	**5**	**10**	**20**
TPC (mg/g) DW	Winter	0.19	0.57	0.95	1.9	3.8
	Spring	0.38	1.14	1.9	3.8	7.6

4. Discussion

Brown seaweeds are relatively low in protein compared to other feed sources, such as alfalfa or soy [40,41]. The highest CP contain in this study was found in winter and spring for both species and is likely caused by N-based fertilisers leaching into coastal areas, as terrestrial crops and grasslands tend to be fertilised during this period [42]. Al-Yaman et al. [43] reported that the phenolic contents in brown seaweeds correlate to their protein content. However, no correlation was observed between CP and phenolics in this study. It is noteworthy that the species of brown seaweeds differed between these two studies.

The NDF results of the seaweeds studied have been found to be comparable to other terrestrial forage, e.g., alfalfa [44,45], corn and wheat [46], which have been reported to be approximately 45%. In the ADF results, no seasonality effect was observed in either seaweed species. The difference was not found to be statistically significant between seasons. Typically, ADF <35% are desirable in livestock diets [47]. In this study, the digestibility of the whole seaweed of *F. serratus* was found to have lower digestibility than *A. nodosum* seaweed, with ADF in the range of 21 to 26% and 15 to 18%, respectively. While a correlation between phenolic and ADF content for *A. nodosum* was observed, this trend was not found in *F. serratus*. The polymer content might also account for the difference in digestibility. In brown seaweeds, the polymers are mainly alginates and sulphated fucoidans [48,49]. In humans, alginate has been reported to be indigestible and is likely to have the same effect in pigs given that our digestive systems are very similar [50]. It is considered beneficial as an anti-obesity fibre that promotes weight loss [51,52], whereas fucoidan has been reported to have no effect on weight loss in humans in a recent obesity study [53]. Therefore, it is likely than fucoidan is more digestible than alginate in brown seaweeds. Fletcher et al. [54] have reported the fucoidan content in *A. nodosum* is significantly higher than *F. serratus* throughout the year Therefore, a higher fucoidan content in *A. nodosum* may have contributed to the significant difference in ADF.

When comparing IVDMD to total phenolic content, the different trend observed between *A. nodosum* and *F. serratus* suggests that seaweed species produce an effect on digestibility. Studies of phenolic compounds in seaweeds have reported that phlorotannin structures vary between species [55–57]. It is likely that phlorotannins with different linkages, functional groups, reactivities and molecular weights interacted and/or bound differently with other macromolecules within the seaweeds, namely polysaccharides and proteins, thus causing a different effect on digestibility.

Polyethylene glycol (PEG) is used to counteract the effect of tannins in the animal feed [6]. This method has been used for terrestrial tannins and with phlorotannins from brown seaweed [35]. Polyvinylpyrrolidone (PVP) and PEG have been used to bind to tannins for the last three decades. However, the same has been performed on conventional terrestrial tannins [6]. The use of PEG and PVP with molecular weights (MW) between 2000 and 35,000 has been tested for their binding abilities with terrestrial tannins [58], with the most commonly used molecular weights between 3500 and 4000 [6]. The FC assay is a colorimetric assay based on the redox potential of the substrate and its ability to complex with the metallic FC reagent [59]. The reaction produces a dark blue colour upon complexation and this can be measured at 755 nm. Therefore, the PEG MW best at counteracting the effect of the phlorotannins should have the largest decrease in absorbance. In the PEG binding study, the most pronounced decrease was observed in *F. serratus* at PEG 2000 MW. This would suggest that PEG 2000 was the best phlorotannin binder for *F. serratus* phlorotannins. However, it is noteworthy that there was an obvious precipitate formed from 1000 to 20,000 MW in both seaweeds tested, which could have caused higher readings in the absorbance due to turbidity and the precipitates absorbing the light at all wavelengths. Further, the FC assay is a metal complex colorimetric reaction and therefore the affinity between the phlorotannins and FC reagent is likely to be very different from that of the affinity between the tannin and protein.

The gel electrophoresis showed that both PEG 2000 and 4000 MW successfully inhibited phlorotannins. Previous studies have reported PEG 3350 MW at blocking phlorotannin [34] and most terrestrial tannins studies have reported between 3500 and 4000 MW to be most effective [6]. Therefore, 4000 MW was used to mask the effect of phenolics in the subsequent study IVDMD study. As PEG counteracts the effect of the phlorotannins, the results would suggest that PEG was only successful at blocking the phlorotannins in *F. serratus* in summer at all ratios. However, at higher phenolic concentrations, no significant difference in digestibility was observed, suggesting that the ratios and ME of PEG used have not been successful at blocking the phlorotannins. The same trend was observed in *A. nodosum*, whereby there was no significant difference in IVDMD between the control and the samples complexed with PEG, suggesting that the PEG ratio and MW used was not sufficient to bind this concentration of phenolics. PEG may have been less effective in the IVDMD study compared to binding study (Figure 2) and gel electrophoresis (Figure 3) due to pH. The effect of pH on the binding efficiency of terrestrial tannins has previously been reported by Makkar et al. [58].

The results of the PEG binding digestibility study (Figure 4) suggest that the phenolics found in *F. serratus* in summer do affect digestibility, whereas no relationship was found when correlating the IVDMD to phenolics (Figure 1). The lack of correlation in the results shown in Figure 1 may be a result of the limitations in the FC assay, as discussed in a recent review by Ford et al. [60], as opposed to a lack of relationship between phenolics and digestibility, thus highlighting the need to improve characterisation methods used to measure phenolics in seaweeds. Further, as discussed above, pH may have affected the correlation. In addition, these studies have investigated the digestibility of the whole seaweed including the polymer fraction and other compounds as previously discussed, as opposed to the specific relationship between the phlorotannins on the animal feed used. The most common method of inclusion to animal feeds is the inclusion (between 1 and 5%) of the whole seaweed feedstock [5], as opposed to the crude (phenolics) or purified (phlorotannins) due to animal feed regulations governing the use of extracts.

The results (Figure 5) clearly show that phlorotannins can negatively affect digestibility of the pig feed. However, when whole seaweed (Figure 6) was added, the effect was significantly reduced. This would suggest that the other compounds (most likely the polysaccharides) which form the seaweed matrix are inhibiting the interaction between the phlorotannins and digestive enzymes, thus counteracting the reduction in digestion. The results also show that higher concentrations of phenolics can be added by using the whole feedstock as opposite to purified phlorotannin extracts before effecting digestibility. However, in spring, 10% seaweed inclusion with ≈3.8 mg/g did not affect digestibility, whereas 20% seaweed inclusion with ≈3.8 mg/g in winter significantly reduced digestibility, thus confirming that other compounds, or a variation in phenolic structure, may be affecting digestibility. In the literature, the most common inclusion rates reported are between 1 and 5% [5,30,41]. Although this would suggest that supplementing whole seaweeds, as opposed to phlorotannin extracts, might be better to counteract the effect of phlorotannins on digestibility, it is noteworthy to consider that the acetone/water yields may extract higher concentrations of phlorotannins from the biomass than the buffer system used in the digestibility studies, thus also masking the effect of the phlorotannins on digestibility. Further work is required to better understand the digestibility of the polysaccharides, types of phlorotannins and other compounds found in seaweeds to determine the overall effect of adding seaweeds to pig feed.

5. Conclusions

The work herein shows a clear correlation between the digestibility of pig feed and phenolics, when the phenolics were purified in phlorotannin. In the two seaweeds studied, *A. nodosum* and *F. serratus*, the addition of phlorotannin extracts caused a significant decrease in IVDMD. This relationship becomes more complex when the whole seaweed plant was added. Further work is required to better understand the inclusion of whole seaweeds in feeds. In addition, consideration needs to be given to the effect of seasonality on the chemical composition of seaweed being added, as phenolics can vary significantly within the same species during the year. Ideally, the seaweed should be harvested when the phenolics are most abundant in order to reduce the overall amount of seaweed that needs to be supplemented. This research has highlighted that supplementation should not just be based on standardised percentage inclusion of seaweeds, but on the chemistry of biomass (e.g., phenolic and biopolymers) to mitigate the effect on digestibility. To validate the use of seaweeds in pig feed, an animal trial is required to gain a better understanding of the in vivo digestibility of seaweed-supplemented pig feeds before seaweeds could be considered for use in commercial pig feed. In addition, further work is required on the digestibility of other seaweed compounds and the bioavailability of the pholorotannins to the animal if whole seaweed is to be supplemented.

Author Contributions: Conceptualisation, all authors; data curation, L.F. (mainly), M.C. and C.C.; formal analysis, L.F., G.S. and P.J.W.; statistical analysis, L.S., J.D. and P.J.W., funding acquisition, P.J.W.; investigation, all authors; project administration, P.J.W.; resources, P.J.W., G.S. and K.T.; supervision, P.J.W. and G.S.; validation, L.F. and P.J.W.; visualisation, L.F.; writing—original draft, L.F.; editing, all; writing—review and editing, L.F. and P.J.W. All authors have read and agreed to the published version of the manuscript.

References

1. Pinteus, S.; Silva, J.; Alves, C.; Horta, A.; Thomas, O.P.; Pedrosa, R. Antioxidant and cytoprotective activities of fucus spiralis seaweed on a human cell in vitro model. *Int. J. Mol. Sci.* **2017**, *18*, 292. [CrossRef]
2. Torres, M.D.; Kraan, S.; Domínguez, H. Seaweed biorefinery. *Rev. Environ. Sci. Biotechnol.* **2019**, *18*, 335–388. [CrossRef]

3. Balina, K.; Romagnoli, F.; Blumberga, D. Seaweed biorefinery concept for sustainable use of marine resources. *Energy Procedia* **2017**, *128*, 504–511. [CrossRef]
4. Corona, G.; Coman, M.M.; Guo, Y.; Hotchkiss, S.; Gill, C.; Yaqoob, P.; Spencer, J.P.E.; Rowland, I. Effect of simulated gastrointestinal digestion and fermentation on polyphenolic content and bioactivity of brown seaweed phlorotannin-rich extracts. *Mol. Nutr. Food Res.* **2017**, *61*, 1700223. [CrossRef]
5. Makkar, H.P.S.; Tran, G.; Heuzé, V.; Giger-Reverdin, S.; Lessire, M.; Lebas, F.; Ankers, P. Seaweeds for livestock diets: A review. *Anim. Feed Sci. Technol.* **2016**, *212*, 1–17. [CrossRef]
6. Makkar, H.P.S. Effects and fate of tannins in ruminant animals, adaptation to tannins, and strategies to overcome detrimental effects of feeding tannin-rich feeds. *Small Rumin. Res.* **2003**, *49*, 241–256. [CrossRef]
7. Chapman, V.J.; Chapman, D.J. *Seaweeds and Their Uses*; Springer: Dordrecht, The Netherlands, 1980; ISBN 978-94-009-5806-7.
8. Corino, C.; Modina, S.C.; Di Giancamillo, A.; Chiapparini, S.; Rossi, R. Seaweeds in pig nutrition. *Animals* **2019**, *9*, 1126. [CrossRef] [PubMed]
9. Stratakos, A.C.; Sima, F.; Ward, P.; Linton, M.; Kelly, C.; Pinkerton, L.; Stef, L.; Pet, I.; Corcionivoschi, N. The in vitro effect of carvacrol, a food additive, on the pathogenicity of O157 and non-O157 Shiga-toxin producing Escherichia coli. *Food Control* **2018**, *84*, 290–296. [CrossRef]
10. Ford, L.; Stratakos, A.C.; Theodoridou, K.; Dick, J.T.A.; Sheldrake, G.N.; Linton, M.; Corcionivoschi, N.; Walsh, P.J. Polyphenols from Brown Seaweeds as a Potential Antimicrobial Agent in Animal Feeds. *ACS Omega* **2020**, *5*, 9093–9103. [CrossRef]
11. Kadam, S.U.; O'Donnell, C.P.; Rai, D.K.; Hossain, M.B.; Burgess, C.M.; Walsh, D.; Tiwari, B.K. Laminarin from Irish brown seaweeds Ascophyllum nodosum and Laminaria hyperborea: Ultrasound assisted extraction, characterization and bioactivity. *Mar. Drugs* **2015**, *13*, 4270–4280. [CrossRef]
12. Maia, M.R.G.; Fonseca, A.J.M.; Oliveira, H.M.; Mendonça, C.; Cabrita, A.R.J. The potential role of seaweeds in the natural manipulation of rumen fermentation and methane production. *Sci. Rep.* **2016**, *6*, 32321. [CrossRef] [PubMed]
13. Wang, Y.; Xu, Z.; Bach, S.J.; McAllister, T.A. Effects of phlorotannins from Ascophyllum nodosum (brown seaweed) on in vitro ruminal digestion of mixed forage or barley grain. *Anim. Feed Sci. Technol.* **2008**, *145*, 375–395. [CrossRef]
14. Philippe, F.X.; Nicks, B. Review on greenhouse gas emissions from pig houses: Production of carbon dioxide, methane and nitrous oxide by animals and manure. *Agric. Ecosyst. Environ.* **2015**, *199*, 10–25. [CrossRef]
15. McHugh, D.J. *A Guide to the Seaweed Industry*; Food and Agriculture Organization of the United Nations: Rome, Italy, 2003.
16. Mac Monagail, M.; Cornish, L.; Morrison, L.; Araújo, R.; Critchley, A.T. Sustainable harvesting of wild seaweed resources. *Eur. J. Phycol.* **2017**, *52*, 371–390. [CrossRef]
17. Jones, R.T.; Blunden, G.; Probert, A.J. Effects of dietary Ascophyllum nodosum on blood parameters of rats and pigs. *Bot. Mar.* **1979**, *22*, 393–402.
18. Michiels, J.; Skrivanova, E.; Missotten, J.; Ovyn, A.; Mrazek, J.; De Smet, S.; Dierick, N. Intact brown seaweed (Ascophyllum nodosum) in diets of weaned piglets: Effects on performance, gut bacteria and morphology and plasma oxidative status. *J. Anim. Physiol. Anim. Nutr.* **2012**, *96*, 1101–1111. [CrossRef]
19. Bach, S.J.; Wang, Y.; McAllister, T.A. Effect of feeding sun-dried seaweed (Ascophyllum nodosum) on fecal shedding of Escherichia coli O157:H7 by feedlot cattle and on growth performance of lambs. *Anim. Feed Sci. Technol.* **2008**, *142*, 17–32. [CrossRef]
20. Gahan, D.A.; Lynch, M.B.; Callan, J.J.; O'Sullivan, J.T.; O'Doherty, J.V. Performance of weanling piglets offered low-, medium- or high-lactose diets supplemented with a seaweed extract from *Laminaria* spp. *Animal* **2009**, *3*, 24–31. [CrossRef]
21. McDonnell, P.; Figat, S.; Odoherty, J.V. The effect of dietary laminarin and fucoidan in the diet of the weanling piglet on performance, selected faecal microbial populations and volatile fatty acid concentrations. *Animal* **2010**, *4*, 579–585. [CrossRef]
22. Kadam, S.U.; Tiwari, B.K.; O'Donnell, C.P. Application of novel extraction technologies for bioactives from marine algae. *J. Agric. Food Chem.* **2013**, *61*, 4667–4675. [CrossRef]
23. Bikker, P.; Stokvis, L.; van Krimpen, M.M.; van Wikselaar, P.G.; Cone, J.W. Evaluation of seaweeds from marine waters in Northwestern Europe for application in animal nutrition. *Anim. Feed Sci. Technol.* **2020**, *263*, 114460. [CrossRef]

24. Bleakley, S.; Hayes, M. Algal Proteins: Extraction, Application, and Challenges Concerning Production. *Foods* **2017**, *6*, 33. [CrossRef] [PubMed]
25. Ruiz, Á.R.; Gadicke, P.; Andrades, S.M.; Cubillos, R. Supplementing nursery pig feed with seaweed extracts increases final body weight of pigs. *Austral J. Vet. Sci.* **2018**, *50*, 83–87. [CrossRef]
26. Schmid, M.; Guihéneuf, F.; Stengel, D.B. Fatty acid contents and profiles of 16 macroalgae collected from the Irish Coast at two seasons. *J. Appl. Phycol.* **2014**, *26*, 451–463. [CrossRef]
27. Chapman, A.R.O.; Craigie, J.S. Seasonal growth in Laminaria longicruris: Relations with dissolved inorganic nutrients and internal reserves of nitrogen. *Mar. Biol.* **1977**, *40*, 197–205. [CrossRef]
28. Parys, S.; Kehraus, S.; Pete, R.; Küpper, F.C.; Glombitza, K.-W.; König, G.M. Seasonal variation of polyphenolics in Ascophyllum nodosum (Phaeophyceae). *Eur. J. Phycol.* **2009**, *44*, 331–338. [CrossRef]
29. Polat, S.; Ozogul, Y. Seasonal proximate and fatty acid variations of some seaweeds from the northeastern Mediterranean coast. *Oceanologia* **2013**, *55*, 375–391. [CrossRef]
30. Belanche, A.; Jones, E.; Parveen, I.; Newbold, C.J. A metagenomics approach to evaluate the impact of dietary supplementation with Ascophyllum nodosum or Laminaria digitata on rumen function in Rusitec fermenters. *Front. Microbiol.* **2016**, *7*, 299. [CrossRef]
31. Lee, S.H.; Han, J.S.; Heo, S.J.; Hwang, J.Y.; Jeon, Y.J. Protective effects of dieckol isolated from Ecklonia cava against high glucose-induced oxidative stress in human umbilical vein endothelial cells. *Toxicol. Vitr.* **2010**, *24*, 375–381. [CrossRef]
32. Özkan Gülzari, S.; Lind, V.; Aasen, I.M.; Steinshamn, H. Effect of supplementing sheep diets with macroalgae species on in vivo nutrient digestibility, rumen fermentation and blood amino acid profile. *Animal* **2019**, *13*, 2792–2801. [CrossRef]
33. Huang, Q.; Liu, X.; Zhao, G.; Hu, T.; Wang, Y. Potential and challenges of tannins as an alternative to in-feed antibiotics for farm animal production. *Anim. Nutr.* **2017**, *4*, 137–150. [CrossRef] [PubMed]
34. Wang, Y.; Alexander, T.W.; Mcallister, T.A. In vitro effects of phlorotannins from Ascophyllum nodosum (brown seaweed) on rumen bacterial populations and fermentation. *J. Sci. Food Agric.* **2009**, *89*, 2252–2260. [CrossRef]
35. Vissers, A.M.; Pellikaan, W.F.; Bouwhuis, A.; Vincken, J.P.; Gruppen, H.; Hendriks, W.H. Laminaria digitata phlorotannins decrease protein degradation and methanogenesis during in vitro ruminal fermentation. *J. Sci. Food Agric.* **2018**, *98*, 3644–3650. [CrossRef] [PubMed]
36. Van Soest, P.J.; Robertson, J.B.; Lewis, B.A. Methods for Dietary Fiber, Neutral Detergent Fiber, and Nonstarch Polysaccharides in Relation to Animal Nutrition. *J. Dairy Sci.* **1991**, *74*, 3583–3597. [CrossRef]
37. Angell, A.R.; Mata, L.; de Nys, R.; Paul, N.A. The protein content of seaweeds: A universal nitrogen-to-protein conversion factor of five. *J. Appl. Phycol.* **2016**, *28*, 511–524. [CrossRef]
38. Campbell, M.; Ortuño, J.; Ford, L.; Davies, D.R.; Koidis, A.; Walsh, P.J.; Theodoridou, K. The effect of ensiling on the nutritional composition and fermentation characteristics of brown seaweeds as a ruminant feed ingredient. *Animals* **2020**, *10*, 1019. [CrossRef]
39. Tiwari, U.P.; Jha, R. Nutrient profile and digestibility of tubers and agro-industrial coproducts determined using an in vitro model of swine. *Anim. Nutr.* **2016**, *2*, 357–360. [CrossRef]
40. Friedman, M.; Brandon, D.L. Nutritional and health benefits of soy proteins. *J. Agric. Food Chem.* **2001**, *49*, 1069–1086. [CrossRef]
41. Evans, F.D.; Critchley, A.T. Seaweeds for animal production use. *J. Appl. Phycol.* **2014**, *26*, 891–899. [CrossRef]
42. The British Survey of Fertiliser Practice: Fertiliser Use on Farm Crops for Crop Year. 2015. Available online: https://www.summerfieldbooks.com/product/the-british-survey-of-fertiliser-practice-fertiliser-use-on-farm-crops-for-crop-year-2015/ (accessed on 1 July 2020).
43. Sabeena, S.F.; Alagarsamy, S.; Sattari, Z.; Al-Haddad, S.; Fakhraldeen, S.; Al-Ghunaim, A.; Al-Yamani, F. Enzyme-assisted extraction of bioactive compounds from brown seaweeds and characterization. *J. Appl. Phycol.* **2020**, *32*, 615–629. [CrossRef]
44. Uddin, M.E.; Larson, R.A.; Wattiaux, M.A. Effects of dairy cow breed and dietary forage on greenhouse gas emissions from manure during storage and after field application. *J. Clean. Prod.* **2020**, *270*, 122461. [CrossRef]
45. Cherney, D.J.; Siciliano-Jones, J.; Pell, A.N. Technical note: Forage in vitro dry matter digestibility as influenced by fiber source in the donor cow diet. *J. Anim. Sci.* **1993**, *71*, 1335–1338. [CrossRef] [PubMed]
46. Soufizadeh, M.; Pirmohammadi, R.; Alijoo, Y.; Behroozyar, H.K. Indigestible neutral detergent fibers:

Relationship between forage fragility and neutral detergent fibers digestibility in total mixed ration and some feedstuffs in dairy cattle. *Vet. Res. Forum* **2018**, *9*, 49–57. [PubMed]

47. Understanding and Using a Feed Analysis Report. Available online: https://extensionpubs.unl.edu/publication/9000016366079/understanding-and-using-a-feed-analysis-report/ (accessed on 1 September 2020).
48. Lu, J.; Yang, H.; Hao, J.; Wu, C.; Liu, L.; Xu, N.; Linhardt, R.J.; Zhang, Z. Impact of hydrolysis conditions on the detection of mannuronic to guluronic acid ratio in alginate and its derivatives. *Carbohydr. Polym.* **2015**, *122*, 180–188. [CrossRef]
49. Øverland, M.; Mydland, L.T.; Skrede, A. Marine macroalgae as sources of protein and bioactive compounds in feed for monogastric animals. *J. Sci. Food Agric.* **2018**, *99*, 13–24. [CrossRef]
50. Furness, J.B.; Cottrell, J.J.; Bravo, D.M. Comparative physiology of digestion. *J. Anim. Sci.* **2015**, *93*, 485–491. [CrossRef]
51. Chater, P.I.; Wilcox, M.D.; Houghton, D.; Pearson, J.P. The role of seaweed bioactives in the control of digestion: Implications for obesity treatments. *Food Funct.* **2015**, *6*, 3420–3427. [CrossRef]
52. Cherry, P.; O'hara, C.; Magee, P.J.; Mcsorley, E.M.; Allsopp, P.J. Risks and benefits of consuming edible seaweeds. *Nutr. Rev.* **2019**, *77*, 307–329. [CrossRef]
53. Hernández-Corona, D.M.; Martínez-Abundis, E.; González-Ortiz, M. Effect of Fucoidan administration on insulin secretion and insulin resistance in overweight or obese adults. *J. Med. Food* **2014**, *17*, 830–832. [CrossRef]
54. Fletcher, H.R.; Biller, P.; Ross, A.B.; Adams, J.M.M. The seasonal variation of fucoidan within three species of brown macroalgae. *Algal Res.* **2017**, *22*, 79–86. [CrossRef]
55. Catarino, M.D.; Silva, A.M.S.; Cardoso, S.M. Fucaceae: A Source of Bioactive Phlorotannins. *Int. J. Mol. Sci.* **2017**, *18*, 1327. [CrossRef] [PubMed]
56. Ferreres, F.; Lopes, G.; Gil-Izquierdo, A.; Andrade, P.B.; Sousa, C.; Mouga, T.; Valentão, P. Phlorotannin extracts from fucales characterized by HPLC-DAD-ESI-MS n: Approaches to hyaluronidase inhibitory capacity and antioxidant properties. *Mar. Drugs* **2012**, *10*, 2766–2781. [CrossRef] [PubMed]
57. Peng, Y.; Hu, J.; Yang, B.; Lin, X.P.; Zhou, X.F.; Yang, X.W.; Liu, Y. Chemical composition of seaweeds. In *Seaweed Sustainability: Food and Non-Food Applications*; 2015; ISBN 9780124199583. Available online: https://www.sciencedirect.com/book/9780124186972/seaweed-sustainability#book-info (accessed on 1 September 2020).
58. Makkar, H.P.S.; Blümmel, M.; Becker, K. Formation of complexes between polyvinyl pyrrolidones or polyethylene glycols and tannins, and their implication in gas production and true digestibility in in vitro techniques. *Br. J. Nutr.* **1995**, *73*, 897–913. [CrossRef] [PubMed]
59. Singleton, V.L.; Orthofer, R.; Lamuela-Raventós, R.M. Analysis of total phenols and other oxidation substrates and antioxidants by means of folin-ciocalteu reagent. *Methods Enzymol.* **1998**, *299*, 152–178.
60. Ford, L.; Theodoridou, K.; Sheldrake, G.N.; Walsh, P.J. A critical review of analytical methods used for the chemical characterisation and quantification of phlorotannin compounds in brown seaweeds. *Phytochem. Anal.* **2019**, *30*, 587–599. [CrossRef]

2

Porcine Digestible Peptides (PDP) in Weanling Diets Regulates the Expression of Genes Involved in Gut Barrier Function, Immune Response and Nutrient Transport in Nursery Pigs

Francesc González-Solé [1], Lourdes Criado-Mesas [2], Carmen Villodre [1], Wellington C. García [1,3], Mercè Farré [4], Elisabet Borda [5], Francisco J. Pérez-Cano [6,7], Josep M. Folch [1,2], David Solà-Oriol [1,*] and José F. Pérez [1]

1 Department of Animal and Food Sciences, Animal Nutrition and Welfare Service, Autonomous University of Barcelona, 08193 Bellaterra, Spain; francesc.gonzalez.sole@uab.cat (F.G.-S.); carmenvitu@gmail.com (C.V.); wncg_100583@hotmail.com (W.C.G.); JosepMaria.Folch@uab.cat (J.M.F.); JoseFrancisco.Perez@uab.cat (J.F.P.)
2 Plant and Animal Genomics, Centre for Research in Agricultural Genomics (CRAG), CSIC-IRTA-UAB-UB, 08193 Bellaterra, Spain; lourdes.criado@cragenomica.es
3 Department of Animal Production, Agrarian University of Ecuador, Guayaquil 090108, Ecuador
4 Department of Mathematics, Area of Statistics and Operations Research, Autonomous University of Barcelona, 08193 Bellaterra, Spain; farre@mat.uab.cat
5 R&D Animal Nutrition Department, Bioiberica S.A.U., 08389 Palafolls, Spain; eborda@bioiberica.com
6 Physiology Section, Department of Biochemistry and Physiology, Faculty of Pharmacy and Food Science, University of Barcelona (UB), 08028 Barcelona, Spain; franciscoperez@ub.edu
7 Nutrition and Food Safety Research Institute (INSA-UB), 08921 Santa Coloma de Gramenet, Spain
* Correspondence: David.Sola@uab.cat

Simple Summary: Porcine digestive peptides (PDP) are a coproduct of the heparin industry obtained from the enzymatic hydrolysis of porcine intestinal mucosa. They have proven to be a valid substitute for other high quality dietary protein sources for piglets, like spray-dried plasma (SDP), but knowledge about their influence on intestinal function is still scarce. This study found that substituting soybean ingredients with PDP and SDP to the diets of weaned piglets increased growth rate at 14 d post-weaning. In addition, the combination of PDP with SDP increased the expression of certain genes related to intestinal function in the jejunum, which suggests that this combination might have functional properties that contribute to improving the intestinal health of the pigs, although more research is needed to confirm it.

Abstract: This study was conducted to investigate the effects of dietary supplementation of porcine digestible peptides (PDP), spray-dried plasma (SDP), or a combination of both, on growth performance and the expression of genes related to intestinal function of weaned pigs. A total of 180 piglets (trial 1) and 198 piglets (trial 2) were used to evaluate the partial substitution of soybean ingredients with 2% SDP or 2% PDP (trial 1), and with 3% SDP or the combination of 1% SDP and 2% PDP (SDP-PDP; trial 2) during the pre-starter period (0–14 days). The gene expression of 56 genes was quantified in a qPCR platform in jejunum and ileum samples obtained from piglets 14 d after weaning (trial 2). Piglets fed SDP, PDP and SDP-PDP had a higher body weight (BW), average daily gain (ADG) and feed efficiency (G:F) than the soybean control on day 14 ($p < 0.05$). In addition, the combination of SDP and PDP upregulated ten genes in jejunum samples ($p < 0.05$) related to intestinal function. More research is needed to confirm that gene expression upregulation by PDP in combination with SDP has an impact on intestinal function and to elucidate its underlying mechanisms.

Keywords: gene expression; high throughput RT-qPCR; intestinal function; piglet; porcine digestible peptides

1. Introduction

Weaning is a stressful period for piglets as they have to deal with a change from sow milk to a less digestible, plant-based dry solid diet that contains many ingredients that the pig has not eaten before [1]. One major consequence of weaning is a reduction in feed intake, which in turn causes a reduction in the villi height in the small intestine [2] and a disruption of the gut microbiota ecosystem with a loss of diversity [3]. After weaning, piglets are more susceptible to gut inflammatory problems and their intestinal function can be affected. This often leads to post-weaning diarrhea and increased mucosal permeability [4]. A strategy for helping piglets to reduce intestinal disturbances during this period is to reduce the inclusion of less-digestible vegetal protein sources in their feed, like soybean ingredients, and substitute them with high-quality digestible animal protein sources. Certain protein sources, like spray-dried plasma (SDP) contain biologically active components that give them physiological or regulatory functions beyond their nutritional value [5]. In particular, SDP has shown potential for modulating the immune response, reducing intestinal mucosa inflammation and maintaining its integrity [6–8].

Porcine digestible peptides (PDP) are a coproduct of the heparin industry and are obtained from the enzymatic hydrolysis of porcine intestinal mucosa. Currently, PDP can be used in postweaning diets as the cheapest alternative for SDP, fish meal and other sources of high-quality protein in terms of preference and digestibility [9,10]. The ability of PDP to increase villus height suggests that they may improve nutrient uptake [11]. However, there is no available literature exploring possible effects of PDP on intestinal function when replacing major soy protein ingredients in the diet. In the present study, in an initial trial we hypothesize that including PDP and SDP in a high crude protein (CP) diet partially substituting a high content of soybean ingredients might improve piglet performance after weaning (trial 1). In a second trial we also hypothesize that partially substituting soybean meal (SBM) in weanling diets with SDP or a combination of SDP and PDP could improve piglet performance and the expression of genes related to intestinal function (trial 2).

2. Materials and Methods

The experimental procedures used in the two trials were approved by the Ethical Committee on Animal Experimentation of the Autonomous University of Barcelona (CEAAH 3817), and are in full compliance with national legislation following the EU-Directive 2010/63/EU for the protection of animals used for scientific purposes.

2.1. Animals, Housing and Diet

2.1.1. Trial 1

The first trial was conducted in the weanling unit of a commercial farm throughout the pre-starter period (0–14 d post-weaning). A total of 180 male and female weaned commercial piglets ((Landrace × Large White) × Piétrain, weaned at 28 d) with a body weight (BW) of 7.5 ± SD 1.15 kg were moved to the nursery unit. These animals were not given previous access to creep feed during lactation. Piglets were distributed into two blocks according to initial BW (heavy piglets: 8.6 ± SD 0.03 kg; light piglets: 6.4 ± SD 0.02 kg). Each block contained 9 pens of 10 animals to which the three experimental treatments were randomly assigned (6 pens or replicates/treatment). Each pen (3.2 m^2 in floor area) had a commercial non-lidded hopper (TR5, Rotecna, Agramunt, Spain) and a nipple waterer to ensure ad libitum feeding and free water access.

Treatments consisted of three different iso-protein pre-starter diets: a control diet (CON) with a high content of soybean ingredients and two extra diets with partial replacement of the extruded soybeans by 2% SDP (AP 820 P, APC Europe S.L., Granollers, Spain) or 2% PDP (Palbio 62SP, Bioiberica S.A.U., Palafolls, Spain). The basal pre-starter diet was formulated to contain 2470 kcal of net energy (NE)/kg, 19.5% CP/kg and 1.28% digestible Lys (Table 1) to meet the requirements for maintenance and growth of newly weaned piglets [12]. Diets were presented in mash form and were fed ad libitum for fourteen consecutive days. No antimicrobials or ZnO were used in the experimental diets.

2.1.2. Trial 2

The second trial was conducted in a different commercial farm throughout the pre-starter (0–14 d post-weaning) and starter period (14–35 d post-weaning). A total of 198 male and female weaned commercial crossing piglets ((Landrace × Large White) × Piétrain, weaned at 21 d) with a BW of 5.7 ± SD 0.60 kg were moved to the nursery unit without transport to be used in the trial. Piglets had access to creep feed during lactation. Animals were distributed into two blocks according to initial BW (heavy piglets: 6.3 ± SD 0.02 kg; light piglets: 5.1 ± SD 0.01 kg) and each block contained 9 pens of 11 animals to which three experimental treatments were randomly assigned (6 replicates for each treatment). Each pen (3 m^2 in floor area) had a commercial non-lidded hopper (TR5, Rotecna) and a nipple waterer to ensure ad libitum feeding and free water access. The different characteristics of this farm compared to the farm of the first trial forced some changes to the experimental design of the experiment.

Treatments consisted of three different iso-protein pre-starter diets: a CON diet with a high content of SBM and two extra diets with partial replacement of the SBM by 3% spray-dried plasma (SDP; AP 820 P, APC Europe S.L.) or a combination of 1% spray-dried plasma and 2% porcine digestible peptides (SDP-PDP; Palbio 62SP, Bioiberica S.A.U.). The basal pre-starter diet was formulated to contain 2470 kcal NE/kg, 20.5% CP/kg and 1.35% digestible Lys (Table 2) and to meet the requirements for maintenance and growth of newly weaned piglets [13]. The pre-starter diets were fed ad libitum for fourteen consecutive days and a common starter diet was also fed ad libitum from 15 to 35 d post-weaning (Table 2). No antibiotics, alternative antimicrobials or ZnO were included in the diets.

2.2. Data and Sample Collection

Piglets were ear tagged and individually weighed at weaning (0 d) and 14 d post weaning in trial 1 and at weaning (0 d), 7 d, 14 d and 35 d post-weaning in trial 2. Feed disappearance from each hopper was measured throughout the experimental period. The average daily feed intake (ADFI), average daily gain (ADG) and feed efficiency (G:F) were calculated for the experimental period.

One piglet per pen was sedated with a combination of zolazepam/tiletamine and xylazine, and was euthanized with a pentobarbital injection on day 14 in trial 2. Portions of 0.5 cm of jejunum and ileum tissues were collected at 30 cm and 1.3 cm from the ileo-cecal valve respectively. Intestinal sections were rinsed in phosphate-buffered saline (PBS) and immediately snap frozen in 1 mL of RNAlater (Deltalab, Rubí, Spain). Samples were stored at −80 °C until analysis.

2.3. Proximate Analysis of Diets

Diet proximate analyses from both trials were performed following the Association of Official Agricultural Chemists methodology: dry matter (AOAC 934.01 [14]), ash (AOAC 942.05 [15]), ether extract (AOAC 2003.05 [15]) and crude protein (AOAC 968.06 [16]). Neutral-detergent fiber was determined according to the method of Van Soest et al. [17].

Table 1. Composition of the experimental diets used on trial 1, % as fed basis.

	Experimental Diets [1]		
Item	**CON**	**SDP**	**PDP**
Ingredient, %			
Maize	36.9	38.7	38.8
Wheat	16.0	16.0	16.0
Extruded Soybeans	15.0	11.4	11.2
Barley	13.0	13.0	13.0
Soybean meal 44% crude protein (CP)	6.8	6.8	6.8
Soybean protein concentrate 56% CP	5.6	5.6	5.6
Sweet milk whey	2.5	2.5	2.5
Spray-dried plasma 80% CP	-	2.0	-
Porcine digestible peptides 62% CP	-	-	2.0
Mono calcium phosphate	1.34	1.37	1.30
Calcium carbonate	0.62	0.64	0.67
L-Lysine HCl	0.55	0.49	0.52
Salt	0.53	0.40	0.16
Vitamin-Mineral premix [2]	0.40	0.40	0.40
DL-Methionine	0.27	0.23	0.57
L-Threonine	0.25	0.21	0.23
L-Valine	0.15	0.11	0.18
L-Tryptophan	0.09	0.08	0.10
Calculated composition			
Net Energy (NE), kcal/kg	2470	2470	2470
Dry Matter, %	89.1	89.0	89.1
Ash, %	5.4	5.3	5.4
Crude Protein, %	19.5	19.7	19.5
Calcium, %	0.650	0.655	0.650
Total P, %	0.678	0.671	0.671
Digestible amino acids			
Lys, %	1.280	1.280	1.280
Met, %	0.509	0.470	0.829
Cys, %	0.222	0.262	0.213
Met+Cys, %	0.768	0.768	1.078
Thr, %	0.832	0.832	0.832
Trp, %	0.282	0.282	0.282
Val, %	0.896	0.896	0.934
Analyzed composition			
Dry Matter, %	89.1	88.8	88.8
Ether Extract, %	5.1	4.2	4.3
Neutral Detergent Fiber, %	10.5	10.2	10.0
Crude Protein, %	17.9	18.2	18.4

[1] Experimental diets: CON: control diet; SDP: diet with 2% spray-dried plasma inclusion; PDP: diet with 2% porcine digestible peptides inclusion. [2] Supplied the following per kg of diet: 7000 IU of vitamin A (acetate); 500 IU of vitamin D3 (cholecalciferol); 250 IU of vitamin D (25-hydroxicholecalciferol); 45 mg of vitamin E; 1 mg of vitamin K3; 1.5 mg of vitamin B1; 3.5 mg of vitamin B2; 1.75 mg of vitamin B6; 0.03 mg of vitamin B12; 8.5 mg of D-pantothenic acid; 22.5 mg of niacin; 0.1 mg of biotin; 0.75 mg of folacin; 20 mg of Fe (chelate of amino acids); 2.5 mg of Cu (sulphate); 7.5 mg of Cu (chelate of glycine); 0.05 mg of Co (sulphate); 40 mg of Zn (oxide); 12.5 mg Zn (chelate of amino acids); 12.5 mg of Mn (oxide); 7.5 of Mn (chelate of glycine); 0.35 mg of I, 0.5 of Se (organic); and 0.1 mg of Se (inorganic).

2.4. Gene Expression Study by qPCR

Gene expression was quantified by RT-qPCR to study the expression of 56 genes in two intestinal tissues in an Open Array Real-Time PCR Platform (Applied Biosystems, Waltham, MA, USA) by the Servei Veterinari de Genètica Molecular at the Veterinary Faculty of the Universitat Autònoma de Barcelona (Spain).

The pre-amplified product was diluted 1:10 with 0.1× Tris-EDTA pH 8.0 and 6 μL was transferred to 384-well plates. These were analyzed in duplicate in Taqman Open Array gene expression plates custom-designed in a QuantStudio 12K Flex Real-Time PCR system (ThermoFisher Scientific, Waltham, MA, USA). One sample was used as an inter-plate control to check the replication of results from different plates.

Table 2. Composition of the experimental diets used on trial 2 (%, as fed basis).

	Experimental Diets [1]			
Item	**CON**	**SDP**	**SDP-PDP**	**Starter**
Ingredients, %				
Maize	29.68	32.53	32.11	29.79
Soybean Meal 47% crude protein (CP)	25.82	20.6	21.34	21.34
Wheat	16	16	16	15
Barley	6.5	6.5	6.5	20
Dextrose	6.5	6.5	6.5	-
Sweet Milk Whey	6.5	6.5	6.5	-
Potato Protein	2.5	2.5	2.5	-
Lard	2.67	2.38	2.31	6.53
Di-calcium phosphate	1.72	1.77	1.67	1.56
Spray-dried plasma (80% CP)	-	3	1	-
Porcine digestible peptides 62% PB	-	-	2	-
Salt	0.47	0.27	0.03	0.48
Vitamin-Mineral Premix [2]	0.40	0.40	0.40	0.40
L-Lysine HCL (78)	0.46	0.39	0.39	0.51
DL-Methionine	0.27	0.24	0.25	0.25
L-Threonine	0.21	0.16	0.17	0.24
Calcium Carbonate	0.11	0.11	0.17	0.55
L-Valine	0.11	0.09	0.08	0.14
L-Tryptophan	0.08	0.07	0.09	0.07
Calculated composition				
Net Energy (NE), kcal/kg	2470	2470	2470	2653
Ether Extract, %	4.69	4.43	4.39	8.61
Neutral Detergent Fiber, %	7.87	7.57	7.61	9.92
Crude Protein, %	20.51	20.50	20.50	18.50
Calcium, %	0.65	0.65	0.65	0.73
Total P, %	0.70	0.68	0.68	0.65
Digestible amino acids				
Lys, %	1.350	1.350	1.350	1.230
Met, %	0.545	0.504	0.539	0.478
Met + Cys, %	0.810	0.810	0.810	0.720
Thr, %	0.878	0.878	0.878	0.780
Trp, %	0.297	0.297	0.297	0.264
Analyzed composition				
Dry Matter, %	90.3	92.3	91.8	89.3
Ether Extract, %	4.6	4.3	4.1	8.4
Neutral Detergent Fiber, %	9.7	7.7	8.2	-
Crude Protein, %	19.7	18.7	19.6	17.7

[1] Experimental diets: CON: control diet; SDP: diet with 3% spray-dried plasma inclusion; SDP-PDP: diet supplemented with a combination of 1% spray-dried plasma and 2% porcine digestible peptides. [2] Supplied the following per kg of diet: 7000 IU of vitamin A (acetate); 500 IU of vitamin D3 (cholecalciferol); 250 IU of vitamin D (25-hydroxicholecalciferol); 45 mg of vitamin E; 1 mg of vitamin K3; 1.5 mg of vitamin B1; 3.5 mg of vitamin B2; 1.75 mg of vitamin B6; 0.03 mg of vitamin B12; 8.5 mg of D-pantothenic acid; 22.5 mg of niacin; 0.1 mg of biotin; 0.75 mg of folacin; 20 mg of Fe (chelate of amino acids); 2.5 mg of Cu (sulphate); 7.5 mg of Cu (chelate of glycine); 0.05 mg of Co (sulphate); 40 mg of Zn (oxide); 12.5 mg Zn (chelate of amino acids); 12.5 mg of Mn (oxide); 7.5 of Mn (chelate of glycine); 0.35 mg of I, 0.5 of Se (organic); and 0.1 mg of Se (inorganic).

2.5. Open Array Design

A list of 56 genes related to intestinal health were selected according to the bibliography [18–31] and included: (1) genes participating in the barrier function (*OCLN*, *ZO1*, *CLDN1*, *CLDN4*, *CLDN15*, *MUC2*, *MUC13* and *TFF3*); (2) genes that play an important role in the immune response, such as pattern recognition receptors, cytokines, chemokines and stress proteins (*TLR2*, *TLR4*, *IL1B*, *IL6*, *IL8*, *IL10*, *IL17A*, *IL22*, *IFNG*, *TNF*, *TGFB1*, *CCL20*, *CXCL2*, *IFNGR1*, *HSPB1*, *HSPA4*, *REG3G*, *PPARGC1A*, *FAXDC2* and *GBP1*); (3) genes coding for enzymes and hormones implicated in the digestion process (*GPX2*, *SOD2*, *ALPI*, *SI*, *DAO1*, *HNMT*, *APN*, *IDO1*, *GCG*, *CCK*, *IGF1R* and *PYY*); (4) genes involved in nutrient transport (*SLC5A1*, *SLC16A1*, *SLC7A8*, *SLC15A1*, *SLC13A1*, *SLC11A2*, *MT1A*, *SLC30A1* and *SLC39A4*); (5) genes involved in stress response (*CRHR1*, *NR3C1* and *HSD11B1*); and (6) four reference genes (*ACTB*, *B2M*, *GAPDH* and *TBP*).

Primers were designed spanning exon-exon boundaries or at different exons using PrimerExpress 2.0 software (Applied Biosystems) for 55 genes. The *IL8* gene primer was pre-designed by the company due to its complexity (Table A1). Possible residual genomic DNA amplification and primer dimer formation were controlled. Finally, a customized open array panel containing 56 genes was obtained.

2.6. RNA Extraction and cDNA Preparation

Total RNA was obtained from 100 mg of frozen intestinal tissues with the RiboPure kit (Ambion, Austin, TX, USA) following the manufacturer's protocol. RNA concentration and purity was calculated with a NanoDrop ND-1000 spectrophotometer (NanoDrop products, Wilmington, DE, USA). RNA integrity was checked with Agilent Bioanalyzer-2100 equipment (Agilent Technologies, Santa Clara, CA, USA) following the producer's protocol. One microgram of total RNA was reverse-transcribed into cDNA in a final volume of 20 μL. The High-Capacity cDNA Reverse Transcription kit (Applied Biosystems) and random hexamer primers were used, and the following thermal profile was applied: 25 °C, 10 min; 37 °C, 120 min; 85 °C, 5 s; 4 °C hold. A total of 25 ng of cDNA sample was pre-amplified using a 2× TaqMan PreAmp Master Mix and a 0.2× Pooled Taqman Gene Expression Custom Assays in a final volume of 10 μL. The thermal cycling conditions for the pre-amplification reactions were 10 min at 95 °C; 14 cycles of 15 s at 95 °C and 4 min at 60 °C; and a final step of 10 min at 99 °C. The pre-amplified cDNA product was stored at −20 °C until use.

2.7. Gene Expression Data Analysis

Gene expression data were collected and analyzed using the ThermoFisher Cloud software 1.0 (Applied Biosystems) applying the $2^{-\Delta\Delta Ct}$ method for relative quantification and using the sample with the lowest expression as a calibrator. Some parameters were adjusted: the maximum cycle relative threshold allowed was 26, amplification score < 1.240, quantification cycle confidence > 0.8 and the maximum standard deviation allowed between duplicates was set at <0.38. Samples that did not fit these requirements or had an inconclusive amplification status were deleted. Relative quantification values were checked for normalization by a $\log_{10}$ transformation, and all the statistical analyses were performed with R 3.4.3 software [32] and Bioconductor [33]. We carried out a one-way ANOVA and calculated the Benjamini–Hochberg false discovery rate (FDR q-value) to control multiple p-values [34], setting an upper bound for the expected proportion of false significant tests, that is, false significant treatment differences in mean expression levels between treatments. Pairwise post hoc treatment comparisons were carried out using Tukey's honest significant difference test [35]. Statistical differences between results for the treatments were identified at p-values and q-values under 0.05 for the ANOVA and Tukey's analysis and for the FDR, respectively.

A principal component analysis (PCA) was performed with samples as cases and gene $\log_{10}$-expressions as variables. The function PCA of the FactoMiner R-library [36] was used for dimension reduction and visualization in the first two principal dimensions. The variables factor map was restricted to genes showing cos2-qualities over 0.45 and significant differences. Finally, the heatmap visualization method was used to obtain double clustering both for genes (with a correlation-based distance:

$$d = (1 - r)/2, \tag{1}$$

where d is the distance and r is the correlation coefficient, and the complete linkage hierarchical clustering) and for samples (with Euclidean distance and Ward's D2 method). These methods were chosen based on the following: the Euclidian distance between two samples adds all the squared differences in the log-expression level of them in each gene, and then the Ward's D2 linkage method uses the Euclidean squared-distance to cluster in a way that minimizes the increment of the variance into the resulting clusters. Correlation based distance is preferred for obtaining the gene clusters because the expression levels in different genes may not be comparable (see Murtagh and Legendre [37] for Ward's method and Everitt [38] for clustering methods). The function heatmap.2 in the R-library gplots was used [39].

2.8. Performance Data Statistical Analysis

Production performance data were analyzed with ANOVA using the generalized linear model procedure of the statistical package SAS (version 9.4, SAS Inst. Inc., Cary, NC, USA). Normality and homoscedasticity were checked with Shapiro–Wilk test using the univariate procedure and Levene's test using the generalized linear model procedure, respectively. Data were analyzed taking the experimental treatment and the block of weight as the main factors. Their corresponding interaction was also included in the model. The statistical unit was the pen of 10 pigs in trial 1 and the pen of 11 pigs in trial 2. The results are presented as least square means taking into account the Tukey adjustment. The level of significance considered was $\alpha = 0.05$.

3. Results

3.1. Growth Performance

The productive performance of piglets during the first two weeks after weaning (trial 1) is summarized in Table 3. Animals fed SDP and PDP showed higher BW, ADG and G:F ($p < 0.05$) than piglets in the control group.

Table 3. Effect of the experimental treatments on growth performance of piglets in trial 1.

	Experimental Diets [3]				
Item [2]	**CON**	**SDP**	**PDP**	**SEM**	***p*-Value** [1]
BW, day 0, g	7505	7517	7534	12.0	0.241
day 14, g	9438 ^{b}	10,381 ^{a}	9990 ^{a}	138.9	0.001
ADFI 0–14 d, g/d	236 ^{b}	295 ^{a}	250 ^{b}	11.3	0.002
ADG 0–14 d, g/d	138 ^{b}	205 ^{a}	175 ^{a}	11.8	0.002
G:F 0–14 d	0.554 ^{b}	0.695 ^{a}	0.700 ^{a}	0.0413	0.034

[1] *p*-values come from the ANOVA test. [2] BW: body weight; ADFI: average daily feed intake; ADG: average daily gain; G:F: feed efficiency. [3] Experimental diets: CON: control diet; SDP: diet with 3% spray-dried plasma inclusion; SDP-PDP: diet supplemented with a combination of 1% spray-dried plasma and 2% porcine digestible peptides. ^{a,b} Different letters in the same row indicate significant statistical mean differences in the two diets (Tukey's test p-value < 0.05).

For trial 2, the productive performance results are shown in Table 4. Piglets of the SDP and SDP-PDP group showed greater BW, ADG and G:F than CON piglets at 14 d ($p < 0.05$). Although a numeric difference in BW was observed at 35 d between the supplemented animals and the CON group, no significant differences in BW, ADG or G:F were observed.

Table 4. Effect of the experimental treatments on growth performance of piglets in trial 2.

	Experimental Diets [3]				
Items [2]	**CON**	**SDP**	**SDP-PDP**	**SEM**	***p*-Value [1]**
BW, day 0, g	5724	5720	5721	46.9	0.859
day 7, g	5956 [b]	6344 [a]	6254 [ab]	88.3	0.023
day 14, g	7165 [b]	7894 [a]	7871 [a]	152.7	0.008
day 35, g	15,270	16,252	15,960	460.3	0.335
ADFI, 0–14 d, g	267	282	283	5.96	0.139
14–35 d, g	483	508	525	25.9	0.532
ADG, 0–14 d, g	103 [b]	155 [a]	153 [a]	10.7	0.007
14–35 d, g	364	398	382	22.0	0.571
G:F, 0–14 d	0.380 [b]	0.548 [a]	0.540 [a]	0.0383	0.046
14–35 d	0.761	0.782	0.732	0.0501	0.840

[1] *p*-values come from the ANOVA test. [2] BW: body weight; ADFI: average daily feed intake; ADG: average daily gain; G:F: feed efficiency. [3] Experimental diets: CON: control diet; SDP: diet with 3% spray-dried plasma inclusion; SDP-PDP: diet supplemented with a combination of 1% spray-dried plasma and 2% porcine digestible peptides. [a,b] Different letters in the same row indicate significant statistical mean differences in the two diets (Tukey's test *p*-value < 0.05).

3.2. Intestinal Mucosa Gene Expression Values

After performing qPCR of the intestinal tissue, 37 genes were detected as expressed in jejunum tissue and 27 genes in ileum tissue from the 54 target genes initially included in the open array panel. The conservation of one tissue sample from the jejunum and seven from the ileum was affected and amplification was not possible. Gene expression results of the ileum tissue were not considered due to the high amount of lost samples of this intestinal section.

The results of the analysis of gene expression in the jejunum intestinal tissue (trial 2) are shown in Table 5. Although only two genes showed statistical differences between groups ($p < 0.05$, $q < 0.05$), the other 8 genes showed a trend of being modified by the diets ($p < 0.05$, $q < 0.2$). The *CLDN15* and *TFF3* genes from the barrier function group and the *SLC11A2/DMT1* gene from the nutrient transport group showed a tendency to higher expression for the SDP-PDP treatment compared to the CON treatment ($p < 0.05$). From the immune response functional group, the SDP-PDP group showed a trend of having higher expression levels in three genes: *GBP1* compared to the SDP group, *IFNGR1* compared to the CON group, and *TLR4* compared to the other two groups ($p < 0.05$). Finally, the SDP-PDP group had a higher expression level than the CON and SDP groups in two enzyme-coding genes (*HNMT* and *APN*) that was only significant for *HNMT* ($q < 0.05$). The gene coding for a protein that metabolizes oxidation products, *SOD2*, also showed a significant increase when comparing the SDP-PDP group with the CON and SDP groups.

A PCA was carried out to determine the correlation in the gene expression values among samples distributed in the three diet groups (Figure 1). The sample identifier numbers are represented in the individual factor map (a) and are colored according to the diet assigned to each sample. The variables factor map (b) shows which genes are correlated along the samples. The 2D representation preserves 60% of the total variance (61.27). The results show a correlation in the gene expression pattern. In this tissue, eight of the nine significant genes (*APN, HNMT, SLC11A2, CLCN15, IFNGR1, TLR4, GBP1* and *SOD2*) are well represented in 2D (long arrows) and all of them fall in the first quadrant. Furthermore, the blue colored samples in the SDP-PDP group fall in the same quadrant or near to it, which indicates medium to high expression levels. Therefore, these plots clearly show a relationship between the highest expression levels in these significant genes and the SDP-PDP diet group. The CON and SDP groups are mixed and do not show high expression values.

Table 5. Effect of the experimental treatments on relative gene expression on jejunum mucosa after pre-starter period. Gene expression values are presented as ratios of cycle relative threshold value for each gene normalized to that of the reference sample. *p*-Values come from the ANOVA test and FDR is the false discovery rate.

Function	Genes [1]	Experimental Diets [2]			Contrast Statistic	*p*-Value	*q*-Value (FDR)
		CON	SDP	SDP-PDP			
Barrier function	CLDN15	0.51 ^{b}	0.81 ^{ab}	1.05 ^{a}	4.984	0.027	0.139
	TFF3	1.15 ^{b}	1.07 ^{ab}	1.83 ^{a}	4.290	0.035	0.145
Immune response	TLR4	1.54 ^{b}	1.31 ^{b}	2.86 ^{a}	8.052	0.005	0.059
	GBP1	2.40 ^{ab}	1.80 ^{b}	4.20 ^{a}	5.711	0.015	0.127
	IFNGR1	0.69 ^{b}	0.74 ^{ab}	1.10 ^{a}	4.661	0.028	0.139
Nutrient transport	SLC11A2/DMT1	0.37 ^{b}	0.71 ^{ab}	1.24 ^{a}	3.862	0.046	0.169
Enzyme/Hormone	HNMT	0.86 ^{b}	0.91 ^{b}	1.56 ^{a}	14.111	4×10^{-4}	0.015
	APN	1.59 ^{b}	1.40 ^{b}	3.40 ^{a}	4.577	0.030	0.139
Oxidation	SOD2	1.04 ^{b}	0.95 ^{b}	1.71 ^{a}	11.030	0.001	0.022

[1] Genes: CLDN15: claudin 15; TFF3: trefoil factor 3; SLC11A2/DMT1: solute carrier family 11 member 2/divalent metal transporter 1 (DMT1); TLR4: Toll like receptor 4; GBP1: guanylate binding protein; IFNGR1: interferon gamma receptor 1; HNMT: histamine N-methyltransferase; APN: alanyl aminopeptidase; SOD2: superoxide dismutase 2. [2] Experimental diets: CON: control diet; SDP: diet with 3% spray-dried plasma inclusion; SDP-PDP: diet supplemented with a combination of 1% spray-dried plasma and 2% porcine digestible peptides (2%). [a,b] Different letters in the same row indicate significant statistical mean differences in the two diets (Tukey's test *p*-value < 0.05).

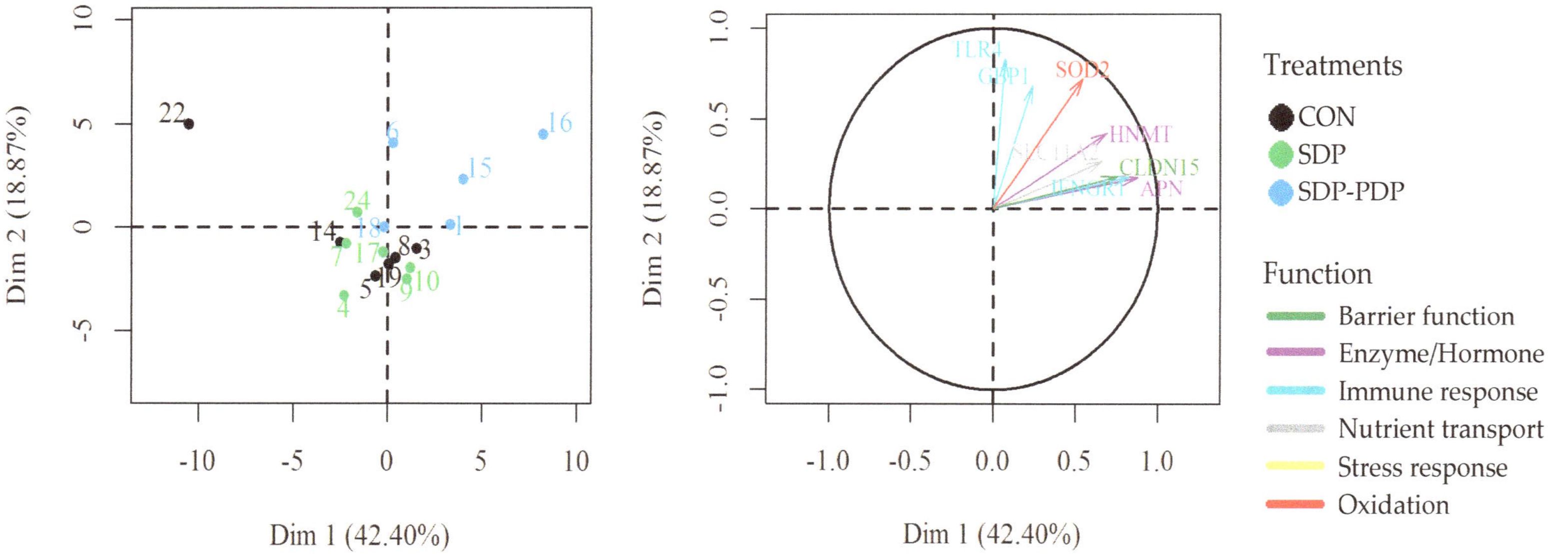

(a) (b)

Figure 1. Principal component analysis (PCA): (**a**) Samples' picture from jejunum (individuals factor map); (**b**) gene expressions arrow diagram from jejunum (variables factor map); CON: control diet; SDP: diet with 3% spray-dried plasma inclusion; SDP-PDP: diet supplemented with a combination of 1% spray-dried plasma and 2% porcine digestible peptides. Samples are labeled with different colors depending on the treatment, and the gene expressions arrows diagram on the right. Only those genes with $p < 0.05$ and having a long arrow have been depicted. Each arrow color indicates a different functional group. Trial 2.

A heatmap was also made in order to observe overall similarities among the gene expression profiles of the animals (Figure 2). On the heatmaps, the samples from the SDP-PDP treatment tend to group in pairs with others in the same group in the first step (the shortest branches). In addition, most SDP-PDP samples fall on one side of the dendrogram, the side with the highest number of red pixels in many genes, indicating a higher level of expression of these genes within the SDP-PDP group. Clustering of genes does not show any evidence of an association by functional group.

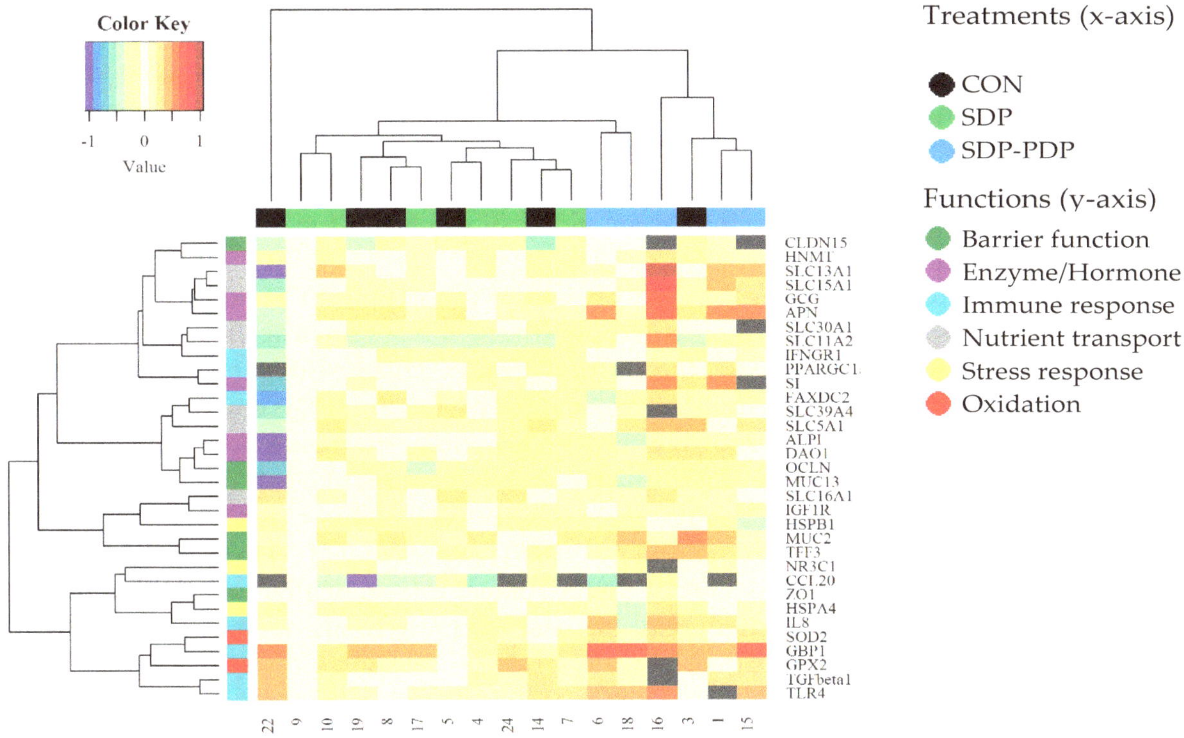

Figure 2. Heatmap representing the gene expression of each sample from jejunum in each gene in trial 2. CON: control diet; SDP: diet with 3% spray-dried plasma inclusion; SDP-PDP: diet supplemented with a combination of 1% spray-dried plasma and 2% porcine digestible peptides. Genes are organized in rows and samples in columns. Samples are labeled with different colors representing every treatment and genes depending on its functional group.

4. Discussion

This study found that substituting soybean ingredients with either 2–3% PDP, SDP or both to the diets of weaned piglets increased growth rate early after weaning. There are two main reasons that could be suggested and contribute to the observed results, such as the reduction of the negative impact of the soybean ingredients and the beneficial effects of the PDP and SDP. Unfortunately, our experimental design does not allow discrimination between them. The diets in the present study were formulated to contain a high inclusion of soybean ingredients, such as soybean meal 44% or 47% CP, extruded soybeans and soybean protein concentrate 56% CP. The β-sheet structures present in the secondary structure of raw legume proteins and the intermolecular β-sheet aggregates derived from heating are negatively correlated with feed digestibility values [40]. The β-sheet structures represent more than 30% of the secondary structure of soybean seeds, while they represent less than 10% in the ingredients of animal origin [41]. Consequently, the flow of protein into the distal parts of the gastrointestinal tract tends to be faster when more soybean ingredients are included, which promotes protein fermentation and selective growth of proteolytic bacteria [41]. Furthermore, soybeans contain anti-nutritional factors, such as antigens, trypsin inhibitors and lectins, which can produce digestive disorders and reduce nutrient availability [42,43]. The consequences of high inclusion of soybean ingredients, aggravated by the situation of post-weaning anorexia, can lead to intestinal disorders, such as post-weaning diarrhea [2,44]. This can negatively affect the animal growth performance. In the present study, the reduction of soy ingredients accounts for 3–5% of the diet, being replaced by either SDP, PDP, or both. In contrast, proteins of animal origin as well as animal protein hydrolysates have higher palatability than vegetal proteins [10,45,46], which can translate into an increased feed intake after weaning.

The substitution of soybean ingredients with SDP improved the productive performance at the end of pre-starter period of both trials. The results reported herein are similar to other studies that showed that SDP improved performance, especially when piglets were challenged with experimental infection and did not receive in-feed medication [47]. Spray-dried plasma is commonly used because it can stimulate feed intake due to its palatability [48], as observed in trial 1. Its beneficial effects are explained by the preservation of blood immunoglobulins, growth factors and bioactive peptides or compounds during the spray-drying process. These components can interact with the gut-associated lymphoid tissue [49], therefore preserving the small intestinal barrier function and reducing intestinal inflammation and damage [6].

Consumption of feed containing the hydrolysates, PDP or SDP-PDP, enhanced the productive performance of piglets compared to their corresponding control group at the end of the pre-starter period, and these groups were equivalent to the SDP group. This result is in accordance with other authors who also obtained similar growth performance by feeding PDP and SDP in early-weaned piglets [50]. Some research data have shown improved growth performance, feed intake and efficiency of animals fed PDP compared to other high-quality protein sources like fish meal [45]. Other studies found that the inclusion of PDP in weanling diets improved villus height of the small intestine compared to some sources of intact protein like SDP [51] or fish meal [11], which can be considered a good indicator of nutrient uptake. Part of the beneficial effects of PDP on growth performance could be due to its content of short-chain peptides that are more easily absorbed by pigs than intact proteins [52] or even free amino acids [53].

This is the first study performed using gene expression to provide some inputs about the effects of PDP on intestinal function. The outcome of the PCA individual factor map of the gene expression shows that the factorial scores of the SDP-PDP group tended to be closer to each other and more separate from the CON and SDP group scores. In addition, the correspondence of most of the arrows representing the significant genes in the PCA variables map and most of the samples in the SDP-PDP group in a similar position indicates that there is a relationship between the higher expression levels in these genes and the differences that the SDP-PDP group showed from the CON and SDP groups. Heatmap representation also helped to visualize a partial clustering of the samples from the SDP-PDP treatment in the jejunum and ileum. In line with this, the statistical ANOVA and Tukey's test showed a stronger effect of the SDP-PDP diet than SDP with respect to the CON diet. Although all results agree, ANOVA and Tukey's test deal with mean treatment values while heatmaps and PCA-plots reflect individual performance. The combination of inferential treatment comparison techniques (ANOVA and Tukey, essentially) and powerful exploratory methods (PCA and heatmap) provide a clearer and dual (individuals and genes) idea of the differences in gene expression.

Proteins coded by the genes analyzed in the current study participate in the barrier function of gut cells, in nutrient transport in the mucosa, in digestion, in the immune response and in the metabolization of oxidation products.

Some trends in the jejunum gene expression due to diets may indicate a potentiation of the epithelial structure by SDP-PDP, because, for example, the TFF3 gene participates in epithelial restitution and maintenance of intestinal mucosa integrity [24]. However, the CLDN15 gene, which expression was also modified, codes for a pore-forming protein [22], which is important for the normal-sized morphogenesis of the small intestine [25].

Diets also had an effect on the expression in the jejunum of genes related to the immune response and metabolization of oxidation products. Again, the SDP-PDP diet showed more changes than the SDP diet alone. Expression of immune response genes TLR4, IFNGR1 and GBP1 showed a trend to increase in the SDP-PDP group, TLR4 compared to the CON and SDP groups; IFNGR1 compared to the CON group and GBP1 compared to the SDP group.

TLR4 is a receptor involved in the recognition of lipopolysaccharide, a major cell wall component of Gram-negative bacteria [19], and IFNGR1 is part of the receptor that mediates the biological effects of IFN-γ [54]. The TLR4 and IFNGR1 genes have been reported to be upregulated in animals under stress and with infectious conditions in order to activate the innate immune response and fight against pathogens properly [24,28,55]. Thus, the tendency for upregulation of TLR4 and IFNGR1 might suggest that piglets fed with SDP-PDP seem to be more prepared for controlling infective processes and other intestinal challenges that can occur during the weaning period. In addition, GBP1 is a GTPase that regulates the inhibition of proliferation and invasion of endothelial cells. It protects against epithelial apoptosis induced by inflammatory cytokines and subsequent loss of the barrier function [27]. Upregulation of GBP1 by the SDP-PDP diet, although without being statistically significant, is probably related to IFNGR1 upregulation because GBP1 expression is strongly induced by IFN-γ [27], although an upregulation of IFN-γ was not observed in this study.

Focusing on nutrient transport, only an up-regulatory trend in the expression of the SLC11A2/DMT1 gene was found in animals fed with the SDP-PDP combination compared to the CON group. The divalent metal transporter (SCL11A2/DMT1) is located on the apical surface of the enterocyte and is involved in the intestinal Fe uptake [18]. SLC11A2/DMT1 gene expression is upregulated in circumstances of low Fe intake [26] and hyperglycemia conditions [56]. We have no evidence of differences in the Fe content or glycemic levels among experimental diets; therefore, the reason why the SDP-PDP treatment influenced the expression of SLC11A2/DMT1 should be researched further.

The enzyme-coding gene HNMT was upregulated and the gene APN showed a trend to increase due to the SDP-PDP diet. Kröger et al. [31] reported that high dietary inclusion of fermentable CP increased the HNMT expression in the colon, which is a histamine-degrading enzyme. They determined that the histamine catabolism activity of HNMT counter-regulated the increased production of this biogenic amine, reducing the fecal score of the piglets fed with a high fermentable CP diet. Considering that all diets had an elevated CP level, an increase in this enzyme could show that SDP-PDP was attenuating the inflammatory effects of histamine more efficiently than other groups. On the other hand, APN is a Zn-dependent enzyme that takes part in the final digestion of peptides [56]. Its upregulation in the small intestine has been documented with products considered beneficial for intestinal health, such as the probiotic *Lactococcus lactis* [23]. Therefore, its increase due to the SDP-PDP diet may also be considered a positive physiological change.

Regarding the antioxidant defense mechanisms, we observed here that the expression of the SOD2 gene was also increased in the SDP-PDP group compared to both the SDP and CON groups. This mitochondrial enzyme is considered the first defense against reactive oxygen species (ROS) formed during normal cell metabolism [30]. Elimination of ROS by SOD2 can be considered as an anti-inflammatory effect due to the important role that ROS plays in triggering and promoting inflammation [57]. As well as the TLR4 gene, expression of SOD2 is stimulated by lipopolysaccharides but cytokines or ROS can also upregulate it [58]. Thus, changes in SOD2 gene expression can be derived from the TLR4 upregulation also induced by the SDP-PDP intervention.

The gene expression results of this trial might indicate that the combination of PDP and SDP had some effects on intestinal function, although most of the differences found among treatments only could be considered as a trend ($p < 0.05$, $q < 0.2$). First of all, the tendency for upregulation of TFF3 and CLDN15 could be considered a potentiation of the epithelial structure of the gut, which would be related to the increase of villi height observed in previous studies with PDP [51]. Secondly, we could speculate that animals fed PDP-SDP might be more prepared for controlling infective processes and defending against other hazardous processes due to the up-regulatory trend in the expression of the immune response genes (TLR4, IFNGR1 and GBP1) and the upregulation of genes related to the degradation of toxic products of metabolism (SOD2 and HNMT). In addition, the trend to increase the expression of the APN gene could suggest an improvement in the digestion of protein. The underlying mechanisms that produced these effects are still unknown and the literature exploring possible functional effects of PDP is still scarce. More research is needed to confirm that gene expression upregulation by PDP in combination with SDP has an impact on intestinal function and to elucidate the underlying mechanisms that are responsible of these effects. Furthermore, it might be interesting to investigate if the effects on gene expression observed in this study are only attributable to the addition of PDP or if it is a synergy of PDP with SDP that is producing the effects, as SDP alone did not show them.

5. Conclusions

The present study suggests that substituting soybean products with animal protein sources like PDP or SDP increases growth performance of weanling piglets at the end of the pre-starter period. In addition, it indicates that PDP can substitute or complement SDP because it showed the same effect on the growth performance of the piglets during this period. Furthermore, changes in the gene expression of the jejunum produced by the SDP-PDP diet suggest that this treatment might be able to produce beneficial effects in the epithelial structure of the gut, in the defensive capacity of the intestine against threats and the digestion of proteins. More research is needed to confirm that gene expression upregulation by PDP in combination with SDP has an impact on intestinal function and to elucidate the underlying mechanisms that are responsible for these effects.

Author Contributions: Conceptualization, D.S.-O., J.F.P. and E.B.; methodology, J.F.P. and D.S.-O.; software, M.F.; validation, J.F.P., F.J.P.-C. and D.S.-O.; formal analysis, M.F.; investigation, F.G.-S., L.C.-M., C.V., W.C.G. and J.M.F.; resources, J.M.F.; data curation, L.C.-M., M.F. and F.G.-S.; writing—original draft preparation, F.G.-S.; writing—review and editing, D.S.-O., J.F.P., F.J.P.-C. and E.B.; visualization, F.G.-S.; supervision, D.S.-O.; project administration, J.F.P.; funding acquisition, E.B. All authors have read and agreed to the published version of the manuscript.

Acknowledgments: The authors thank the ownership of the commercial farm for housing animals and providing production records.

Appendix A

Table A1. List of primers used for the analyses of gene expression of the 56 genes by Open Array Real-Time PCR custom designed.

Gene	Name	Primer Forward (5′-3′)	Primer Reverse (5′-3′)	Probe (5′-3′)
OCLN	Occludin	CAGGTGCACCCTCCAGATTG	CAGGCCTATAAGGAGGTGGACTT	TGACATCAGCCATGTCAT
ZO1	Zonula occludens 1	GCTATGTCCAGAATCTCGGAAAA	TGCTTCTTTCAATGCTCCATACC	TCACCATCTTTTTACAACTAC
CLDN1	Claudin-1	CTTCGACTCCTTGCTGAATCTGA	CTTCCATGCACTTCATACACTTCAT	ACAGCACTTTGCAAGC
CLDN4	Claudin-4	CCTCCGTGCTGTTCCTCAA	GAGGCACAAGCCCAGCAA	CCTTGTGGCACTTTG
CLDN15	Claudin-15	GCTATCTCCTGGTATGCCTTCAA	GGGACTTCCACACTCCTTGGT	ACTTCTTCGACCCCTTGTA
MUC2	Mucin 2	AAGGACGACACCATCTACCTCACT	GGCCAGCTCGGGAATAGAC	CATGGTCAGCACCCCG
MUC13	Mucin 13	CAGTGGAGTTGGCTGTGAAAAC	ATCAAGTTCTGTTCTTCCACATTCTTG	TCCTCTCATTAAGATCAAAC
TFF3	Trefoil factor 3	AGAACCTGCCCGTGACCAT	CACACTGGTTCGCCGACAG	AGGCCAGGATGTTCT
TLR2	Toll-like receptor 2	CTCTCGTTGCGGCTTCCA	AAGACCCATGCTGTCCACAAA	CAAGGTCAACTCTCTG
TLR4	Toll-like receptor 4	CATCCCCACATCAGTCAAGATACT	TCAATTGTCTGAATTTCACATCTGG	ACAGCAATAGCTTCTCCA
IL1B	Interleukin 1 beta	GGTGACAACAATAATGACCTGTTATTTG	GCTCCCATTTCTCAGAGAACCA	ATGAAGTGCTGCACCC
IL6	Interleukin 6	CCAATCTGGGTTCAATCAGGAG	ACAGCCTCGACATTTCCCTTATT	AGATATACCTGGACTACCTC
IL8	Interleukin 8	GGAAAAGTGGGTGCAGAAGGT	GAGAATGGGTTTTTGCTTGTTGT	TACAGATATTTTTGAAGAGAACT
IL10	Interleukin 10	TGAGGCTGCGGCGCT	GAGCTTGCTAAAGGCACTCTTCA	AACAAGAGCAAGGCCGT
IL17A	Interleukin 17	CCAGACGGCCCTCAGATTAC	ATCTTCCTTCCCTTCAGCATTG	CCATGGACTCTCCAACG
IL22	Interleukin 22	TGTTCCCCAACTCTGATAGATTCC	GTTGTTCACATTTCTCTGGATATGCT	AGCTAAGCCAATGCCGTAT
IFNG	Interferon gamma	TGACTTTGTGTTTTTCTGGCTCTT	CACTCTCCTCTTTCCAATTCTTCAA	ATCCTAAAGGACTATTTTAAT
TNF	Tumor necrosis factor alpha	CACCACGCTCTTCTGCCTACT	GACGGGCTTATCTGAGGTTTGA	CAAGGACTCAGATCATCGT
TGFB1	Transforming growth factor beta 1	GCGGCAGCTCTACATTGACTT	GACCTTGCTGTACTGAGTGTCTAGG	CCATGCCAATTTCTGCCT
CCL20	Chemokine (C-C motif) ligand 20	GACCATATTCTTCACCCCAGATTT	CACACACGGCTAACTTTTTCTTTG	ATCAATGCAATCATCTTT

Table A1. *Cont.*

Gene	Name	Primer Forward (5′-3′)	Primer Reverse (5′-3′)	Probe (5′-3′)
CXCL2	Chemokine (C-X-C motif) ligand 2	CATGGTGAAGAAAATCATCGAGAA	GCCAGTAAGTTTCCTCCATCTCTCT	AACAAGAGCAGTGCCAAC
IFNGR1	Interferon gamma receptor 1	CATGTTACCCAAATCTTTGCTGTCT	CAGTATGCACGCTTGAAATTGTC	ATATATATCACCCATCACCTACC
HSPB1	Heat shock protein 27	CGAGGAGCTGACGGTCAAG	GCAGCGTGTATTTTCGAGTGAA	ACGGCTTCATTTCCCGGT
HSPA4	Heat shock protein 70	TCAATTGCCTGCGATTAATGAA	GAATGCCCCATGTCTACAAAAAC	CAGTTGCTCTTGCATATG
REG3G	Regenerating-islet derived protein 3 gamma	TGCCTGATGCTCCTGTCTCA	GGCATAGCAGTAGGAAGCATAGG	CCAAGGTGAAGATTC
PPARGC1A	Peroxisome proliferative activated receptor gamma, coactivator 1 alpha	CTCTGGAACTGCAGGCCTAA	TGGAGAAGCCCTAAAAGGGTTAT	ACCCACAACTCCTCCT
FAXDC2	Fatty acid hydrolase domain containing 2	CCATGACTACCACCATCTCAAGTT	GCAGGATCGTGTGTCTCTCGTA	TGTTCAAGCAGACCAAG
GBP1	Guanylate binding protein 1	AGAATCCATCACAGCAGACGAGTA	CGGATACAGAGTCGAGGCAGGTTAA	TCAAGCTTAAGAAGGGTACCAG
GPX2	Glutathione peroxidase 2	CAACCAATTTGGACATCAGGAG	GGGTAAAGGTGGGCTGGAAT	AGATCCTGAACAGCCTCA
SOD2	Superoxide dismutase	GGGTTGGCTCGGTTTCAA	CATGCTCCCACACGTCGAT	CTGCAAGGAACAACAGGTCT
ALPI	Intestinal alkaline phosphatase	ATGTCTTCTCTTTTGGTGGCTACA	GGAGGTATATGGCTTGAGATCCA	AAGCTCCGTTTTTGGCCT
SI	Sucrase-isomaltase	CGACCCCTTTTGCATGAGTT	AAGGCTGGACCCCATAGGAA	TTTAATGAAAAGCCAACCTG
DAO1	Diamine oxidase	GGAACCAACAGACCTTCAACTATCTC	TTCGGAATCCCAGGACCAT	CCGGACCCTTACTGGAAA
HNMT	Histamine N-methyltransferase	TGTTGAACCAAGTGCTGAACAAAT	ACTTTATGTTCTCGAGGTTTGATGTCTT	ACCAAGTACAAAGAGCTT
APN	Aminopeptidase-N	AGGGCAACGTCAAAAAGGTG	GTCAAAGCATGGGAAGGATTTC	ACACAGATGCAGTCTACAG
IDO1	Indoleamine 2,3 dioxygenase	TTGGCAAATTGGAAGAAAAAGG	CCGGAAATGAGAAGAGAATATCCAT	CCAGTGGGCCCATGACTTAC
GCG	Glucagon	AGGCGTGCCCAGGATTTT	CATCGTGACGTTTGGCAATG	CACCAAGAGGAACAAGAA
CCK	Cholecystokinin	CAGCAGGCTCGAAAAGCAC	AATCCATCCAGCCCATGTAGTC	CAGCCACAGAATAAGTGA

Table A1. *Cont.*

Gene	Name	Primer Forward (5′-3′)	Primer Reverse (5′-3′)	Probe (5′-3′)
IGF1R	Insulin-like growth factor 1 receptor	CCGACGCGGCAACAAC	TCAGGAAGGACAAGGAGACCAA	CTACGTGAAGATCCGCCA
PYY	Peptide tyrosine tyrosine	CAGAGGTATGGAAACGTGACA	CCTTCTGGCCACGACTTGAC	CAAACTGCTCTTCCCTGAA
SLC5A1	Solute carrier family 5 (sodium/glucose cotransporter) member 1	GGCCATCTTTCTCTTACTGGCA	TCCCACTTCATGAAAAGCAAAC	TTTATACGGATACCTTGCAGAC
SLC16A1	Monocarboxylate transporter 1	CCTTGTTGGACCTCAGAGATTCTC	CCAGTATGTGTATTTATAGTCTCCGTATATG	CCTACCACTTTTAGGTCGTC
SLC7A8	Solute carrier family 7 (amino acid transporter light chain, L System) member 8	TGTCGCTTATGTCACTGCAATGT	GACAGGGCGACGGAAATG	CTGTGACTTTTGGAGAGAA
SLC15A1	Solute carrier family 15 (oligopeptide transporter) member 1	GGTTATCCCTTGAGCATCTTCTTC	AGTGCTCTCATTCCATAGTAGGAAAA	TCAACGAGTTCTGTGAAAG
SLC13A1	Solute carrier family 13 (sodium/sulfate symporters) member 1	GGTACCTCCACCAACTTGATCTTC	ATCCAAAGTTGATGCAGTGACAAT	ATTTCAATATGCGCTACCC
SLC11A2	Solute carrier family 11 (proton-coupled divalent metal ion transporter) member 2	GTCTTTGCCGAAGCGTTTTTT	ACCACGCCCCCTTTGTAGA	CCAACCAGCAGGTGGT
MT1A	Metallothionein 1A	TGAATCCGCGTTGCTCTCT	CAGGAGCAGCAGCTCTTCTT	ACGTGCAAAACCTGCAGA
SLC30A1	Solute carrier family 30 (zinc transporter) member 1	AATTGGACCGGACAGATCCA	TCTCTGATAAGATTCCCATTCACTTG	AAAAGTCCAGAAGTGATGC
SLC39A4	Solute carrier family 39 (zinc transporter) member 4	ATCTTTGGGCTCTTGCTCCTT	GCAGCCCCAGCACCTTAG	CTGCTACCCACTACGTCA
CRHR1	Corticotropin releasing hormone receptor 1	CAGGGCCCCATGATATTGG	CCGGAGTTTGGTCATGAGGAT	CTGATCAACTTTATCTTCC

Table A1. *Cont.*

Gene	Name	Primer Forward (5′-3′)	Primer Reverse (5′-3′)	Probe (5′-3′)
NR3C1	Glucocorticoid receptor	GGCAATACCAGGATTCAGGAACT	CCATGAGAAACATCCATGAATACTG	TGACCAAATGACCCTCCT
HSD11B1	Hydroxysteroid (11-beta) dehydrogenase 1	GGTCAGAAGAAACTCTCAAGAAGGTG	GCGAAGGTCATGTCCTCCAT	TCTTCAGCACACTACGTTG
GAPDH	Glyceraldehyde-phosphate-dehydrogenase	TTCGTCAAGCTCATTTCCTGGTA	TCCTCGCGTGCTCTTGCT	AATTTGGCTACAGCAACAG
ACTB	Actin, beta	CAAGGACCTCTACGCCAACAC	TGGAGGCGCGATGATCTT	CACCACCATGTACCCAGG
TBP	TATA-box binding protein	CAGAATGATCAAACCGAGAATTGT	CTGCTCTGACTTTAGCACCTGTTAA	TTTGTCTCTGGAAAAGTTGT
B2M	Beta-2-microglobulin	TCACTCCTAACGCTGTGGATCA	CGGTTAGTGGTCTCGATCCC	AGCACGTGACTCTCGATA

References

1. Lallès, J.P.; Bosi, P.; Smidt, H.; Stokes, C.R. Weaning—A Challenge to Gut Physiologists. *Livest. Sci.* **2007**, *108*, 82–93. [CrossRef]
2. Domeneghini, C.; Di Giancamillo, A.; Arrighi, S.; Bosi, G. Gut-Trophic Feed Additives and Their Effects upon the Gut Structure and Intestinal Metabolism. State of the Art in the Pig, and Perspectives towards Humans. *Histol. Histopathol.* **2006**, *21*, 273–283. [CrossRef] [PubMed]
3. Gresse, R.; Chaucheyras-Durand, F.; Fleury, M.A.; Van de Wiele, T.; Forano, E.; Blanquet-Diot, S. Gut Microbiota Dysbiosis in Postweaning Piglets: Understanding the Keys to Health. *Trends Microbiol.* **2017**, 851–873. [CrossRef] [PubMed]
4. Pluske, J.R. Feed- and Feed Additives-Related Aspects of Gut Health and Development in Weanling Pigs. *J. Anim. Sci. Biotechnol.* **2013**, *4*, 1. [CrossRef]
5. Bhat, Z.F.; Kumar, S.; Bhat, H.F. Bioactive Peptides of Animal Origin: A Review. *J. Food Sci. Technol.* **2015**, *52*, 5377–5392. [CrossRef] [PubMed]
6. Bosi, P.; Casini, L.; Finamore, A.; Cremokolini, C.; Merialdi, G.; Trevisi, P.; Nobili, F.; Mengheri, E. Spray-Dried Plasma Improves Growth Performance and Reduces Inflammatory Status of Weaned Pigs Challenged with Enterotoxigenic Escherichia Coli K88. *J. Anim. Sci* **2004**, *82*, 1764–1772. [CrossRef]
7. Pérez-Bosque, A.; Miró, L.; Polo, J.; Russell, L.; Campbell, J.; Weaver, E.; Crenshaw, J.; Moretó, M. Dietary Plasma Proteins Modulate the Immune Response of Diffuse Gut-Associated Lymphoid Tissue in Rats Challenged with Staphylococcus Aureus Enterotoxin B. *J. Nutr.* **2018**, *138*, 533–537. [CrossRef]
8. Pérez-Bosque, A.; Amat, C.; Polo, J.; Campbell, J.M.; Crenshaw, J.; Russell, L.; Moretó, M. Spray-Dried Animal Plasma Prevents the Effects of Staphylococcus Aureus Enterotoxin B on Intestinal Barrier Function in Weaned Rats. *J. Nutr.* **2018**, *136*, 2838–2843. [CrossRef]
9. Figueroa, J.; Solà-Oriol, D.; Guzmán-Pino, S.A.; Chetrit, C.; Borda, E.; Pérez, J.F. The Use of Porcine Digestible Peptides and Their Continuity Effect in Nursery Pigs. *J. Anim. Sci.* **2016**, *94*, 1531–1540. [CrossRef]
10. Solà-Oriol, D.; Roura, E.; Torrallardona, D. Feed Preference in Pigs: Effect of Selected Protein, Fat, and Fiber Sources at Different Inclusion Rates. *J. Anim. Sci.* **2011**, *89*, 3219–3227. [CrossRef]
11. Xin, G.; QiuGang, M.; Cheng, J.; ShaoChen, G.; Keller, G. Evaluation of DPS Supplement in Early-Weaned Pig Diets. *Acta Zoonutrimenta Sin.* **2001**, *13*, 15–19.
12. FEDNA. *Necesidades Nutricionales Para Ganado Porcino*, 2nd ed.; FEDNA: Madrid, Spain, 2013.
13. National Research Council. *Nutrient Requirements of Swine*, 11th ed.; National Academies Press: Washington, DC, USA, 2012.
14. AOAC. *Official Methods of Analysis of the AOAC*, 15th ed.; Association of Official Analytical Chemists: Washington, DC, USA, 1990.
15. AOAC International. *Official Methods of Analysis of AOAC International*, 18th ed.; Official Methods of Analysis of AOAC International: Gaithersburg, MI, USA, 2007.
16. AOAC International. *Official Methods of Analysis of AOAC International*, 18th ed.; Official Methods of Analysis of AOAC International: Washington, DC, USA, 2005.
17. Van Soest, P.J.; Robertson, J.B.; Lewis, B.A. Methods for Dietary Fiber, Neutral Detergent Fiber, and Nonstarch Polysaccharides in Relation to Animal Nutrition. *J. Dairy Sci.* **1991**, *74*, 3583–3597. [CrossRef]
18. Garrick, M.D.; Dolan, K.G.; Horbinski, C.; Ghio, A.J.; Higgins, D.; Porubcin, M.; Moore, E.G.; Hainsworth, L.N.; Umbreit, J.N.; Conrad, M.E.; et al. DMT1: A Mammalian Transporter for Multiple Metals. *BioMetals* **2003**, *16*, 41–54. [CrossRef] [PubMed]
19. Takeda, K.; Kaisho, T.; Akira, S. Toll-like Receptors. *Annu. Rev. Immunol.* **2003**, *21*, 335–376. [CrossRef]
20. Liu, P.; Pieper, R.; Rieger, J.; Vahjen, W.; Davin, R.; Plendl, J.; Meyer, W.; Zentek, J. Effect of Dietary Zinc Oxide on Morphological Characteristics, Mucin Composition and Gene Expression in the Colon of Weaned Piglets. *PLoS ONE* **2014**, *9*, e91091. [CrossRef]
21. Smith, S.H.; Wilson, A.D.; Van Ettinger, I.; Macintyre, N.; Archibald, A.L.; Ait-Ali, T. Down-Regulation of Mechanisms Involved in Cell Transport and Maintenance of Mucosal Integrity in Pigs Infected with Lawsonia Intracellularis. *Vet. Res.* **2014**, *45*. [CrossRef]
22. Khan, N.; Asif, A.R. Transcriptional Regulators of Claudins in Epithelial Tight Junctions. *Mediat. Inflamm.* **2015**, *2015*. [CrossRef]

23. Xu, S.; Wang, D.; Zhang, P.; Lin, Y.; Fang, Z.; Che, L.; Wu, D. Oral Administration of *Lactococcus lactis*-Expressed Recombinant Porcine Epidermal Growth Factor Stimulates the Development and Promotes the Health of Small Intestines in Early-Weaned Piglets. *J. Appl. Microbiol.* **2015**, *119*, 225–235. [CrossRef]
24. Taupin, D.; Podolsky, D.K. Erratum: Trefoil Factors: Initiators of Mucosal Healing. *Nat. Rev. Mol. Cell Biol.* **2003**, *4*, 721–732. [CrossRef]
25. Tamura, A.; Kitano, Y.; Hata, M.; Katsuno, T.; Moriwaki, K.; Sasaki, H.; Hayashi, H.; Suzuki, Y.; Noda, T.; Furuse, M.; et al. Megaintestine in Claudin-15-Deficient Mice. *Gastroenterology* **2008**, *134*. [CrossRef]
26. Hansen, S.L.; Trakooljul, N.; Liu, H.-C.; Moeser, A.J.; Spears, J.W. Iron Transporters Are Differentially Regulated by Dietary Iron, and Modifications Are Associated with Changes in Manganese Metabolism in Young Pigs. *J. Nutr.* **2009**, *139*, 1474–1479. [CrossRef] [PubMed]
27. Schnoor, M.; Betanzos, A.; Weber, D.A.; Parkos, C.A. Guanylate-Binding Protein-1 Is Expressed at Tight Junctions of Intestinal Epithelial Cells in Response to Interferon-Gamma and Regulates Barrier Function through Effects on Apoptosis. *Mucosal Immunol.* **2009**, *2*, 33–42. [CrossRef] [PubMed]
28. Badia, R.; Lizardo, R.; Martinez, P.; Badiola, I.; Brufau, J. The Influence of Dietary Locust Bean Gum and Live Yeast on Some Digestive Immunological Parameters of Piglets Experimentally Challenged with Escherichia Coli1. *J. Anim. Sci.* **2012**, *90*, 260–262. [CrossRef] [PubMed]
29. Zhou, C.; Liu, Z.; Jiang, J.; Yu, Y.; Zhang, Q. Differential Gene Expression Profiling of Porcine Epithelial Cells Infected with Three Enterotoxigenic Escherichia Coli Strains. *BMC Genom.* **2012**, *13*, 330. [CrossRef] [PubMed]
30. Chmielewska, M.; Łosiewicz, K.; Podlasz, P.; Wasowicz, K. The Expression of Mitochondrial, Cytoplasmic and Extracellular Superoxide Dismutase in the Colonic Wall of Pigs Suffering from Swine Dysenteria. *Pol. J. Vet. Sci.* **2013**, *16*, 463–467. [CrossRef]
31. Kröger, S.; Pieper, R.; Schwelberger, H.G.; Wang, J.; Villodre Tudela, C.; Aschenbach, J.R.; Van Kessel, A.G.; Zentek, J. Diets High in Heat-Treated Soybean Meal Reduce the Histamine-Induced Epithelial Response in the Colon of Weaned Piglets and Increase Epithelial Catabolism of Histamine. *PLoS ONE* **2013**, *8*, e80612. [CrossRef]
32. R Development Core Team. *R: A Language and Environment for Statistical Computing. [2.12.2]*; R Foundation for Statistical Computing: Vienna, Austria, 2013.
33. Gentleman, R.C.; Carey, V.J.; Bates, D.M.; Bolstad, B.; Dettling, M.; Dudoit, S.; Ellis, B.; Gautier, L.; Ge, Y.; Gentry, J.; et al. Bioconductor: Open Software Development for Computational Biology and Bioinformatics. *Genome Biol.* **2004**, *5*. [CrossRef] [PubMed]
34. Benjamini, Y.; Hochberg, Y. Controlling the False Discovery Rate. *J. R. Stat. Soc. B* **1995**, 289–300. [CrossRef]
35. Tukey, J.W. Comparing Individual Means in the Analysis of Variance. *Biometrics* **1949**, *5*, 99. [CrossRef]
36. Lê, S.; Josse, J.; Husson, F. FactoMineR: An R Package for Multivariate Analysis. *J. Stat. Softw.* **2008**, *25*, 1–18. [CrossRef]
37. Murtagh, F.; Legendre, P. Ward's Hierarchical Agglomerative Clustering Method: Which Algorithms Implement Ward's Criterion? *J. Classif.* **2014**, *31*, 274–295. [CrossRef]
38. Everitt, B. *Cluster Analysis*; Heinemann Educational Books: London, UK, 1974. [CrossRef]
39. Warnes, G.; Bolker, B.; Bonebakker, L.; Gentleman, R.; Liaw, W.; Lumley, T.; Maechler, M.; Magnusson, A.; Moeller, S.; Schwartz, M.; et al. *Gplots: Various R Programming Tools for Plotting Data. R Package Version 3.0.1.1*; Science Open, Inc.: Berlin, Germany, 2015.
40. Carbonaro, M.; Maselli, P.; Nucara, A. Relationship between Digestibility and Secondary Structure of Raw and Thermally Treated Legume Proteins: A Fourier Transform Infrared (FT-IR) Spectroscopic Study. *Amino Acids* **2012**, *43*, 911–921. [CrossRef] [PubMed]
41. Pieper, R.; Villodre Tudela, C.; Taciak, M.; Bindelle, J.; Pérez, J.F.; Zentek, J. Health Relevance of Intestinal Protein Fermentation in Young Pigs. *Anim. Heal. Res. Rev.* **2016**, *17*, 137–147. [CrossRef] [PubMed]
42. Li, D.F.; Nelssen, J.L.; Reddy, P.G.; Blecha, F.; Klemm, R.D.; Giesting, D.W.; Hancock, J.D.; Allee, G.L.; Goodband, R.D. Measuring Suitability of Soybean Products for Early-Weaned Pigs with Immunological Criteria. *J. Anim. Sci.* **1991**, *69*, 3299–3307. [CrossRef] [PubMed]
43. Salgado, P.; Freire, J.P.B.; Mourato, M.; Cabral, F.; Toullec, R.; Lallès, J.P. Comparative Effects of Different Legume Protein Sources in Weaned Piglets: Nutrient Digestibility, Intestinal Morphology and Digestive Enzymes. *Livest. Prod. Sci.* **2002**, *74*, 191–202. [CrossRef]

44. Prohászka, L.; Baron, F. The Predisposing Role of High Dietary Protein Supplies in Enteropathogenic *E. coli* Infections of Weaned Pigs. *Zent. Veterinärmedizin R. B* **1980**, *27*, 222–232. [CrossRef]
45. Solà-Oriol, D.; Figueroa, J.; Borda, E.; Chetrit, C.; Pérez, J.F. The Effect of Different Animal and Vegetable Protein Sources on the Feed Intake and Weight Gain of Piglets. *J. Anim. Sci.* **2010**, *88*, 92.
46. Martínez-Puig, D.; Anguita, M.; Baucells, F.; Borda, E.; Pérez, J.; Chetrit, C. Evidence of a Preference in Piglets for an Animal Protein Hydrolysate. In Proceedings of the Joint Annual Meeting ADSA-PSA-AMPA-ASAS, San Antonio, TX, USA, 8–12 July 2007; Journal of Animal Science: San Antonio, TX, USA, 2007; Volume 85, p. 310.
47. Torrallardona, D. Spray Dried Animal Plasma as an Alternative to Antibiotics in Weanling Pigs. *Asian-Australas. J. Anim. Sci.* **2010**, *23*, 131–148. [CrossRef]
48. Ermer, P.M.; Miller, P.S.; Lewis, A.J. Diet Preference and Meal Patterns of Weanling Pigs Offered Diets Containing Either Spray-Dried Porcine Plasma or Dried Skim Milk. *J. Anim. Sci.* **1994**, *72*, 1548–1554. [CrossRef]
49. Pérez-Bosque, A.; Polo, J.; Torrallardona, D. Spray Dried Plasma as an Alternative to Antibiotics in Piglet Feeds, Mode of Action and Biosafety. *Porc. Health Manag.* **2016**, *2*, 1–10. [CrossRef]
50. Myers, A.J.; Goodband, R.D.; Tokach, M.D.; Dritz, S.S.; DeRouchey, J.M.; Nelssen, J.L. The Effects of Porcine Intestinal Mucosa Protein Sources on Nursery Pig Growth Performance. *J. Anim. Sci.* **2014**, *92*, 783–792. [CrossRef]
51. Borda, E.; Martínez-Puig, D.; Pérez, J. An Hydrolyzed Protein Concentrate (Palbio 62) Increases Feed Intake and Villus Height in Early Weaning Pigs. In *Book of Abstracts of the 56th Annual Meeting of the EAAP*; Wageningen Academic Publishers: Uppsala, Sweden, 2005.
52. Gilbert, E.R.; Wong, E.A.; Webb, K.E. Board-Invited Review: Peptide Absorption and Utilization: Implications for Animal Nutrition and Health. *J. Anim. Sci.* **2008**, *86*, 2135–2155. [CrossRef] [PubMed]
53. Rérat, A.; Nunes, C.S.; Mendy, F.; Roger, L. Amino Acid Absorption and Production of Pancreatic Hormones in Non-Anaesthetized Pigs after Duodenal Infusions of a Milk Enzymic Hydrolysate or of Free Amino Acids. *Br. J. Nutr.* **1988**, *60*, 121–136. [CrossRef] [PubMed]
54. Xing, F.; Jiang, C.; Liang, S.; Kang, L.; Jiang, Y. Genomic Structure and Characterization of MRNA Expression Pattern of Porcine Interferon Gamma Receptor 1 Gene. *Int. J. Immunogenet.* **2010**, *37*, 477–485. [CrossRef] [PubMed]
55. Bank, U.; Bohr, U.R.M.; Reinhold, D.; Lendeckel, U.; Ansorge, S.; Malfertheiner, P.; Täger, M. Inflammatory Bowel Diseases: Multiple Benefits from Therapy with Dipepridyl- and Alanyl-Aminopeptidase Inhibitors. *Front. Biosci.* **2008**, 3699–3713. [CrossRef]
56. Zhao, L.; Bartnikas, T.; Chu, X.; Klein, J.; Yun, C.; Srinivasan, S.; He, P. Hyperglycemia Promotes Microvillus Membrane Expression of DMT1 in Intestinal Epithelial Cells in a PKCα-Dependent Manner. *FASEB J.* **2019**, *33*, 3549–3561. [CrossRef]
57. Li, C.; Zhou, H.M. The Role of Manganese Superoxide Dismutase in Inflammation Defense. *Enzym. Res.* **2011**, *2011*, 387176. [CrossRef]
58. Visner, G.A.; Dougall, W.C.; Wilson, J.M.; Burr, I.A.; Nick, H.S. Regulation of Manganese Superoxide Dismutase by Lipopolysaccharide, Interleukin-1, and Tumor Necrosis Factor. Role in the Acute Inflammatory Response. *J. Biol. Chem.* **1990**, *265*, 2856–2864.

3

Rumen Methanogenesis, Rumen Fermentation and Microbial Community Response to Nitroethane, 2-Nitroethanol and 2-Nitro-1-Propanol: An In Vitro Study

Zhenwei Zhang, Yanlu Wang, Xuemeng Si, Zhijun Cao, Shengli Li and Hongjian Yang *

State Key Laboratory of Animal Nutrition, College of Animal Science and Technology, China Agricultural University, Beijing 100193, China; qingyibushuo@163.com (Z.Z.); yanluwang@yeah.net (Y.W.); sxmswun@126.com (X.S.); caozhijun@cau.edu.cn (Z.C.); lisheng0677@163.com (S.L.)

* Correspondence: yang_hongjian@sina.com

Simple Summary: The present study comparatively investigates the inhibitory difference of nitroethane (NE), 2-nitroethanol (NEOH), and 2-nitro-1-propanol (NPOH) on in vitro rumen fermentation, microbial populations, and coenzyme activities associated with methanogenesis. The results showed that both NE and NEOH were more effective in reducing ruminal methane (CH_4) production than NPOH. This work provides evidence that NE, NEOH, and NPOH were able to inhibit methanogen population and dramatically decrease methyl-coenzyme M reductase gene expression and the content of coenzymes F_{420} and F_{430} with different magnitudes in order to reduce ruminal CH_4 production.

Abstract: Nitroethane (NE), 2-nitroethanol (NEOH), and 2-nitro-1-propanol (NPOH) were comparatively examined to determine their inhibitory actions on rumen fermentation and methanogenesis in vitro. Fermentation characteristics, CH_4 and total gas production, and coenzyme contents were determined at 6, 12, 24, 48, and 72 h incubation time, and the populations of ruminal microbiota were analyzed by real-time PCR at 72 h incubation time. The addition of NE, NEOH, and NPOH slowed down in vitro rumen fermentation and reduced the proportion of molar CH_4 by 96.7%, 96.7%, and 41.7%, respectively ($p < 0.01$). The content of coenzymes F_{420} and F_{430} and the relative expression of the *mcr*A gene declined with the supplementation of NE, NEOH, and NPOH in comparison with the control ($p < 0.01$). The addition of NE, NEOH, and NPOH decreased total volatile fatty acids (VFAs) and acetate ($p < 0.05$), but had no effect on propionate concentration ($p > 0.05$). Real-time PCR results showed that the relative abundance of total methanogens, *Methanobacteriales*, *Methanococcales*, and *Fibrobacter succinogenes* were reduced by NE, NEOH, and NPOH ($p < 0.05$). In addition, the nitro-degradation rates in culture fluids were ranked as NEOH (−0.088) > NE (−0.069) > NPOH (−0.054). In brief, the results firstly provided evidence that NE, NEOH, and NPOH were able to decrease methanogen abundance and dramatically decrease *mcr*A gene expression and coenzyme F_{420} and F_{430} contents with different magnitudes to reduce ruminal CH_4 production.

Keywords: nitrocompounds; methanogenesis; rumen fermentation; microbial community; coenzyme

1. Introduction

Nitrocompounds are classified into aromatic compounds containing nitro groups in the aromatic ring and aliphatic–aromatic compounds containing nitro groups only in the aliphatic side chain. Among the naturally occurring aliphatic nitrocompounds, 3-nitro-1-propanol (3-NPOH) and 3-nitro-1-propionic acid (NPA) are regarded as nitrotoxins that can accumulate to toxic levels in certain forages (e.g., *Astragalus*); their toxic and metabolic properties were well investigated in ruminants and monogastric animals nearly a century ago [1]. Over the last decades, certain commercially available short-chain aliphatic nitrocompounds have been chemically synthesized and are known to be potent methane-inhibiting compounds, since methane (CH_4) from livestock is increasingly considered as a significant greenhouse gas [2].

Among these commercially available nitrocompounds, Latham et al. [2] summarized the supplemental effects of nitroethane (NE), 2-nitroethanol (NEOH), and 2-nitro-1-propanol (NPOH) in comparison with 3-NPOH and NPA on CH_4 and volatile fatty acid (VFAs) production during in vitro rumen incubations, finding that the yield of VFAs was almost not affected, whereas CH_4 production consistently decreased by up to 58%–99%. Regarding in vivo studies in sheep [3] and feedlot steer [4–6], CH_4-reducing activity varied depending on dose levels and depletion of nitrocompounds. More recently, in the study of Hristov et al. [7], as much as a 64-fold increase in hydrogen (H_2) production was observed in 3-nitrooxypropanol-treated cattle, which was still only about 3% of the H_2 spared from ruminal production. Despite the fact that CH_4 emission is negatively associated with energy retention and greenhouse gas production, rumen archaea play an important ecological role in methanogenesis, however, few studies have examined the microbial response to nitrocompounds. In many of the aforementioned studies, the inhibited archaea populations or methanogenesis-associated enzymes were not characterized though methyl-coenzyme M reductase (*mcr*). The coenzymes F_{420} and F_{430} are known in nature as key enzymes involved in CH_4 formation from H_2 and CO_2 by archaea [8–10]. Although the naturally occurring nitrocompounds 3-NPOH and NPA are recognized to be metabolized by rumen microorganisms to aminopropanol and β-alanine, a nonessential amino acid may be utilized by ruminant animals and potentially used as a source of carbon, nitrogen, and energy, making it an attractive candidate [1,2]. To avoid unknown risks of toxic intermediates in feeding practice trials regarding commercially available NE, NEOH, and NPOH, the present study comparatively investigated the inhibitory difference of each nitrocompound on in vitro rumen fermentation, microbial populations, and coenzyme activities associated with methanogenesis.

2. Materials and Methods

The donor animals and experimental procedures were approved by the requirements of Beijing Municipal Council on Animal Care according to the protocol of CAU20171014-1.

2.1. Nitrocompounds

Nitroethane (NE), 2-nitroethanol (NEOH), and 2-nitro-1-propanol (NPOH) were purchased commercially (Sigma Aldrich, Inc., LA, USA). Structures of these nitrocompounds are shown in Figure 1. Their analytical grades were 99%, 90%, and 98%, respectively.

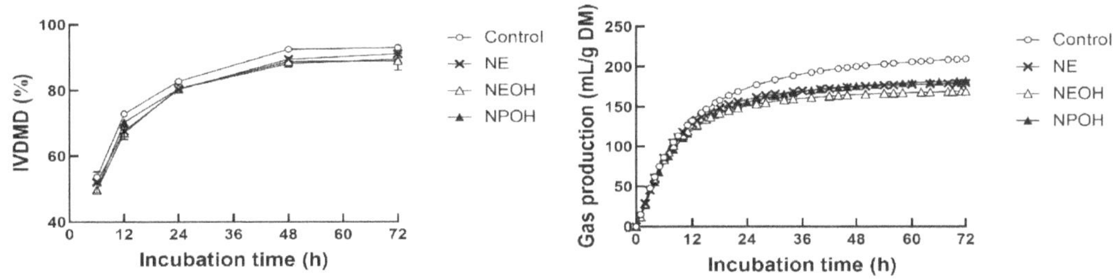

(a) (b)

Figure 1. Effect of NE, NEOH, and NPOH addition to culture fluids on in vitro dry matter disappearance (IVDMD (**a**) and cumulative gas production (**b**) of grain-rich feed incubated with rumen fluids obtained from lactating dairy cows. NE: nitroethane; NEOH: 2-nitroalcohol; NPOH: 2-nitro-1-propanol. Statistical analyses showed that the effects of nitrocompounds on IVDMD and gas production were significant at $p < 0.01$.

2.2. Animals

Five multiparous and rumen-cannulated Holstein lactating dairy cows (540 ± 25.3 kg body weight; 100 ± 8.5 days in milk, 33.0 ± 0.78 kg/d milk yield; mean ± SD) were selected as the donors of rumen fluid. Cows were maintained on a total mixed ration (calculated as % dry matter basis) of alfalfa hay (16.32%), whole corn silage (24.61%), 1 kg corn meal (3.95%), extruded corn (19.60%), soybean meal (14.38%), soybean hull (4.09%), extruded soybean (3.43%), whole cottonseed (8.25%), trace mineral, and vitamin premix (5.37%), according to the Chinese Feeding Standard of Dairy Cow (NY/T 34 2004).

2.3. In Vitro Experiment

In vitro fermentations in anaerobic glass bottles (volume capacity of 120 mL) incubated with rumen fluids were performed following the previous description of Zhang and Yang [11]. The treatments included the control (no additive treatment), 10 mmol/L of NE, 10 mmol/L of NEOH, and 10 mmol/L of NPOH. Corn meal and alfalfa hay (500 mg; 80:20, w/w) were used as the fermentation substrates.

Rumen fluids were collected before morning feeding from each rumen-cannulated donor cow into a pre-warmed thermos flask at 39 °C. After filtering through 4 layers of cheesecloth and mixed in equal proportion, 25 mL of rumen fluids were incubated into anaerobic glass bottles with 50 mL buffered medium (pH 6.8) [12]. The batch cultures were performed at 39 °C in both automated and manual systems. In the automated system, five bottles per treatment were connected to the gas inlets of an automated gas recording system (AGRS) and continuously incubated for 72 h to continuously record cumulative gas production (GP). In the manual system, five bottles per treatment were connected to pre-emptied air bags to collect fermentation gas samples and removed at 6, 12, 24, 48, and 72 h of incubation. The batch cultures were repeated and completed in three consecutive runs. One milliliter of gas sample was drawn out of the air bags using a syringe to measure the CH_4 concentration according to the gas chromatographic method.

2.4. Sampling

After 6, 12, 24, 48, and 72 h of incubation in the manual system, the contents of each bottle were filtered through a nylon bag (8 × 12 cm; 42 μm pore size) and dried at 105 °C to determine the in vitro dry matter disappearance (IVDMD). Then, the culture fluids (6 × 1.0 mL) were sampled into DNase-free polypropylene tubes and stored at −80 °C for later analysis of VFA, nitrocompounds, microbial populations, *mcr*A (methyl coenzyme-M reductase subunit A) gene expression, coenzyme F_{420} content, and coenzyme F_{430} content.

2.5. Measurement of VFA, Coenzyme, and Nitrocompound Contents

The culture fluids (1.0 mL) from each of the 5 aforementioned time stamps were mixed with 300 μL metaphosphoric acid solution (25%, w/v) for 30 min and centrifuged at 10,000× *g* for 15 min at 4 °C. Supernatants (0.5 mL) were injected into gas chromatography to determine the concentrations of acetate, propionate, isobutyrate, butyrate, isovalerate, and valerate [11].

Following the description of Reuter et al. [13], coenzyme F_{420} was determined and expressed as fluorescence intensity. Assays were performed at 37 °C anaerobically in the dark. Culture fluid samples (1.0 mL) were stirred continuously and boiled at 95 °C in water bath for 30 min. Fluid aliquots were then centrifuged at 10,000× *g* for 10 min, and a volume of 500 μL from supernatants was mixed with 1 mL of isopropanol. Subsequently, the mixture was precipitated for 2 h and centrifuged at 10,000× *g* again for 15 min. Finally, the fluorescence intensity of the supernatants was measured at 420 nm by the fluorescence spectrophotometer (Thermo Fisher Scientific Co., Ltd., shanghai, China). Coenzyme F_{430} content was examined via the ultraviolet/visible spectrum by determining the loss of absorbance [14]. Briefly, the culture fluids were quenched with equal volumes of methanol and centrifuged aerobically at 6153× *g* for 20 min in dim light. The precipitate was discarded and the supernatants were determined colorimetrically using a spectrophotometer at 430 nm (Laspec Technology Co., Ltd., shanghai, China). Coenzyme F_{430} content was expressed as the relative absorbance of coenzyme F_{430} at 430 nm.

The contents of NE, NEOH, and NPOH were determined colorimetrically by a spectrophotometer (Laspec Technology Co., Ltd., shanghai, China) [15]. The culture fluids (1 mL) were firstly centrifuged at 10,000× *g* for 15 min. Supernatants (50 μL) were then diluted with 2 mL of distilled water and mixed with 100 μL of NaOH (0.65 M) and 100 μL of diazotized p-nitroaniline. Finally, the absorbances of the culture fluids were measured at a wavelength of 405 nm.

2.6. Analysis of *mcr*A Gene Expression

Total RNA of culture fluid was extracted by RNeasy Mini kit (Tiangen Biotech, Beijing, China) using an RNase-Free DNase Set (Qiagen). The detailed procedures for analysis of *mcr*A gene expression were described by Guo et al. [16]. The specific primer sets for the *mcr*A gene (F: 5′-TTCG GTGG ATCD CARA GRGC-3′, R: 3′-GBAR GTCG WAWC CGTA GAAT CC-5′) and the 16S rRNA gene (F: 5′-CGGC AACG AGCG CAAC CC-3′, R: 3′-CCAT TGTA GCAC GTGT GT AG CC-5′) were applied as described by Denman et al. [17] and Denman and McSweeney [18]. The $2^{-\Delta\Delta Ct}$ method was used for the expression analysis of the *mcr*A gene, with the 16S rRNA set as the reference gene.

2.7. Microbial Population Analysis with Real-Time PCR

Total DNA of culture fluid (1 mL) was extracted with the FastDNA kit and FastPrep instrument (Tiangen®Biotech, Beijing, China) by a bead-beating method as described by Denman and McSweeney [18]. According to the real-time PCR method [18], enumeration of total bacteria, total methanogens, *Methanobacteriales*, *Methanococcales*, *Methanomicrobiales*, protozoa, fungi, *Ruminococcus flavefaciens*, *Ruminococcus albus*, and *Fibrobacter succinogenes* was measured using the Bio-Rad Multicolor Real-Time PCR Detection System (Bio-rad Company, CA, USA) with qReal Master Mix SYBR®Green (Tiangen®Biotech, Beijing, China). The primer sets for the detection and enumeration of the microbial populations were described by Denman and McSweeney [18], Zhou et al. [19], and Yu et al. [20]. The abundance of the microbial population was expressed as a proportion of total estimated rumen bacterial (16S rDNA) according to the following equation: relative quantification = $2^{-(\text{CT target - CT total bacteria})}$, where CT represents the threshold cycle.

2.8. Data Calculation and Statistical Analysis

Cumulative gas production data from the AGRS were fitted according to an exponential model as described by France [21]. In addition, the average gas production rate (AGPR) (mL/h) and hydrogen recovery (2Hrec) were also calculated following the description by Zhang et al. [22].

The data for gas production, IVDMD, fermentation gas composition, coenzyme contents, and VFAs were subjected to analysis of variance with the MIXED model procedure of SAS (Statistical Analysis for Windows, SAS Institute Inc., Cary, NC, USA). The model was applied as $Y_{ijk} = \mu + R_i + N_j + T_k + (N \times T)_{jk} + e_{ijk}$,, where Y_{ijk} is the dependent variable, μ represents the overall mean, R_i is the effect of the experimental run, N_j is the effect of the nitrocompound treatment, T_k is the effect of the incubation time, $N \times T$ are the interactions between the nitrocompounds and the incubation time, and e_{ijk} is the residual. The data for microbial abundance analysis were applied according to the model $Y_{ijk} = \mu + R_i + N_j + e_{ijk}$, where Y_{ijk} is the dependent variable, μ represents the overall mean, R_i is the effect of the experimental run, N_j is the effect of the nitrocompound treatment, and e_{ijk} is the residual. Least square means and standard error (SEM) were calculated, and treatment differences were estimated using a multiple comparisons test (Tukey/Kramer). Correlation analyses between variables were performed using the CORR procedure of SAS. Significance was declared at $p < 0.05$.

3. Results

3.1. IVDMD and Gas Production Kinetics

As the incubation time increased, both IVDMD (Figure 1a) and gas production (Figure 1b) continuously increased. Addition of NE, NEOH, and NPOH slowed down the fermentation process and caused IVDMD to decline (Figure 1a, $p < 0.01$).

Regarding the kinetics of gas production, NE, NEOH, and NPOH addition decreased asymptotic gas production (A) (Table 1, $p < 0.01$), while NE and NEOH addition increased the fractional gas production rate (c, h^{-1}). Neither NE nor NEOH addition altered AGPR, but NPOH decreased AGPR compared to the control.

Table 1. Effect of NE, NEOH, and NPOH addition (10 mmol/L) to culture fluids on gas production kinetics and fermentation gas composition during 72 h incubation.

Items [1]	Treatment [2]				SEM	p [3]		
	Control	NE	NEOH	NPOH		Treatment	Time	Trt × Time
Gas production kinetics								
A, mL/g DM	199 ^{a}	178 ^{b}	170 ^{c}	173 ^{bc}	1.96	<0.01	-	-
c, h^{-1}	0.09 ^{b}	0.11 ^{a}	0.12 ^{a}	0.10 ^{b}	0.02	<0.01	-	-
$T_{1/2}$, h	3.1	3.1	3.0	3.1	0.02	0.07	-	-
AGPR, mL/h	13.3 ^{a}	13.9 ^{a}	13.9 ^{a}	11.9 ^{b}	0.29	<0.01	-	-
Fermentation gas composition, %								
CH_4	15.1 ^{a}	0.5 ^{c}	0.5 ^{c}	8.8 ^{b}	0.39	<0.01	<0.01	<0.01
H_2	0.3 ^{d}	9.8 ^{b}	10.5 ^{a}	2.0 ^{c}	0.23	<0.01	<0.01	<0.01
CO_2	84.6 ^{b}	89.4 ^{a}	89.0 ^{a}	89.2 ^{a}	0.36	<0.01	<0.01	0.01

$^{a–d}$ Means with different superscripts within a row are significantly different ($p < 0.05$); [1] A: asymptotic gas production; c: fractional gas production rate; $T_{1/2}$: the time when half of A occurred; AGPR: average gas production rate. [2] NE: nitroethane; NEOH: 2-nitroalcohol; NPOH: 2-nitro-1-propanol. [3] Interaction effect between treatment and incubation time.

3.2. Fermentation Gas Composition

H_2 accumulation in the NE, NEOH, and NPOH groups was far greater than in the control group (Table 1, $p < 0.01$). The molar proportions of CO_2 were increased (5.7%, 5.2%, and 5.4%) by NE, NEOH, and NPOH compared to the control, whereas the molar proportions of CH_4 were notably decreased (96.7%, 96.7%, and 41.7%, Table 1). An interaction between the nitrocompound treatment and the incubation time was observed for gas composition ($p < 0.01$). The molar proportion of CH_4 constantly increased with increasing incubation time in the control and NPOH groups, but it was at a pretty low level in both the NE and NEOH groups ($p < 0.01$, Figure 2a). In contrast, molar H_2 production in the NE and NEOH groups continuously increased with increasing incubation time and was always far greater than the control and NPOH throughout the incubation ($p < 0.01$, Figure 2b). In addition, as the incubation time increased, the molar proportion of CO_2 gradually decreased in all groups, and it was greater in the nitrocompound-treated cultures than the control ($p < 0.01$, Figure 2c).

3.3. Coenzymes Related to CH_4 Production

Compared to the control, both coenzyme F_{420} fluorescence intensity and F_{430} ultraviolet (UV) absorbance declined in the NE, NEOH, and NPOH groups (Table 2, $p < 0.01$). In addition, NE, NEOH, and NPOH addition decreased *mcr*A gene expression by 83.1%, 79.7%, and 53.5%, respectively.

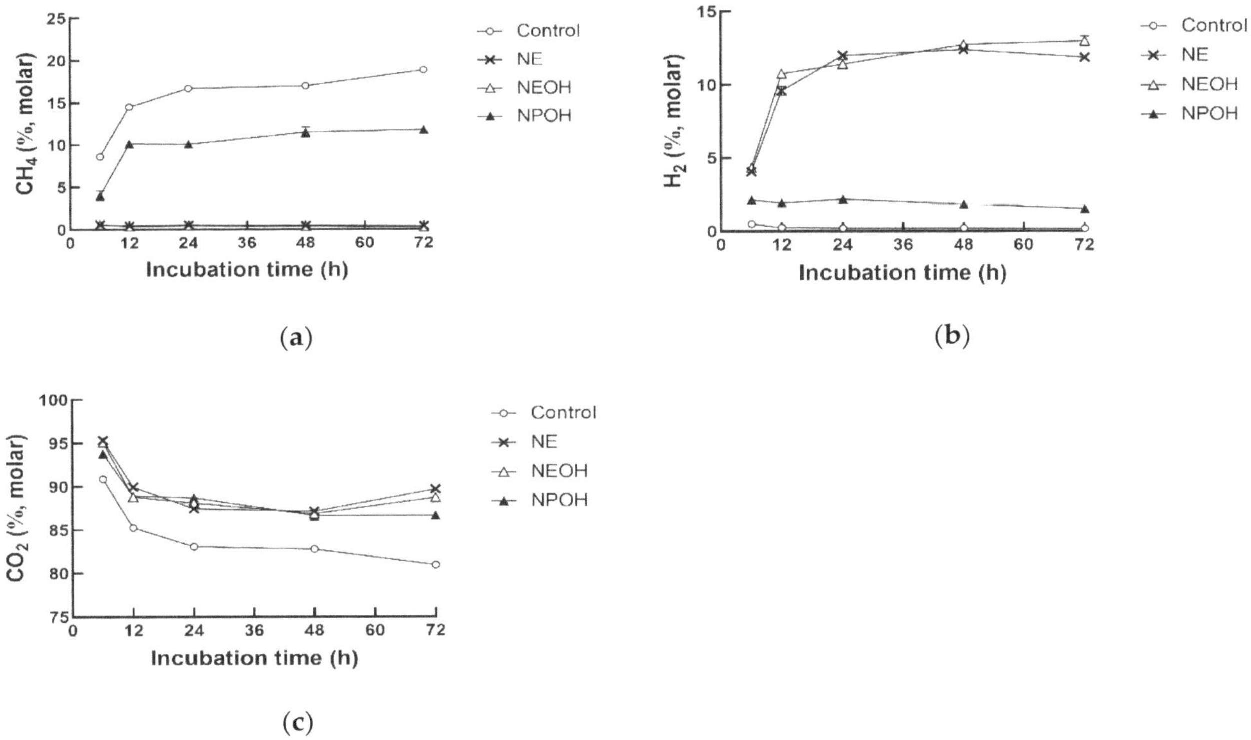

Figure 2. Effect of NE, NEOH, and NPOH addition to culture fluids on CH_4 (**a**), H_2 (**b**), and CO_2 (**c**) production of grain-rich feed incubated with rumen fluids obtained from lactating dairy cows. NE: nitroethane; NEOH: 2-nitroalcohol; NPOH: 2-nitro-1-propanol. Statistical analyses showed the effects of nitrocompounds on CH_4, H_2, and CO_2, with correlations significant at $p < 0.01$.

Table 2. Effect of NE, NEOH, and NPOH addition (10 mmol/L) on coenzyme content, *mcr*A gene expression, and volatile fatty acid (VFA) production in fermentation fluids across different incubation times (6, 12, 24, 48, and 72 h).

Items [1]	Treatment [2]				SEM	p [3]		
	Control	NE	NEOH	NPOH		Treatment	Time	Trt × Time
Coenzyme content								
F_{420}	16.7 [a]	11.9 [c]	11.4 [c]	15.2 [b]	0.23	<0.01	<0.01	<0.01
F_{430}	0.59 [a]	0.48 [b]	0.44 [c]	0.50 [b]	0.02	<0.01	<0.01	0.01
*mcr*A expression	1 [a]	0.11 [c]	0.12 [c]	0.42 [b]	0.02	<0.01	<0.01	<0.01
Total VFA, mmol/L	102.8 [a]	93.7 [c]	94.7 [bc]	97.4 [b]	1.64	<0.01	<0.01	<0.01
Acetate, mmol/L	60.9 [a]	52.1 [c]	53.1 [c]	54.7 [b]	1.31	<0.01	<0.01	0.02
Propionate, mmol/L	24.0	23.1	24.2	23.9	0.53	0.33	<0.01	0.02
Butyrate, mmol/L	12.8 [bc]	13.5 [a]	12.5 [c]	13.2 [ab]	0.30	<0.01	<0.01	<0.01
BCVFA, mmol/L	5.1	5.0	4.9	5.6	0.31	0.14	<0.01	0.14
2Hrec, %	67.6 [a]	47.2 [c]	48.5 [c]	59.3 [b]	0.87	<0.01	<0.01	<0.01

[a–d] Means with different superscripts within a row are significantly different ($p < 0.05$); [1] F_{420}: F_{420} fluorescence intensity; F_{430}: UV absorbance of coenzyme F_{430}; BCVFA: branch-chained volatile fatty acids; 2Hrec: hydrogen recovery. [2] NE: nitroethane, NEOH: 2-nitroalcohol; NPOH: 2-nitro-1-propanol. [3] Interaction effect between treatment and incubation time.

An interaction between the nitrocompound treatment and the incubation time was evident for *mcr*A gene expression and coenzyme F_{420} and F_{430} contents ($p < 0.01$). As the incubation time increased, *mcr*A gene expression in the NPOH group relative to the control peaked at 12 h during the 72 h incubation, and it was continuously far lower in the NE and NEOH groups than in the control and NPOH groups (Figure 3a). Coenzyme F_{420} fluorescence intensity gradually declined in both the NE and NEOH groups with increasing incubation time, and it was continuously lower in the nitrocompound-treated cultures than the control (Figure 3b). The F_{430} ultraviolet (UV) absorbance continuously decreased in the NE and NEOH groups with increasing incubation time (Figure 3c), but it fluctuated during 72 h incubation in the NPOH group. In addition, the coenzyme F_{430} content was constantly lower in the nitrocompound-treated cultures than the control.

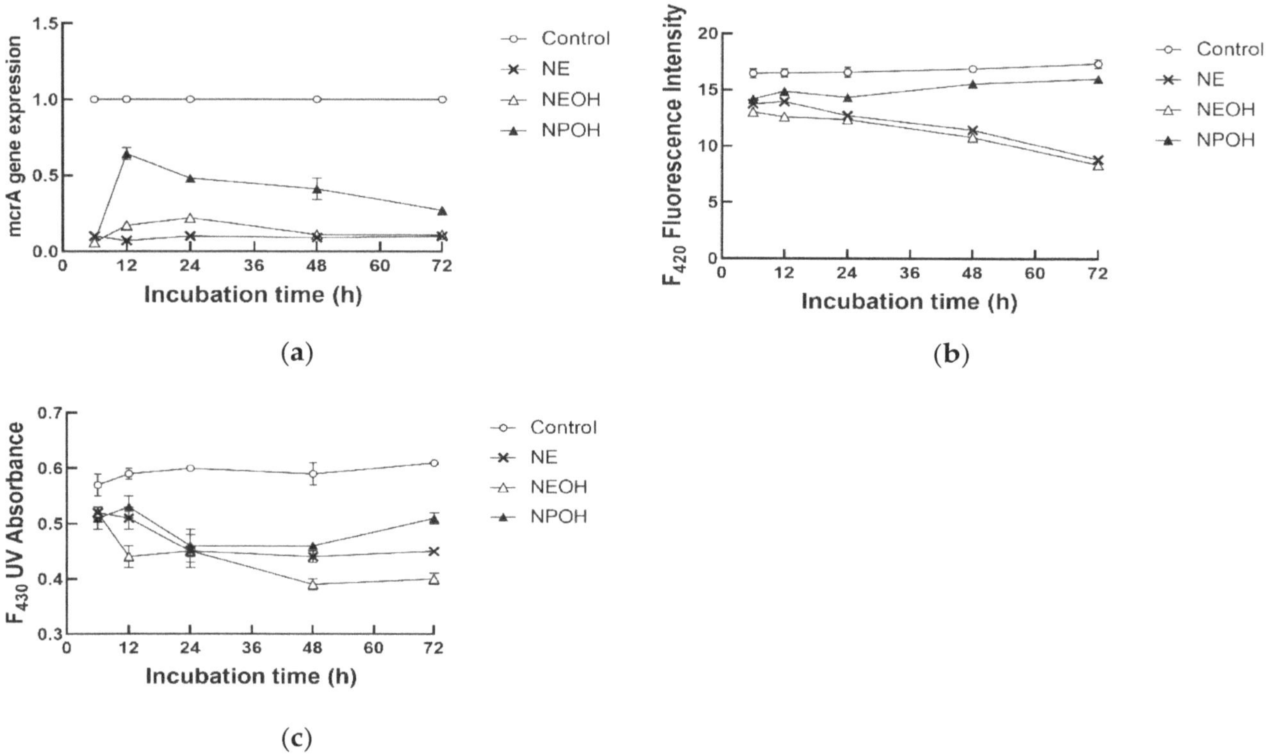

Figure 3. Effect of NE, NEOH, and NPOH addition to culture fluids on *mcr*A gene expression (**a**) and the contents of coenzyme F_{420} (**b**) and coenzyme F_{430} (**c**) of grain-rich feed incubated with rumen fluids obtained from lactating dairy cows. NE: nitroethane; NEOH: 2-nitroalcohol; NPOH: 2-nitro-1-propanol. Statistical analyses showed that the effects of nitrocompounds were significant at $p < 0.01$.

3.4. Fermentation Characteristics

Concentrations of total VFAs and acetate were lower in the NE, NEOH and NPOH groups than in the control (Table 2, $p < 0.01$), whereas the concentration of butyrate increased with NE addition (Table 2, $p < 0.01$). NE, NEOH, and NPOH addition had no significant influence on propionate and branched-chain VFAs (BCVFAs). Compared to the control, NE, NEOH, and NPOH addition decreased 2Hrec by 30.2%, 28.3%, and 12.3%, respectively (Table 2, $p < 0.01$). An interaction between the nitrocompound treatment and the incubation time was observed for total VFA, acetate, propionate, and butyrate. The concentrations of total VFAs, acetate, propionate, butyrate, and BCVFA were continuously increased in all groups with increasing incubation time (Figure 4). In addition, concentrations of acetate were continuously lower in nitrocompound-treated cultures than in the control, whereas the concentrations of propionate, butyrate, and BCVFAs fluctuated during 72 h incubation.

3.5. Microbial Populations

Real-time PCR results showed that NE, NEOH, and NPOH addition decreased the relative populations of total methanogens, *Methanobacteriales*, *Methanomicrobiales*, and *Fibrobacter succinogenes* (Table 3, $p < 0.05$). Compared to the control, the relative populations of total methanogens decreased by 49.2%, 36.9%, and 41.5% in the NE, NEOH, and NPOH groups, and the populations of *Methanobacteriales* decreased by 46.1%, 35.9%, and 17.9%, respectively. In addition, the relative populations of fungi tended to decrease according to nitrocompound treatment ($p = 0.09$). Compared with control, the populations of *R. albus* increased by 50.0% and 50.0% in the NEOH and NPOH groups, and the populations of

R. flavefaciens decreased by 84.5% and 70.7% in the NE and NEOH groups. However, the populations of protozoa were not affected ($p = 0.29$).

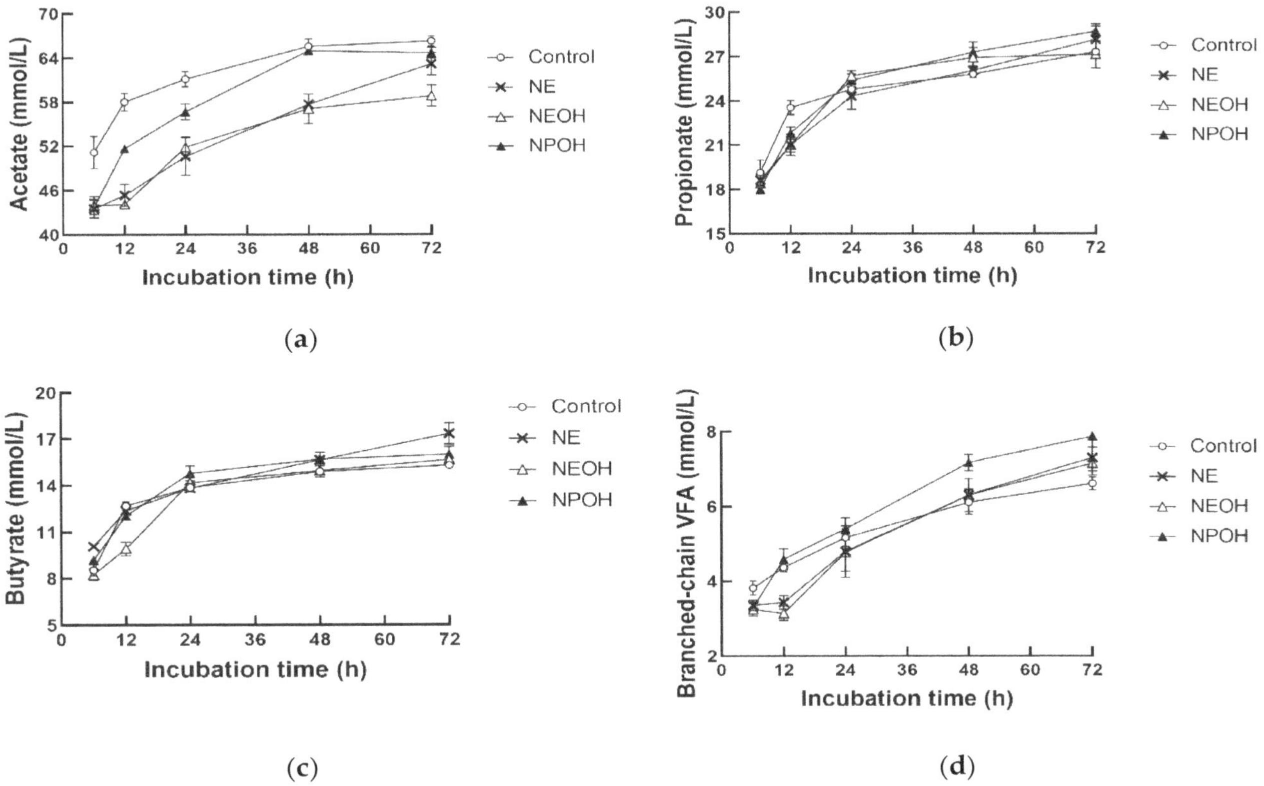

Figure 4. Effect of NE, NEOH, and NPOH addition to culture fluids on acetate (**a**), propionate (**b**), butyrate (**c**), and branched-chain VFA (**d**) of grain-rich feed incubated with rumen fluids obtained from lactating dairy cows. NE: nitroethane; NEOH: 2-nitroalcohol; NPOH: 2-nitro-1-propanol. Statistical analyses showed that the effects of nitrocompounds on acetate and butyrate were significant at $p < 0.01$, but the effects of nitrocompounds on propionate and branched-chain VFAs were not significant.

Table 3. Effects of NE, NEOH, and NPOH addition (10 mmol/L) to culture fluids on microorganism relative populations after 72 h incubation.

Items	Treatment					
	Control	NE	NEOH	NPOH	SEM	*p*
Total methanogen	0.65 [a]	0.33 [b]	0.41 [b]	0.38 [b]	0.041	0.04
Methanobacteriales	0.39 [a]	0.21 [c]	0.25 [bc]	0.32 [b]	0.013	0.02
Methanomicrobiales, $\times 10^{-3}$	14.2 [a]	4.6 [c]	9.5 [b]	5.1 [c]	0.985	0.01
Methanococcales, $\times 10^{-3}$	0.44 [bc]	0.56 [ab]	0.65 [a]	0.38 [c]	0.041	0.03
Fungus, $\times 10^{-3}$	1.82 [a]	0.37 [b]	1.25 [ab]	1.14 [ab]	0.240	0.09
Protozoa, $\times 10^{-3}$	0.28	0.29	0.43	0.22	0.068	0.29
Ruminococcus albus	0.12 [b]	0.10 [b]	0.18 [a]	0.18 [a]	0.012	<0.01
Ruminococcus flavefaciens	0.58 [a]	0.09 [b]	0.17 [b]	0.43 [a]	0.056	0.01
Fibrobacter succinogenes, $\times 10^{-3}$	3.33 [a]	1.39 [b]	0.97 [b]	0.56 [b]	0.232	<0.01

[a–c] Means with different superscripts within a row are significantly different ($p < 0.05$). NE: nitroethane; NEOH: 2-nitroalcohol; NPOH: 2-nitro-1-propanol.

3.6. Correlation between CH_4 Production and mcrA Gene Expression/Coenzyme Contents

CH_4 production was positively correlated with *mcr*A gene expression (Table 4; r = 0.88, $p < 0.05$), coenzyme F_{420} content (Table 4; r = 0.74, $p < 0.01$), and coenzyme F_{430} content (Table 4; r = 0.24, $p < 0.05$). In addition, there was a positive correlation between *mcr*A gene expression and the coenzyme

F_{420} content ($r = 0.78$, $p < 0.01$), and between *mcr*A gene expression and the coenzyme F_{430} content ($r = 0.22$, $p < 0.05$). A positive correlation was also observed between the coenzyme F_{420} content and the coenzyme F_{430} content ($r = 0.26$, $p < 0.05$).

Table 4. Correlations between CH_4 production, *mcr*A gene expression, coenzyme F_{420} content, and coenzyme F_{430} content, regardless of the type of nitrocompound added.

Items	*mcrA*	F_{420}	F_{430}
methane production	0.88 *	0.74 **	0.24 *
mcrA		0.78 **	0.22 *
F_{420}			0.26 *

Significance: * p <0.05, ** p <0.01.

3.7. Disappearance of Nitrocompounds

Concentrations of NE, NEOH, and NPOH decreased with increasing incubation time (Figure 5). The disappearance of nitrocompounds was fitted to a linear model, and the nitrocompound degradation rate was ranked as NEOH (−1.20) > NE (−0.80) > NPOH (−0.78).

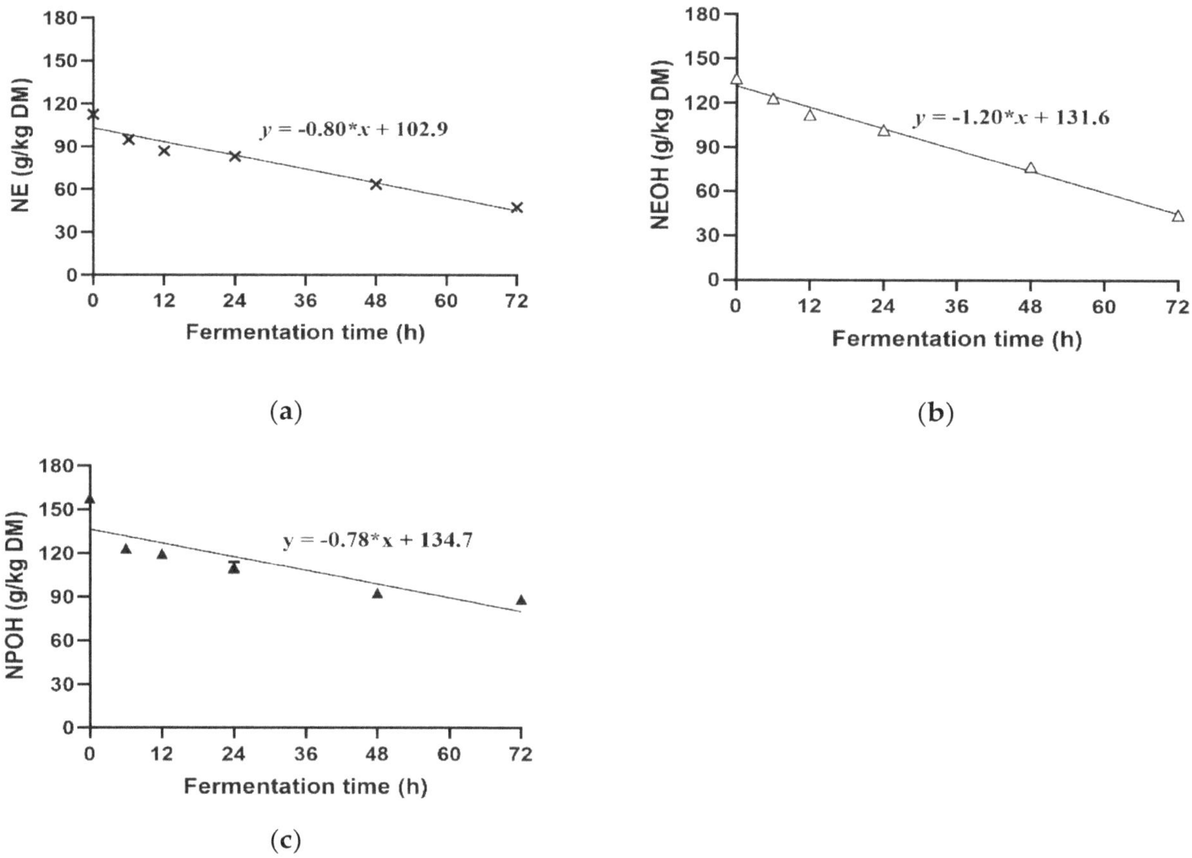

Figure 5. Variation of NE (**a**), NEOH (**b**), and NPOH (**c**) in culture fluids during 72 h incubation of grain-rich feed incubated in batch cultures of mixed rumen microorganisms. DM: dry matter; NE: nitroethane; NEOH: 2-nitroalcohol; NPOH: 2-nitro-1-propanol.

4. Discussion

4.1. Effects of Nitrocompounds on IVDMD and Gas Production

A majority of CH_4 inhibition strategies tend to compromise fermentative efficiency, resulting in the reduction of certain digestive processes [11,23]. In the present study, the addition of NE, NEOH, and NPOH caused notable reductions of IVDMD, gas production, and total VFAs, indicating that the activity of the microbes responsible for the degradation of substrates was inhibited by these nitrocompounds. The inclusive level of nitrocompounds was determined according to previous studies [11,24], however, a remarkable reduction in total VFA production occurred in the NE, NEOH, and NPOH groups, indicating that the fermentative bacterial population may be sensitive to a 10 mM dose level of nitrocompounds.

4.2. *Effects of Nitrocompounds on CH_4 Production*

The antimethanogenic activity of NE, NEOH, and NPOH was previously observed. For instance, more than 90% of CH_4 production was inhibited by NE, NEOH, and NPOH at a concentration of 9~24 mM after 24 h incubation [5,24]. The addition of NE, NEOH, and NPOH in the present study caused reductions in CH_4 production of 96.7%, 97.2%, and 39.5%, respectively, which agreed with the results of earlier studies [24,25]. In addition, both NE and NEOH were more effective in reducing ruminal CH_4 production than NPOH.

The nitro group is strongly electron-withdrawing. Because of this property, nitrocompounds may be able to serve as electron acceptors within ruminal microbes to reduce CH_4 production. Early studies done by Anderson et al. [26] revealed that nitrocompounds could serve as electron acceptors within ruminal microbes to inhibit CH_4 production. However, nitrocompounds also directly inhibit ruminal methanogenesis [27]. The first author in a review summarized the anti-methanogenic roles of nitrocompounds and their potential inhibitory action modes in terms of VFA production, hydrogen accumulation, formate oxidation and ferredoxin-linked hydrogenase activity [28]. Notable accumulation of H_2 in NE-, NEOH-, and NPOH-treated cultures and CH_4 production inhibition were observed in the present study. Similarly, Anderson et al. [25] also found that accumulations of H_2 were higher in nitrocompound-supplemented cultures than in controls after 24 h of incubation. In the unperturbed rumen, H_2 is usually present at approximately 1 μmol/L (0.1 kPa); however, it often increases to concentrations that inhibit hydrogenase activity (1 kPa), while ruminal CH_4 production is inhibited due to declined H_2 consumption by methanogens [23]. Considering that hydrogenases are reversible enzymes that can catalyze either the production or oxidation of H_2, the authors in the present study speculated that nitrocompounds may have inhibited H_2-oxidation hydrogenase activity as well, which was in accordance with Angermaier and Simon [29], who reported that NEOH inhibited ferredoxin-linked hydrogenase uptake activity. Likewise, Anderson et al. [30] also noted that NE, NEOH, NPOH, and NPA each inhibited oxidation of H_2 or formate when these reductants were added in excess (60 mM) to mixed cultures of rumen microbial populations, thereby implicating a possible mechanism of activity against ruminal methanogenesis.

Consistent with the different levels of CH_4 inhibition efficiency, NE and NEOH were shown to be nearly equally effective in promoting H_2 accumulation in vitro, with both of them promoting H_2 accumulation more effectively than NPOH. The molar proportions of H_2 in total fermentation gas production were 9.8%, 10.5%, and 2.0% in the NE, NEOH, and NPOH groups, respectively. However, the decreasing extent of 2Hrec in the present study was far less than that of CH_4 production. The fate of the remaining H_2 is not known with certainty, however, the consumption of reducing equivalents may occur during anabolic processes, including cell growth, intracellular polyhydroxyalkonoate, or extracellular polysaccharide production [25].

4.3. *Effects of Nitrocompounds on VFA Production*

In order to compensate for the disruption of electron flow to ruminal methanogenesis, some CH_4 inhibitors, such as sodium sulfite, organic halides, and monensin, cause notable increases in propionate production during fermentation, which is frequently accompanied by decreased acetate [23]. However, NE, NEOH, and NPOH addition in the present study had little effect on propionate produced by ruminal populations. Therefore, the results indicate that the reducing equivalents produced during ruminal fermentation were not necessarily directed toward increased production of propionate. In addition, the concentrations of acetate in the present study were decreased in the nitrocompound-treated cultures, along with the reduction of CH_4 production. It is well recognized that acetate accompanies H_2 production, and the latter can be used for CH_4 formation by methanogenic archaea, with CH_4 being positively associated with the acetate to propionate ratio [31].

4.4. Effects of Nitrocompounds on Methane Production and Related Changes in Microbial Populations and Coenzyme Contents

Methanogens are a unique group of ruminal microbes that generate CH_4 as a stoichiometric end-product of their metabolism. Methanogen populations are generally closely associated with CH_4 production [32]. In the present study, NE, NEOH, and NPOH addition decreased total methanogens by 49.2%, 36.9%, and 41.5%, respectively, and the authors speculated that NE, NEOH, and NPOH likely exerted a direct inhibition of ruminal methanogenesis via direct suppression of methanogens. *Methanobacteriales* are the predominant populations and constitute the major portion of methanogen community in ruminants, being the second most prevalent archaea in the rumen ecology [33]. Presently, both the relative abundances of *Methanobacteriaceae* and *Methanomicrobiales* were reduced with NE, NEOH, and NPOH addition, but different sensitivity responses to these nitrocompounds were observed. *Methanobacteriaceae* reduction varied in the order of NE (−46.2%) > NEOH (−35.9%) > NPOH (−17.9%), whereas *Methanomicrobiales* reduction varied in the order of NE (−67.6%) > NPOH (−64.1%) > NEOH (−33.1%).

Rumen protozoa constitute only a small portion of ruminal microorganisms, however, they play important roles in feed degradation and making energy and protein available to the hosts. The RT-PCR results showed no apparent decrease in the protozoal population in the NE-, NEOH- and NPOH-treated cultures, indicating that there were no adverse effects of the nitrocompounds on the rumen protozoa populations.

Few studies have determined microbial responses to nitrocompounds. Anderson et al. [26] noted that 3-NPOH and NPA modestly inhibited total culturable anaerobes from bovine rumens, but inhibited microbes were not characterized. Rumens harbor different types of bacteria, which are most actively involved in plant fiber degradation. It is well recognized that the depression of cellulolysis can decrease the rate and extent of neutral detergent fiber digestion [34], and the inhibition to cellulolytic microorganisms (*R. flavefaciens, F. succinogenes*) by nitrocompounds may explain why decreased IVDMD was observed in the cultures supplemented with NE, NEOH, or NPOH. Fibrolytic bacteria families, including *R. flavefaciens, R. albus*, and *F. succinogenes*, have vbeen shown to release many fibrolytic enzymes and promote H_2, acetate and formic acid production for methanogen utilization [35]. Real-time PCR results showed that the relative populations of the methanogens *R. Flavefaciens* and *F. succinogenes* decreased with nitrocompound addition, suggesting that the mutual-aid interaction between methanogens and fibrolytic bacteria might be one reason why *R. Flavefaciens* and *F. succinogenes* decreased along with methanogen inhibition by NE, NEOH, and NPOH. The rumen protozoa produce fermentation end-products similar to those made by the bacteria, particularly acetate, butyrate, and H_2. They utilize large amounts of starch at one time and can store it in their bodies. The corn-rich substrate applied in the present study may have been adequate to maintain the growth of protozoa, which may explain why the nitrocompounds had no negative effect on the abundance of protozoa. In the present study, the abundance of *R. albus* increased with NEOH and NPOH addition, while anaerobic rumen fungi decreased with nitrocompound supplementation. This phenomenon could be attributed to the antagonistic association between ruminal fungi and cellulolytic bacteria [36]. In summary, although nitrocompounds could change the relative abundance of some microbial populations, the differences in diversity and metabolic activity in response to NE, NEOH, and NPOH need further investigation in order to determine the maximal inhibitory effect on CH_4 production with minimal adverse effects on rumen fermentation.

The hydrogenotrophic, methylotrophic, and acetoclastic pathways are the three major pathways for ruminal CH_4 production. The biochemical reactions and enzyme profiles involved in methanogenesis are well identified and described [37]. Methyl-coenzyme M reductase (*mcr*) is a key enzyme responsible for catalyzing the CH_4-producing step in the process of methanogenesis. As a gene encoding the alpha-subunit of *mcr*, *mcr*A was evolutionarily highly conserved, probably due to functional constraints on the catalytic activity of *mcr* [35]. Recently, the determination of *mcr*A gene expression was accepted for *mcr* activity measurement [16]. A positive correlation between decreased *mcr*A gene expression and decreased molar CH_4 proportions was found in the present study, suggesting that NE, NEOH, and NPOH inhibited CH_4 production by decreasing *mcr* activity. In addition, the activity of *mcr* is dependent mainly on the unique nickel-containing tetrapyrrole known as coenzyme F_{430} [38]. As reported by Gunsalus and Wolfe [39], coenzyme F_{430} is a yellow nonfluorescent compound released from *mcr*, with an absorption maximum at 430 nm on its UV-Vis absorption spectrum. In the present study, NE, NEOH, and NPOH addition also reduced the content of coenzyme F_{430}. Coenzyme F_{420} is of importance for methanogenesis and can act as an indicator for methanogenic activity [8,40]. The coenzyme F_{420} content in this study also decreased in the nitrocompound-treated cultures and was accompanied by CH_4 reduction, with the anti-methanogenic activity ranked as NEOH > NE > NPOH. A possible reason for these results might be due to the toxic action of nitrocompounds to methanogens [24,27]; this needs further clarification.

4.5. Degradation of NE, NEOH, and NPOH

Decreased concentrations of NE, NEOH, and NPOH were observed with increasing incubation time in all of the nitrocompound-treated cultures, thus confirming the presence of competent nitrocompound-degrading microbes within the incubations. A previous study by Anderson et al. [26] revealed that most of the rumen microorganisms tolerated nonlethal concentrations of naturally occurring nitrocompounds. In addition, the degradation rate of nitrocompounds can be enhanced via exposure to nitropropionic acid-containing forages. This phenomenon could be ascribed the improvement of nitro-degrading activity or enrichment in numbers of nitro-degrading microorganisms. Until now, *Denitrobacterium detoxificans* was recognized as a unique nitrocompound-reducing microbe which oxidizes reducing substrates, including H_2, formate, and lactate, to reduce nitrocompounds to their respective amines and minor nitrites [41]. Consequently, *D. detoxificans* has the potential to outcompete ruminal methanogens for available reductants [15]. In the present study, the nitrocompound disappearance rate was ranked as NEOH (−1.20) > NE (−0.80) > NPOH (−0.78), suggesting that the rumen microbes presented divergent metabolic capabilities regarding nitrocompound degradation, thus partially explaining why NE, NEOH, and NPOH showed different competition for methanogenesis-produced reductants.

5. Conclusions

Along with a dramatic increase in H_2 accumulation, both NE and NEOH were shown to be more effectivene in inhibiting methanogenesis than NPOH. Although nitrocompound addition decreased acetate and total VFA production, it had no negative effect on propionate. In addition, NE, NEOH, and NPOH addition decreased the population abundance of total methanogens, *Methanobacteriales*, and *Methanomicrobiales*, also causing decreases in *mcr*A gene expression and coenzyme F_{420} and F_{430} contents. The results provided evidence that NE, NEOH, and NPOH could reduce methanogen populations and dramatically decrease *mcr*A gene expression and coenzyme F_{420} and F_{430} contents with different magnitudes to reduce overall ruminal CH_4 production.

Author Contributions: Conceptualization, Z.Z. and H.Y.; Data curation, H.Y.; formal analysis, Z.Z., Y.W., X.S., and H.Y.; investigation, Z.Z., Y.W., and X.S.; writing—original draft preparation, Z.Z.; writing—review and editing, Z.Z., Z.C., S.L. and H.Y.; funding acquisition, H.Y.; project administration, H.Y. All authors have read and agreed to the published version of the manuscript.

References

1. Anderson, R.C.; Majak, W.; Rassmussen, M.A.; Callaway, T.R.; Beier, R.C.; Nisbet, D.J.; Allison, M.J. Toxicity and metabolism of the conjugates of 3-nitropropanol and 3-nitropropionic acid in forages poisonous to livestock. *J. Agric. Food Chem.* **2005**, *53*, 2344–2350. [CrossRef] [PubMed]
2. Latham, E.A.; Anderson, R.C.; Pinchak, W.E.; Nisbet, D.J. Insights on alterations to the rumen ecosystem by nitrate and nitrocompounds. *Front. Microbiol.* **2016**, *7*, 228. [CrossRef] [PubMed]
3. Anderson, R.C.; Carstens, G.E.; Miller, R.K.; Callaway, T.R.; Schultz, C.L.; Edrington, T.S.; Harvey, R.B.; Nisbet, D.J. Effect of oral nitroethane and 2-nitropropanol administration on methane-producing activity and volatile fatty acid production in the ovine rumen. *Bioresour. Technol.* **2006**, *97*, 2421–2426. [CrossRef] [PubMed]
4. Gutierrez-Bañuelos, H.; Anderson, R.C.; Carstens, G.E.; Slay, L.J.; Ramlachan, N.; Horrocks, S.M.; Callaway, T.R.; Edrington, T.S.; Nisbet, D.J. Zoonotic bacterial populations, gut fermentation characteristics and methane production in feedlot steers during oral nitroethane treatment and after the feeding of an experimental chlorate product. *Anaerobe.* **2007**, *13*, 21–31. [CrossRef]
5. Gutierrez-Banuelos, H.; Anderson, R.C.; Carstens, G.E.; Tedeschi, L.O.; Pinchak, W.E.; Cabrera-Diaz, E.; Krueger, N.A.; Callaway, T.R.; Nisbet, D.J. Effects of nitroethane and monensin on ruminal fluid fermentation characteristics and nitrocompound-metabolizing bacterial populations. *J. Agric. Food Chem.* **2008**, *56*, 4650–4658. [CrossRef]
6. Brown, E.G.; Anderson, R.C.; Carstens, G.E.; Gutierrez-Bañuelos, H.; McReynolds, J.L.; Slay, L.J.; Callaway, T.R.; Nisbet, D.J. Effects of oral nitroethane administration on enteric methane emissions and ruminal fermentation in cattle. *Anim. Feed Sci. Technol.* **2011**, *166–167*, 275–281. [CrossRef]
7. Hristov, A.N.; Oh, J.; Giallongo, F.; Frederick, T.W.; Harper, M.T.; Weeks, H.L.; Brancob, A.F.; Moatec, P.J.; Deightonc, M.H.; Williamsc, R.O.S.; et al. An inhibitor persistently decreased enteric methane emission from dairy cows with no negative effect on milk production. *Proc. Natl. Acad. Sci. USA* **2015**, *112*, 10663–10668. [CrossRef]
8. Hendrickson, E.L.; Leigh, J.A. Roles of coenzyme F420-reducing hydrogenases and hydrogen- and F420-dependent methylenetetrahydromethanopterin dehydrogenases in reduction of F420 and production of hydrogen during methanogenesis. *J. Bacteriol.* **2008**, *190*, 4818. [CrossRef]
9. Wongnate, T.; Sliwa, D.; Ginovska, B.; Smith, D.; Wolf, M.W.; Lehnert, N.; Raugei, S.; Ragsdale, S.W. The radical mechanism of biological methane synthesis by methyl-coenzyme M reductase. *Science* **2016**, *352*, 953–958. [CrossRef]
10. Zheng, K.; Ngo, P.D.; Owens, V.L.; Yang, X.P.; Mansoorabadi, S.O. The biosynthetic pathway of coenzyme F430 in methanogenic and methanotrophic archaea. *Science* **2016**, *354*, 339–342. [CrossRef]
11. Zhang, D.F.; Yang, H.J. Combination effects of nitrocompounds, pyromellitic diimide, and 2-bromoethanesulfonate on *in vitro* ruminal methane production and fermentation of a grain-rich feed. *J. Agric. Food Chem.* **2012**, *60*, 364–371. [CrossRef] [PubMed]
12. Menke, K.H.; Steingass, H. Estimation of the energetic feed value obtained by chemical analysis and in vitro gas production using rumen fluid. *Anim. Res. Dev.* **1988**, *28*, 7–55.
13. Reuter, B.W.; Egeler, T.; Schneckenburger, H.; Schoberth, S.M. *In vivo* measurement of F420 fluorescence in cultures of *Methanobacterium thermoautotrophicum*. *J. Bacteriol.* **1986**, *4*, 325–332. [CrossRef]
14. Ellefson, W.L.; Whitman, W.B.; Wolfe, R.S. Nickel-containing factor F430: Chromophore of the methylreductase of *Methanobacterium*. *Proc. Natl. Acad. Sci. USA* **1982**, *79*, 3707–3710. [CrossRef] [PubMed]
15. Ochoa-Garcia, P.A.; Arevalos-Sanchez, M.M.; Ruiz-Barrera, O.; Anderson, R.C.; Maynez-Perez, A.O.; Rodriguez-Almeida, F.A.; Chavez-Martinez, A.; Gutierrez-Banuelos, H.; Corral-Luna, A. *In vitro* reduction of methane production by 3-nitro-1-propionic acid is dose-dependent. *J. Anim. Sci.* **2019**, *97*, 1317–1324. [CrossRef]
16. Guo, Y.Q.; Liu, J.X.; Lu, Y.; Zhu, W.Y.; Denman, S.E.; McSweeney, C.S. Effect of tea saponin on methanogenesis, microbial community structure and expression of *mcr*A gene, in cultures of rumen micro-organisms. *Lett. Appl. Microbiol.* **2008**, *47*, 421–426. [CrossRef]
17. Denman, S.E.; Tomkins, N.W.; McSweeney, C.S. Quantitation and diversity analysis of ruminal methanogenic populations in response to the antimethanogenic compound bromochloromethane. *FEMS Microbiol. Ecol.* **2007**, *62*, 313–322. [CrossRef]

18. Denman, S.E.; McSweeney, C.S. Development of a real-time PCR assay for monitoring anaerobic fungal and cellulolytic bacterial populations within the rumen. *FEMS Microbiol. Ecol.* **2006**, *58*, 572–582. [CrossRef]
19. Zhou, M.; Hernandez-Sanabria, E.; Guan, L.L. Assessment of the microbial ecology of ruminal methanogens in cattle with different feed efficiencies. *Appl. Environ. Microbiol.* **2009**, *75*, 6524–6533. [CrossRef]
20. Yu, Y.; Lee, C.; Kim, J.; Hwang, S. Group-specific primer and probe sets to detect methanogenic communities using quantitative real-time polymerase chain reaction. *Biotechnol. Bioeng.* **2005**, *89*, 670–679. [CrossRef]
21. France, J.; Dijkstra, J.; Dhanoa, M.S.; Lopez, S.; Bannink, A. Estimating the extent of degradation of ruminant feeds from a description of their gas production profiles observed *in vitro*: Derivation of models and other mathematical considerations. *Br. J. Nutr.* **2000**, *83*, 143–150. [CrossRef] [PubMed]
22. Zhang, Z.W.; Wang, Y.L.; Wang, W.K.; Cao, Z.J.; Li, S.L.; Yang, H.J. The inhibitory action mode of nitrocompounds on in vitro rumen methanogenesis: A comparison of nitroethane, 2-nitroethanl and 2-nitro-1-propanol. *J. Agric. Sci.* **2019**, *9*, 1–9. [CrossRef]
23. Van Nevel, C.J.; Demeyer, D.I. Control of rumen methanogenesis. *Environ. Monit. Assess.* **1996**, *42*, 73–97. [CrossRef] [PubMed]
24. Anderson, R.C.; Callaway, T.R.; Kessel, J.A.S.V.; Soo, J.Y.; Edrington, T.S.; Nisbet, D.J. Effect of select nitrocompounds on ruminal fermentation; an initial look at their potential to reduce economic and environmental costs associated with ruminal methanogenesis. *Bioresour. Technol.* **2003**, *90*, 59–63. [CrossRef]
25. Anderson, R.C.; Huwe, J.K.; Smith, D.J.; Stanton, T.B.; Krueger, N.A.; Callaway, T.R.; Edrington, T.S.; Harvey, R.B.; Nisbet, D.J. Effect of nitroethane, dimethyl-2-nitroglutarate and 2-nitro-methyl-propionate on ruminal methane production and hydrogen balance *in vitro*. *Bioresour. Technol.* **2010**, *101*, 5345–5349. [CrossRef]
26. Anderson, R.C.; Rasmussen, M.A.; Allison, M.J. Metabolism of the plant toxins nitropropionic acid and nitropropanol by ruminal microorganisms. *Appl. Environ. Microbiol.* **1993**, *59*, 3056–3061. [CrossRef]
27. Anderson, R.C.; Rasmussen, M.A. Use of a novel nitrotoxin-metabolizing bacterium to reduce ruminal methane production. *Bioresour. Technol.* **1998**, *64*, 89–95. [CrossRef]
28. Zhang, Z.W.; Cao, Z.J.; Wang, Y.L.; Wang, Y.J.; Yang, H.J.; Li, S.L. Nitrocompounds as potential methanogenic inhibitors in ruminant animals: A review. *Anim. Feed Sci. Technol.* **2018**, *236*, 107–114. [CrossRef]
29. Angermaier, L.; Simon, H. On the reduction of aliphatic and aromatic nitro compounds by Clostridia, the role of ferredoxin and its stabilization. *Hoppe-Seyler's Z. Physiol. Chem.* **1983**, *364*, 961–975. [CrossRef]
30. Anderson, R.C.; Krueger, N.A.; Stanton, T.B.; Callaway, T.R.; Edrington, T.S.; Harvey, R.B.; Jung, Y.S.; Nisbet, D.J. Effects of select nitrocompounds on *in vitro* ruminal fermentation during conditions of limiting or excess added reductant. *Bioresour. Technol.* **2008**, *99*, 8655–8661. [CrossRef]
31. Soltan, Y.A.; Hashem, N.M.; Morsy, A.S.; El-Azrak, K.M.; Sallam, S.M. Comparative effects of Moringa oleifera root bark and monensin supplementations on ruminal fermentation, nutrient digestibility and growth performance of growing lambs. *Anim. Feed Sci. Technol.* **2018**, *235*, 189–201. [CrossRef]
32. Machmüller, A. Medium-chain fatty acids and their potential to reduce methanogenesis in domestic ruminants. *Agric. Ecosyst. Environ.* **2006**, *112*, 107–114. [CrossRef]
33. Janssen, P.H.; Kirs, M. Structure of the archaeal community of the rumen. *Appl. Environ. Microbiol.* **2008**, *74*, 3619–3625. [CrossRef] [PubMed]
34. Wina, E.; Muetzel, S.; Hoffmann, E.; Makkar, H.P.S.; Becker, K. Saponins containing methanol extract of sapindus rarak affect microbial fermentation, microbial activity and microbial community structure in vitro. *Anim. Feed Sci. Technol.* **2005**, *121*, 159–174. [CrossRef]
35. Vogels, G.D.; Hoppe, W.F.; Stumm, C.K. Association of methanogenic bacteria with rumen ciliates. *Appl. Environ. Microbiol.* **1980**, *40*, 608–612. [CrossRef]
36. Dehority, B.A.; Tirabasso, P.A. Antibiosis between ruminal bacteria and ruminal fungi. *Appl. Environ. Microbiol.* **2000**, *66*, 2921–2927. [CrossRef]
37. Lessner, D.J. Methanogenesis Biochemistry. In *eLS*; John Wiley & Sons, Ltd.: Hoboken, NJ, USA, 2009.
38. Hallam, S.J.; Peter, R.G.; Christina, M.P.; Paul, M.R.; Edward, F.D. Identification of methyl coenzyme M reductase A (mcrA) genes associated with methane-oxidizing archaea. *Appl. Environ. Microbiol.* **2003**, *69*, 5483–5491. [CrossRef]

39. Gunsalus, R.P.; Wolfe, R.S. Chromophoric factors F342 and F430 of Methanobacterium thermoautotrophicum. *FEMS Microbiol. Ecol.* **1978**, *3*, 191–193. [CrossRef]
40. Schönheit, P.; Keweloh, H.; Thauer, R.K. Factor F420 degradation in *Methanobacterium thermoautotrophicum* during exposure to oxygen. *FEMS Microbiol. Ecol.* **2010**, *12*, 347–349. [CrossRef]
41. Anderson, R.C.; Rasmussen, M.A.; Jensen, N.S.; Allison, M.J. Denitrobacterium detoxificans gen. nov.; sp. nov.; a ruminal bacterium that respires on nitrocompounds. *Int. J. Syst. Evol. Microbiol.* **2000**, *50*, 633–638. [CrossRef]

4

Rumen In Vitro Fermentation and In Situ Degradation Kinetics of Winter Forage Brassicas Crops

José Daza [1], Daniel Benavides [2], Rubén Pulido [3], Oscar Balocchi [1], Annick Bertrand [4] and Juan Keim [1,*]

1 Animal Production Institute, Faculty of Agricultural Sciences, Universidad Austral de Chile, Valdivia PO Box 567, Chile; jose_daza11@hotmail.com (J.D.); obalocch@uach.cl (O.B.)
2 Graduate School, Faculty of Agricultural Sciences, Universidad Austral de Chile, Valdivia PO Box 567, Chile; dabepez89@hotmail.com
3 Animal Science Institute, Faculty of Veterinary Sciences, Universidad Austral de Chile, Valdivia PO Box 567, Chile; rpulido@uach.cl
4 Soils and Crops Research and Development Centre, Agriculture and Agri-Food Canada, Québec City, QC G1V 2J3, Canada; Annick.Bertrand@AGR.GC.CA
* Correspondence: juan.keim@uach.cl;

Simple Summary: Winter brassica crops such as kales and swedes are used to supply feed in times of seasonal shortage. However, to the best of our knowledge, there is little information about the fermentation characteristics of these forages in the rumen. This study assessed the nutrient concentration, in vitro fermentation and in situ rumen degradation characteristics of *Brassica oleracea* (L.) ssp. *acephala* (kales) and *Brassica napus* (L.) ssp. *napobrassica* (swedes). The kales and swedes both showed different nutrient concentrations and fermented fast and extensively in the rumen. However, in vitro fermentation of swedes resulted in lower acetate and greater proportions of butyrate and propionate. Varieties of swedes showed more differences in terms of degradation and fermentation in the rumen compared to kale varieties.

Abstract: The aim of the present study was to evaluate the nutritional value, the rumen in vitro fermentation, and the in situ degradation of *Brassica oleracea* (L.) ssp. *acephala* (kales) and *Brassica napus* (L.) ssp. *napobrassica* (swedes) for winter use. Five varieties of each brassica were used in three field replicates and were randomized in a complete block nested design. All forage varieties were harvested at 210 days post-sowing to analyze the chemical composition, in vitro gas production, volatile fatty acid (VFA) production and in situ dry matter (DM) and crude protein (CP) degradability. Kales presented higher DM and neutral detergent fiber (NDF) content ($p < 0.01$), whereas swedes showed higher CP, metabolizable energy (ME), glucose, fructose, total sugars, NFC, and nonstructural carbohydrate (NSC) content ($p < 0.01$). The kale and swede varieties differed in their CP and sugar concentrations, whereas the kale varieties differed in their DM and raffinose content. The rates of gas production were higher for swedes than for kales ($p < 0.01$). No differences between the brassica species ($p > 0.05$) were observed in the total VFA production, whereas kales had a higher proportion of acetate and swedes had higher proportions of butyrate ($p < 0.05$). Only the swede varieties showed differences in VFA production ($p < 0.05$). The soluble fraction "a", potential and effective in situ DM degradability were higher in swedes ($p < 0.01$), but kales presented greater DM and CP degradation rates. Differences were observed between brassica species in the chemical composition, degradation kinetics, and ruminal fermentation products, whereas differences among varieties within species were less frequent but need to be considered.

Keywords: kale; swede; volatile fatty acids; degradation rates

1. Introduction

Brassicas such as kales (*Brassica oleracea* (L.) ssp. *acephala*) and swedes (*Brassica napus* (L.) ssp. *napobrassica*) are used for ruminant feed during winter [1], which is a season with low pasture growth in humid temperate regions [2]. These forages can offer high dry matter (DM) production and nutritional quality in a short time, which is related to high metabolizable energy (ME), water-soluble carbohydrates (WSC), and low neutral detergent fiber (NDF) content [3,4]. Winter brassicas have been used successfully in sheep [4], dry cows [5], and lactating dairy cows [6]. In addition, forage brassicas have an environmental advantage; they reduce the amount of enteric methane (CH_4) per unit of DM intake compared to ryegrass pasture [4,7]. Although nutrient concentrations in brassicas have been widely described, the nutritive value of these forages depends on the quantity of nutrients available to the animal, which is determined by fermentation processes [8] and the presence of secondary compounds such as glucosinolates and S-methyl-cysteine sulfoxide that are present in brassicas [9], thus meaning animal responses can be affected.

Complementary evaluation methods, such as the ruminal digestibility of nutrients or products of ruminal fermentation and metabolism, have been suggested to determine the real nutritive value of forages [10,11]. The ruminal in situ incubation technique is considered a reference method to estimate degradation parameters, when adjusted to suitable nonlinear models [12]. These parameters are used by feeding evaluation models to estimate nutritive value, nutrient supply, and animal performance [8]. On the other hand, the in vitro gas production technique (IVGPT) allows the determination of fermentation kinetics [13]; estimates of DM, protein, and fiber degradation; ruminal volatile fatty acid (VFA) content; and microbial protein synthesis [14]. The popularity of in vitro gas production (GP) stems mainly from the ability to exercise experimental control, the capacity to nondestructively screen a large number of substrates, the kinetic information obtained, and relatively low costs [10]. Thus, IVGPT offers a unique tool for researchers to address a wide range of nutritional issues in ruminants [15].

Whereas degradation kinetics and ruminal fermentation of summer brassica species (rape and turnip) and varieties [16] have been reported in the literature, few reports exist on the effect of winter brassica species (kales and swedes) and varieties on the in situ degradation kinetics and fermentation end products. For example, Sun et al. [4] have observed that sheep fed swedes showed modified VFA profiles in their rumen fluid and lowered methane yield in contrast with those fed kales or perennial ryegrass. Keogh et al. [1] have reported no effects on the rumen VFA concentration from increases in the dietary proportion of kales in the diets of dry cows. Valderrama and Anrique [17] have reported DM and crude protein (CP) degradation kinetics of kale leaves; however, to the best of our knowledge, such data have not been reported for swedes. Moreover, the nutritive value of brassicas varies among species and varieties within species [3,16], and, therefore, information is still lacking about rumen fermentation and the kinetics of winter brassica species such as kales and swedes.

Hence, the aim of this study was to determine the nutritive value of forage brassica species (kales and swedes) and varieties for winter use, based on their nutrient concentration, in vitro ruminal fermentation, and in situ rumen degradation kinetics.

2. Materials and Methods

All animal procedures were performed in accordance with the UK Animals (Scientific Procedures) Act and associated guidelines, and approved by the Animal Ethics Committee of the Austral University of Chile (approval number 144/2013).

2.1. Site and Experimental Design

This experiment was carried out at the Agricultural Research Station (39°47′ S, 73°13′ W) of the Austral University of Chile on a Typic Hapludand soil with an initial water pH of 5.8, Olsen-P of 19.1 mg/kg, exchangeable potassium of 214 mg/kg, and aluminum saturation of 3.1% (measured for the first 20 cm of the soil profile).

Prior to soil preparation and sowing, the weeds were controlled chemically with glyphosate at a dosage of 2025 g/ha of active ingredient. Two brassica species were evaluated (kales and swede), and five varieties were sown for each species: Caledonian (K1), Elba (K2), Sovereign (K3), Regal (K4), and Coleor (K5) for the kales and Major Plus (S1), Aparima Gold (S2), Highlander (S3), Dominion (S4), and Invitation (S5) for swedes. The plot sizes were 6 m by 4 m, with three replicates for each variety, and plots were arranged in field blocks. The varieties were established in October 2014 at a seed dosage of 4.0 (kales) and 1.5 kg/ha (swedes). Fertilizers were applied at sowing to correct any soil nutrient deficiencies. A fertilizer mixture (7 % N-30 % P_2O_5 -12 % K_2O) at doses of 500 kg/ha (35 kg N/ha, 150 kg P_2O_5/ha, and 60 kg K_2O/ha) and 46 kg/ha boronatrocalcite were applied at sowing. After emergence, weeds were controlled chemically by applying Lontrel 3A (clopyralid 475 g of active ingredient (a.i.)/L) and Tordon 24 k (picloram 240 g a.i./L) at doses of 300 and 200 cc/ha respectively. Five weeks after sowing, when the plants had two or three leaves, 125 kg N/ha (urea) was applied, and in January 2015, applications of 750 cc/ha of Aramo (tepraloxydim 200 g a.i./L) and 200 cc/ha of karate (50 g a.i./L Lambda-cyhalothrin) were made.

During the trial, five cuts were made, with an approximate interval of 30 days between cuts, with the first harvest occurring at 90 days after plant emergence. In each cut, 4 m^2 of each crop was harvested. The kale varieties were cut to 20 cm above the ground level, the swede varieties were collected manually, and soil attached to the roots was removed. Plants were weighed and then separated into the main components (leaf and stem for kales and leaf and bulb for swedes). The samples were then dried in a forced-air oven at 60 °C for 48 h for determination of the dry matter (DM).

Samples for nutrient concentrations and in vitro and in situ incubations were harvested at 210 days post-sowing, and the plants were separated into their morphological components (leaf and stems for kale varieties and leaf and bulbs for swede varieties) before being chopped and then being frozen at −20 °C. Later, they were lyophilized (Virtis 10-45 MR-BA, Gardiner, New York, NY, USA) and then ground (Wiley mill, 158 Arthur H. Thomas, Philadelphia, Pennsylvania, PA, USA) to 5 mm for in situ incubations and to 1 mm for nutrient concentrations analyses and in vitro gas production. For the in vitro and in situ incubations, samples of brassica species were composed of a leaf to stem ratio of 35:65 for kales and a leaf to bulb ratio 30:70 for swedes. The ratios were the average proportions of organ components at harvest obtained in this study.

2.2. *In Situ Incubations*

Three dry Holstein-Friesian cows (one for each block) fitted with ruminal cannulas (4′ Pliable Rumen Cannula w/Stopper and U Bolt, Ankom Technologies, Macedon, New York, NY, USA) were used. At the time of the experiment, the ruminal pH (6.55 ± 0.32) of each cow was measured. Cows were offered grass silage (7.5 kg DM), summer turnips (4.5 kg DM) and commercial concentrate (2.0 kg DM). Samples of each variety were incubated in duplicate (~4 g DM) in Dacron bags (10 cm by 20 cm; pore size of 40–60 μm) and sealed. Up to 20 bags were deposited inside a lingerie bag (30 cm by 40 cm in size). Brassica samples from each block were incubated in a different cow and a control sample (commercial concentrate) was incubated in each cow to evaluate cow-to-cow variation.

Prior to incubation in the rumen, the bags were soaked in warm water (40 °C) for 20 min. Nine incubation times were considered: 0, 2, 4, 8, 10, 12, 14, 24, and 48 h. The samples corresponding to 0 h were not introduced into the rumen and were used to determine the soluble fraction. After the incubation, the bags containing the residue were removed from the rumen and were washed under running cold water until no further color appeared; then, they were frozen at −20 °C for 24 h to stop fermentative activity. Thereafter, the bags were defrosted, thawed in water at 4 °C, and washed with a commercial washing machine for 30 min at a "normal" wash setting. Finally, residues were oven-dried at 60 °C for 48 h. The residues were weighed, the DM was calculated by placing the samples in an oven at 105 °C for 12 h, and the CP concentration was determined to calculate nutrient loss.

A correction for small particle loss was made as follows. The samples of each brassica variety (1.5 g) were weighed in a beaker. Then, 40 mL of tap water was added and the mix was stored at room temperature (20 °C) for 1 h; afterward, the mix was filtered through a nitrogen-free filter paper and washed eight times with 20 mL of water. The residues were oven-dried and analyzed individually. Degradation parameters (a, b, and c), potential degradation (PD) and effective degradability (ED) were corrected according to Hvelplund and Weisbjerg [18].

2.3. *In Vitro Incubations*

Duplicates (1 g) of each sample were incubated in 160 mL glass bottles. Each brassica variety, a control (commercial concentrate) sample and two blanks (bottles without substrate) were used. Subsequently, 85 mL of Goering-Van Soest medium and 4 mL of reducing agent (NaOH 2.5 mM and cysteine-HCl 2.5 mM) were added at 39 °C under continuous gasification (CO_2) to maintain anaerobic conditions and the bottles were covered with rubber stoppers and aluminum seal.

The inoculum was extracted from two dry Holstein-Friesian cows with ruminal cannulas with a live weight (PV) of 560 ± 20 kg and a ruminal pH of 6.6 ± 0.53; at the time of the extraction, the animals were offered the same diet as in the in situ trial. Rumen fluid was obtained before the cows were fed in the morning and was stored in a thermos flask to preserve the temperature until being transferred to the laboratory. Once there, the fluid was filtered through four layers of cheesecloth while being maintained at a temperature of 39 °C and under a constant flow of CO_2. Rumen fluids from the two donor cows were mixed in equal proportions and then inoculated (10 mL) into the bottles. After inoculation, the bottles were placed in a water bath at 39 °C under continuous horizontal movement at 50 rpm.

Once the rumen fluid was inoculated, the initial gas was extracted from the bottles. The gas pressure in the headspace of the bottles above atmospheric pressure was measured manually with a pressure transducer (PCE Instruments, Tobarra, Albacete, Spain) at 2, 3, 4, 5, 6, 8, 10, 12, 18, 24, 36, and 48 h, and the volume of gas produced was measured by extraction using syringes connected through a three-way Luer valve from the bottles until the visual display of the transducer read zero, and once the volume of gas produced was recorded, it was eliminated. Fermentations were stopped after 48 h by placing the bottles on ice. Each field block (five varieties of kales and five varieties of swedes) was incubated at different runs. Thus, the first block was incubated at a first run, those corresponding to block number two were incubated at the second run, and samples from block three were incubated at the third run. A control standard (commercial concentrate) was incubated at each incubation run to control the day-to-day variation.

Once the in vitro incubation was finished, the samples were kept on ice to stop fermentative processes and residue duplicates from each sample were collected and then centrifuged at 15,000× *g* and 4 °C. After centrifugation, 0.9 mL of the supernatant was extracted to determine the VFA concentrations with a GG-2010 gas chromatograph (Shimadzu Corporation, Kyoto, Japan).

2.4. *Analyses*

The dry matter content was measured by weighing the samples before and after drying with a forced-air oven, initially at 60 °C for 48 h and then at 105 °C for 12 h. The CP concentration was determined by combustion (Leco Model FP-428, Nitrogen Determinator, Leco Corporation, St Joseph, Minnesota, MI, USA) based on the DUMAS method (nitrogen × 6.25); digestible organic matter on a dry matter basis (DOMD) was measured according to Tilley and Terry [19]; neutral detergent fiber (aNDF) was measured by using a heat stable amylase [20]; and ash and ether extract (EE) were analyzed according [21] (Methods ID 942.05 and ID 920.39 for ash and EE respectively). Sugars (raffinose, sucrose, glucose, and fructose) were analyzed by Waters ACQUITY ultrahigh-performance liquid chromatography (UPLC, Waters, Milford, Massachusetts, MA, USA), and starch quantification was determined by colorimetric detection of non-soluble residues after enzymatic digestion with amyloglucosidase according to Pelletier et al. [22]. The sum of sugars and starch yielded the content of total nonstructural carbohydrates (NSC). An estimation of the combined organic acids plus neutral detergent soluble fiber (OA + NDSF) was calculated according to Hall et al. [23], where OA + NDSF =

NFC − NSC. Non-fibrous carbohydrates were calculated as follows: NFC = 100 − CP − aNDF − EE − ash + neutral detergent insoluble crude protein (NDICP).

2.5. Calculations

The in situ disappearance of DM and CP was determined using the non-linear model described by Ørskov and McDonald [12] to determine the potential degradation according to the exponential model

$$\mathrm{PD} = a + b \times (1 - e^{-kt})$$

where a is the soluble fraction (fraction washed out at $t = 0$; this value resulted from the incubation of 0 h bags corrected for particle loss and fixed into the model), b is the insoluble but potentially degradable fraction, k is the degradation rate (per hour), and t is the time (h).

The effective degradability (ED) was calculated assuming a fractional passage rate (kp) of 2%, 5%, and 8% per hour according to the following equation:

$$\mathrm{ED} = a + b \times c/(c + \mathrm{kp})$$

These parameters were corrected for the losses of small particles which are degraded in a similar way to the particles remaining in the bag, as reported by Hvelplund and Weisbjerg [18].

After correcting for gas production of the blanks, the obtained GP data were adjusted to the generalized Michaelis-Menten model without a lag phase [11], as seen in the equation

$$\mathrm{GP} = \mathrm{A} \times [T^n/(T^n + K^n)]$$

where GP is the gas production at time T, A is the asymptote of GP (mL), n is the determined value of the shape of the curve, and K is the time taken to produce half of A.

The following parameters were calculated according to Groot et al. [24] and France et al. [11], i.e.,

fermentation rate at half-life (C) = $n/(2 \times K)$
maximal fermentation rate (MDR) = $(n - 1)^{((n-1)/n)/\mathrm{k}}$
time to ferment x% of the substrate (t_x) = $K \times ((\mathrm{X}/(1 - \mathrm{X}))^{(1/\mathrm{n})})$

where X = 0.25, 0.75, and 0.90 of A.

2.6. Statistical Analyses

Parameters of in vitro GP, in situ degradation kinetics, VFA content and nutrient concentration were averaged for analytical replicates and analyzed with the MIXED procedure of SAS (SAS Institute, Cary, North Carolina, NC, USA).

Data were analyzed under a nested design, with three replicates organized in complete randomized blocks. Varieties were nested within species and the random effect of the field replicate was included as a block. When significant differences ($p < 0.05$) were found, the Tukey-Kramer multiple-comparison test was used in the LSMEANS procedure statement in SAS.

3. Results

3.1. Nutrient Concentration and Sugar Profile

Kales had greater concentrations of DM, EE (+4 g/kg), and aNDF (+123 g/kg) than swedes ($p < 0.01$; Table 1). Swedes showed greater CP (+25 g/kg; $p < 0.01$) and total sugar concentrations (+75 g/kg; $p < 0.01$) than kales, as well as individual sugars, such as glucose and fructose, NSC, OA + NDSF and DOMD. Raffinose and sucrose concentrations were greater in kales ($p < 0.01$). The ash and starch concentrations did not vary between the species ($p > 0.05$).

The kale varieties differed in their DM, CP, EE, raffinose, glucose, fructose, sugars, starch, and NSC concentrations ($p < 0.01$). Coleor had greater concentrations of CP (+30 g/kg) than Regal; Elba and Sovereign greater EE than Coleor; Coleor greater raffinose than Coledonian and Sovereign; and Regal and Coledonian greater concentrations of glucose and fructose than Sovereign. Finally, Regal showed greater concentrations of total sugars than Elba and Sovereign; Sovereign and Coleor greater concentrations of starch than Coledonian; and Regal the greatest NSC concentration of all varieties (231 g/kg), whereas Elba and Sovereign had the lowest (190 g/kg and 194 g/kg, respectively).

Furthermore, the swede varieties differed in their CP, glucose, fructose, sugars, starch, and NSC concentrations ($p < 0.01$). The concentrations of ash, aNDF, sucrose, OA + NDSF, and DOMD for varieties of both species were not different ($p > 0.05$). Invitation showed greater concentrations of CP (+30 g/kg) than Aparima Gold. Major Plus and Highlander showed the greater concentration of glucose, fructose, and sugars (280 g/kg and 288 g/kg, respectively, compared to 233 g/kg, 241 g/kg, and 234 g/kg in Aparima Gold, Dominion and Invitation). For starch, Aparima Gold and Dominion had greater concentrations (32 g/kg and 26 g/kg, respectively) than Highlander (7 g/kg). Finally, the concentrations of NSC in Major Plus and Highlander (291 g/kg and 295 g/kg, respectively) were greater than that for Invitation.

3.2. In Situ Degradation Parameters

Dry matter degradation parameters differed between brassica species (Table 2) except for the insoluble but potentially degradable fraction "b" ($p > 0.05$). Swedes had a higher soluble fraction "a" compared to kales (591 g/kg and 499 g/kg, respectively; $p < 0.01$) but a lower degradation rate "c" (0.25 h^{-1} and 0.34 h^{-1}, respectively; $p < 0.01$). However, swedes showed greater PD (+90 g/kg) and ED with ruminal passage rates of 2%, 5%, and 8% per hour compared to kales ($p < 0.01$). No significant differences were found in the in situ degradation parameters for the kale varieties ($p > 0.05$). Within the swede varieties, fraction "a" was greater for Major Plus (634 g/kg) compared to Aparima Gold (545 g/kg), Dominion (581 g/kg), and Invitation (586 g/kg; $p < 0.01$).

For the in situ CP degradation parameters, a species effect was observed on the fractional degradation rate "c", with kales having a faster degradation rate compared to swedes (0.48 h^{-1} and 0.36 h^{-1}, respectively; $p < 0.01$), whereas no brassica species effect was observed on the other in situ CP degradation parameters with average values of 559 g/kg, 380 g/kg, and 939 g/kg for "a", "b", and PD, respectively ($p > 0.05$). Fractions "a" and "b" were affected by the kale varieties: Coleor presented a higher "a" than Regal and Caledonian (+159 g/kg and +139 g/kg, respectively; $p < 0.01$). On the other hand, Caledonian and Regal had a greater "b" fraction compared to Coleor ($p < 0.05$). Effective degradability at 8% per hour of Sovereign (904 g/kg) and Coleor (900 g/kg) were higher than that of Regal (839 g/kg; $p < 0.05$). No effects for the swede varieties were found for any of the CP degradation parameters ($p > 0.05$).

Table 1. Nutrient concentration (g/kg dry matter) of five varieties of kale (from K1 to K5) and swedes (from S1 to S5).

Parameter	Kales	Swedes	[1] SEM	Kales					Swedes					[1] SEM	p Value	
				K1	K2	K3	K4	K5	S1	S2	S3	S4	S5		Species	Variety
DM (g/kg sample)	122	74	0.3	108 ^{b}	123 ^{ab}	135 ^{a}	107 ^{b}	136 ^{a}	71	81	72	71	77	0.6	<0.01	<0.01
Ash	79	77	0.2	88	83	76	73	74	74	75	73	81	82	0.5	NS	NS
CP	111	136	0.4	114 ^{ab}	111 ^{ab}	116 ^{ab}	88 ^{b}	128 ^{ab}	127 ^{ab}	123 ^{b}	124 ^{ab}	151 ^{ab}	153 ^{a}	0.9	<0.01	<0.01
EE	13	9	0.04	13 ^{ab}	14 ^{a}	14 ^{a}	13 ^{ab}	10 ^{b}	10	9	9	9	9	0.09	<0.01	<0.05
aNDF	300	176	0.7	271	309	328	297	293	169	196	165	169	182	1.6	<0.01	NS
Raffinose	7	2	0.4	6 ^{b}	7 ^{ab}	5 ^{b}	7 ^{ab}	10 ^{a}	2	1	2	2	2	0.9	<0.01	<0.01
Sucrose	59	15	4.6	47	52	75	50	70	11	17	12	16	22	10.4	<0.01	NS
Glucose	63	139	3.1	74 ^{ab}	59 ^{bc}	44 ^{c}	83 ^{a}	56 ^{bc}	157 ^{a}	127 ^{b}	156 ^{a}	123 ^{b}	127 ^{b}	6.8	<0.01	<0.01
Fructose	51	99	3.0	58 ^{ab}	50 ^{abc}	35 ^{c}	69 ^{a}	42 ^{bc}	109 ^{a}	88 ^{b}	115 ^{a}	100 ^{ab}	82 ^{b}	6.6	<0.01	<0.01
Sugars	180	255	5.3	185 ^{ab}	168 ^{b}	160 ^{b}	209 ^{a}	179 ^{ab}	280 ^{a}	233 ^{b}	288 ^{a}	241 ^{b}	234 ^{b}	11.9	<0.01	<0.01
Starch	25	19	3.7	14 ^{b}	21 ^{ab}	34 ^{a}	22 ^{ab}	31 ^{a}	11 ^{bc}	32 ^{a}	7 ^{c}	26 ^{ab}	20 ^{abc}	8.4	NS	<0.05
NSC	205	274	7.1	199 ^{ab}	190 ^{b}	194 ^{b}	231 ^{a}	210 ^{ab}	291 ^{a}	265 ^{ab}	295 ^{a}	267 ^{ab}	253 ^{b}	15.8	<0.01	<0.05
OA + NDSF	292	327	5.4	315	293	272	298	284	328	331	332	322	321	12.2	<0.01	NS
DOMD	727	831	0.7	740	715	698	732	749	841	823	841	828	824	1.7	<0.01	NS

^{a,b,c} within a row, different letters represent the significant differences at p value < 0.05. [1] SEM standard error of the mean; DM, dry matter; CP, crude protein; EE, ether extract; aNDF, neutral detergent fiber with a heat stable amylase; NSC, non-structural carbohydrates, representing the sum of sugars and starch; OA + NDSF, organic acids and neutral detergent soluble fiber; DOMD, digestible organic matter on DM basis. Kale varieties: K1, Caledonian; K2, Elba; K3, Sovereign; K4, Regal; K5, Coleor. Swede varieties: S1, Major Plus; S2, Aparima; S3, Highlander; S4, Dominion; S5, Invitation.

Table 2. In situ ruminal digestion kinetics, potential degradability (PD) and effective degradability (ED, considering a passage rate of 2%, 5%, and 8% per hour) of five varieties of kale (from K1 to K5) and swedes (from S1 to S5).

Parameter	Kales	Swedes	[1] SEM	Kales					Swedes					[1] SEM	p Value	
				K1	K2	K3	K4	K5	S1	S2	S3	S4	S5		Species	Variety
Dry matter degradation parameters																
a	499	591	0.6	519	499	479	504	491	634 [a]	545 [b]	612 [ab]	581 [b]	586 [b]	1.3	<0.01	<0.01
b	370	368	0.6	363	374	374	376	363	349	387	352	367	384	1.4	NS	NS
c	0.34	0.25	0.02	0.35	0.32	0.37	0.30	0.35	0.22	0.29	0.28	0.22	0.22	0.05	<0.01	NS
PD	869	959	0.6	882	873	853	881	854	982	931	964	964	969	1.9	<0.01	NS
ED, 2%/h	846	931	0.8	860	849	831	855	835	952	907	940	919	936	1.7	<0.01	NS
ED, 5%/h	817	895	0.7	831	819	802	823	809	915	875	910	881	896	1.5	<0.01	NS
ED, 8%/h	793	867	0.7	807	794	779	797	786	886	848	885	851	864	1.5	<0.01	NS
Crude protein degradation parameters																
a	550	568	1.4	494 [bc]	554 [abc]	596 [ab]	474 [c]	633 [a]	580	532	520	594	615	3.2	NS	<0.01
b	383	377	1.4	429 [a]	387 [ab]	354 [ab]	431 [a]	313 [b]	384	398	390	361	353	3.1	NS	<0.05
c	0.48	0.36	0.02	0.53	0.42	0.53	0.43	0.47	0.34	0.40	0.41	0.31	0.32	0.05	<0.01	NS
PD	933	945	1.1	923	941	950	906	946	964	930	911	955	968	2.5	NS	NS
ED, 2%/h	918	927	1.0	909	924	937	888	933	943	912	898	935	947	2.3	NS	NS
ED, 5%/h	897	900	0.9	887	900	920	862	916	915	886	873	906	921	2.1	NS	NS
ED, 8%/h	878	877	0.9	867 [ab]	879 [ab]	904 [a]	839 [b]	900 [a]	890	864	852	882	898	2.0	NS	<0.05

[a, b, c] within a row, different letters represent the significant differences at p value < 0.05. [1] SEM standard error of the mean; a, soluble fraction; b, insoluble but potentially degradable fraction; c, fractional disappearance rate constant at which b is degraded. Kale varieties: K1, Caledonian; K2, Elba; K3, Sovereign; K4, Regal; K5, Coleor. Swede varieties: S1, Major Plus; S2, Aparima; S3, Highlander; S4, Dominion; S5, Invitation.

3.3. *In Vitro Fermentation Products*

The in vitro GP parameters were affected by the brassica species, whereas for varieties within species no effect was observed (Table 3). The GPs at 24 h and 48 h and "A" were higher for swedes compared to kales (256 mL/g, 285 mL/g, and 285 mL/g against 233 mL/g, 260 mL/g, and 261 mL/g DM, respectively; $p < 0.01$). Fermentation rate parameters "C" and "MDR" (h^{-1}) were slightly faster for swedes (0.15 h^{-1}) compared to kales (0.14 h^{-1}; $p < 0.05$). However, no differences were observed for the time to fermentation of 25%, 50%, and 75% of the substrate (h), whereas 90% of the substrate was fermented 1.1 h earlier for swedes ($p < 0.05$).

Table 3. In vitro gas production kinetics of five varieties of kales (from K1 to K5) and swede (from S1 to S5).

Parameter	Kales	Swedes	[1] SEM	*p* Value	
				Species	Variety
24 h GP	232	256	3.0	<0.01	NS
48 h GP	260	285	3.1	<0.01	NS
A	261	285	3.3	<0.01	NS
K	6.1	6.1	0.08	NS	NS
C	0.14	0.15	0.01	<0.05	NS
MDR	0.14	0.15	0.01	<0.05	NS
t25	3.2	3.2	0.04	NS	NS
t75	11.8	11.5	0.2	NS	NS
t90	22.6	21.5	0.5	<0.05	NS

[1] SEM standard error of the mean; 24 h GP (mL/g DM), gas production after 24 h of incubation; 48 h GP (mL/g DM), gas production after 48 h of incubation; A, asymptotic gas production (mL/g DM); K, time to ferment 50% of the substrate (h); C, degradation rate at half-life (per h); MDR, maximal degradation rate (per h); t25, t75, and t90, time to ferment 25%, 75%, and 90% of the substrate, respectively (h).

For the total volatile fatty acids (tVFA) concentrations, the propionate and branched-chain VFA (BCVFA) molar proportions of tVFAs showed no effects for brassica species ($p > 0.05$; Table 4). The acetate molar proportions of tVFAs (+48 mmol/mol) and A:P ratio were greater for kales, whereas the fermentation of swedes increased the molar proportions of butyrate (+33 mmol/mol). Kale varieties showed no differences for tVFA and the relative proportions of the different VFAs, whereas the tVFAs and the relative proportion of acetate and butyrate were affected by swede varieties. Major Plus showed a higher concentration of tVFAs (61.3 mM) compared to Dominion (45.5 mM), and Major Plus, Aparima Gold, and Invitation had higher relative proportions of acetate (555 mmol/mol, 564 mmol/mol, and 572 mmol/mol, respectively) compared to Dominion, which showed a higher relative proportion of butyrate (162 mmol/mol) than Aparima Gold and Invitation (128 mmol/mol and 120 mmol/mol, respectively).

Table 4. In vitro rumen volatile fatty acids (VFA) of five varieties of kale (from K1 to K5) and swede (from S1 to S5).

Parameter	Kales	Swedes	[1] SEM	Kales					Swedes					[1] SEM	*p* Value	
				K1	K2	K3	K4	K5	S1	S2	S3	S4	S5		Species	Variety
VFA concentration																
Total VFA (mM)	52.5	51.2	1.7	52.6	49.6	49.1	56.9	54.5	61.3 [a]	49.1 [ab]	49.6 [ab]	45.5 [b]	50.4 [ab]	3.8	NS	<0.05
Acetate (mmol/mol)	597	549	4.2	595	591	597	597	602	555 [ab]	564 [ab]	537 [bc]	520 [c]	572 [a]	7.2	<0.05	<0.01
Propionate (mmol/mol)	245	258	1.9	245	249	243	243	244	255	254	266	258	257	4.3	NS	NS
Butyrate (mmol/mol)	106	139	2.3	107	104	106	111	103	141 [ab]	128 [b]	143 [ab]	162 [a]	120 [b]	4.9	<0.01	<0.01
BCVFA (mmol/mol)	52	54	1.5	53	55	54	48	50	48	54	54	60	52	2.7	NS	NS
A:P ratio	2.4	2.1	0.04	2.4	2.4	2.5	2.5	2.5	2.2	2.2	2.0	2.0	2.2	0.09	<0.01	NS

[a, b, c] within a row, different letters represent the significant differences at p value < 0.05. [1] SEM standard error of the mean; BCVFA, branched-chain VFA (isobutyrate + isovalerate + valerate); A:P ratio, acetate to propionate ratio. Kale varieties: K1, Caledonian; K2, Elba; K3, Sovereign; K4, Regal; K5, Coleor. Swede varieties: S1, Major Plus; S2, Aparima; S3, Highlander; S4, Dominion; S5, Invitation.

4. Discussion

4.1. Nutrient Concentration

The current study reports the nutritive value of five varieties of two winter forage brassica species that are used to complement the feeding base used in grazing systems when pasture availability is not sufficient to fulfill livestock requirements. Forage brassicas have a higher demonstrated DOMD than perennial pastures; Sun et al. [4], for example, have reported that the DOMD of kales and swedes (883 g/kg and 918 g/kg DM) is 25% higher than that in perennial ryegrass pastures (703 g/kg DM). However, DOMD values observed by the current study are lower than those previously reported for kales and swedes (727 g/kg and 831 g/kg DM).

Sun et al. [4] and Westwood and Mulcock [3] have observed that kales contain higher values of DM, EE, and aNDF than swedes, which is similar to the effect observed in the present study. Regarding CP, swedes showed a higher concentration (136 g/kg) compared to kales (111 g/kg). Westwood and Mulcock [3] have reported similar values for swedes (137 g/kg) but lower values for kales (97 g/kg); however, they have observed that CP is highly variable among kale varieties, ranging from 63 g/kg to 138 g/kg, which is in accordance with our study where the CP concentration of the kale varieties ranged from 88 g/kg to 128 g/kg. These differences are mainly associated with the leaf to stem ratio, as forage brassicas concentrate more CP in their leaves [25]. For example, Valderrama and Anrique [17] have reported CP concentrations of 225 g/kg in kale leaves.

The content of sugars was similar to that reported by Sun et al. [4], with swedes having a higher concentration of sugars compared to kales. Winter brassicas have higher sugars compared to grass-based permanent pastures and perennial ryegrass, where values range from 73 g/kg to 118 g/kg when harvested during winter and early spring, the times when winter brassicas are used [26,27].

The amount of raffinose and sucrose were higher in kales; however, these amounts were not enough to compensate for the higher concentration of fructose and glucose in swedes. Importantly, varieties of both kales and swedes differed in the concentrations of fructose and glucose. In the case of kales, the Regal concentrated almost twofold the amounts of glucose and fructose compared to Sovereign, whereas for swedes, the Major Plus and Highlander concentrated approximately 20 g/kg and 30 g/kg more than the other three varieties. This approach is important, as it considers the type and amount of sugars demonstrated to affect fermentation in the rumen and the end products such as the VFAs [28].

Additionally, the concentration of OA + NDSF was evaluated, and this was higher for swedes (327 g/kg DM) than for kales (291 g/kg DM). This parameter was evaluated in summer brassicas by Keim et al. [16], who found it to be 313 g/kg and 314 g/kg DM for turnip and rape, respectively. Neutral detergent soluble fiber is mainly composed of galactans, β-glucans, soluble hemicelluloses, and pectic substances [23], while organic acids (OA) are mainly carbohydrate derivatives and may contain lactate, citric acid cycle components and secondary plant compounds such as oxalate and shikimate [23]. Previous studies have reported concentrations of pectins between 77 g/kg and 129 g/kg DM for brassicas in general [29]. Likewise, Sun et al. [4] have reported 80 g/kg and 69 g/kg DM for kales and swedes, respectively. In addition to sugars, these components of the NDSF could have an effect on ruminal fermentation and VFA concentrations, and although NDSFs are readily and extensively broken down in the rumen, they do not mimic the pH-lowering effect of starch because they generally produce little or no lactate, and their fermentation ceases at low pH [30]. On the other hand, organic acids do not support microbial growth [23] and therefore have little impact on rumen metabolism.

4.2. In Situ Degradation Parameters

Brassica forages are characterized by their high concentration of readily fermentable carbohydrates and high digestibility [9], which is in accordance with the high soluble fraction and fast degradation rates observed in our study for both species. To the best of our knowledge, limited data exist from reports on degradation kinetics of kales [17] and no published literature has been found regarding the degradation kinetics of swedes. The greater soluble fraction observed in swedes is in accordance with its greater concentration of sugars and nonstructural carbohydrates compared to kales. Both, kales and swedes showed a fast degradation rate and extensive degradation of DM and CP compared with other forages used for livestock feeding during winter such as grass pastures, silages, and hay [31,32], and they showed a similar potential degradability but faster degradation rate compared to other winter fodder crops, such as fodder beets [33], and summer brassicas, such as turnips and forage rape [16]. The faster degradation rates observed in our study compared to those observed by Keim et al. [16] in summer brassicas and the values reported by Valderrama and Anrique [17] for kales might be explained by the composition of the diet offered to the cannulated cows. In this experiment, cows were offered brassicas in their diet, and, therefore, rumen microbes might have been adapted to degrade these kind of ingredients, which would have increased the fermentation rate [34].

Even though kales showed a faster degradation rate of crude protein than swedes, due to the similar soluble (a) and insoluble but potentially degradable (b) fractions, PD and ED were similar among species. The crude protein ED of kales was similar to the values reported by Valderrama and Anrique [17] when calculated at a passage rate of 2% per hour and was slightly greater with passage rates of 5% and 8% per hour because of the fast degradation rate observed in our study (0.48 h^{-1}). Importantly, the CP fractions "a" and "b" differed among the kale varieties and need to be considered for the ration formulation models. This variation in degradation characteristics associated with genetic differences has been reported previously by Sun et al. [35] with perennial ryegrass. The varieties with greater CP concentration (Coleor) also showed the highest soluble fraction of crude protein.

4.3. In Vitro Fermentation

Higher 24 h, 48 h, and asymptotic GP in swedes reflects greater digestibility compared to kales, since total GP is an indicator of forage digestibility [36] and has been related to DM degradability [37], which is in accordance with the greater DM potential and effective degradability observed for swedes compared with kales. This difference may be attributed to the chemical composition, as forages with lower aNDF content and higher NSC, such as swedes, present greater GP [38]. As observed for DM potential and effective in situ degradability, the in vitro GP parameters were not affected by varieties within species, as a good correlation between the GP measurements and in situ degradability has been found [39]. Contrary to what was observed with in situ DM degradation rates (faster for kales than swedes), in vitro gas production rates (C and MDR) were faster for swedes. This is because in vitro gas production rates take into account the gas produced from the soluble fraction, whereas the in situ degradation rate comes only from the insoluble but potentially degradable fraction. Nevertheless, C and MDR values of both swedes and kales (0.14 and 0.15, respectively) demonstrated faster fermentation compared to other typical feedstuffs used for ruminant feeding during winter, such as concentrate (0.11 h^{-1} for both c and MDR), hay (0.03 and 0.05 h^{-1} for c and MDR), silage (0.07 and 0.08 h^{-1} for c and MDR) and pasture (0.09 and 0.10 h^{-1} for c and MDR) [40].

Fast fermentation of both kales and swedes is reflected by parameters t25, k, t75, and t90, where, for example, 90% of both species ferment in less than 23 h, mainly due to the presence of readily fermentable carbohydrates such as sugars and NDSF. Hall et al. [41] have shown that fermentation of NDSF is faster than fermentation from NDF, indicating that forages high in NDSF tend to exhibit a rapid fermentation.

In contrast to the GP, the total VFA production was similar among species. This finding could be because the digested OM is fermented to VFA and GP or converted to microbial protein and therefore total VFA and GP are not always well correlated [42]. Total VFA production was different across swede varieties and related to the in situ DM soluble fraction; that is, the Major Plus variety presented the greater soluble fraction and tVFA production, whereas Dominion showed the lowest values for both tVFA and DM soluble fraction. The acetate and acetate to propionate ratio were greater for kale, which is related to the high aNDF and lower NSC concentrations in kales compared to swedes, as has been reported previously [27,42]. In addition, the highest proportion of butyrate found in swedes coincides with the higher sugar concentration, since the fermentation of sugars generally increases the amount of butyrate in the rumen [28]. The greater acetate and lower butyrate proportions of tVFA for kales compared with swedes are in agreement with in vivo studies with sheep, where acetate was reduced from 515 mmol/mol to 412 mmol/mol, while butyrate was increased from 118 mmol/mol to 176 mmol/mol for sheep fed swedes compared with those fed kales [4].

Kale varieties had no effect on VFA production and the relative proportions of VFAs, whereas differences in the acetate and butyrate relative proportions of tVFA were observed for varieties of swedes. The varieties with greater butyrate relative proportions of tVFA were those with greater sucrose and fructose concentrations as well as a higher DM soluble fraction.

4.4. Implications

Although nutrient concentrations of winter brassicas and variations among varieties have been widely described [3] and their use in sheep [4], dry cows [5] and lactating dairy cows [6] have been reported, to the best of our knowledge few studies have evaluated the rumen fermentation processes of winter brassicas.

The main products of rumen fermentation are VFAs, and among these, propionate is a substrate for gluconeogenesis and is the main source of glucose in the animal, whereas the non-glucogenic acetate and butyrate are sources for long-chain fatty acid synthesis [43]. Glucogenic and lipogenic nutrient supply and VFA profile have been associated with animal energy balance. It has been suggested that in lactating cows increased energy intake is channeled largely through increases in rumen production of butyric and propionic acids and their yield of ATP to the host animal [43,44]. From an environmental point of view, production of acetate and butyrate liberates hydrogen, whereas propionate serves as a net hydrogen sink [45]. Consequently, diets that increase propionate and decrease acetate in the rumen are often associated with a reduction in ruminal CH_4 production [46].

Therefore, the results we obtained for in vitro fermentation end products may lead us to infer some animal responses. However, these extrapolations must be done carefully, because the mechanisms governing microbial efficiency and VFA molar proportions in vitro are not necessarily valid in vivo. For example, in the rumen itself, feed and microbial biomass are subject to passage and VFA subject to passage and absorption [10], processes that do not occur under in vitro conditions.

One of our main findings is that swedes increased the relative molar proportion of butyrate at the expense of acetate compared with kales, however there are differences among varieties of swedes that must be considered and depend basically on their sugar concentration and type of sugars. There is generally a positive association of feeding sugars, such as sucrose, and increased milk and milk fat production, which may be due to the greater molar proportion of butyrate produced from those sugars [44,47]. However, to the best of our knowledge no studies have reported milk production responses of dairy cows fed swedes. In vivo results reported by Sun et al. [4] are similar to our study, with total rumen VFA concentrations being similar for sheep fed swedes or kales and greater butyrate and lower relative molar proportions of acetate found for sheep fed swedes. Complementarily, dry matter intake (DMI) was greater when kales were offered but energy intake was greater when feeding swedes resulting in lower methane emissions; however, no animal responses were reported. Conversely, Keogh et al. [6] have observed no differences in body weight and body condition score change between pre-calving dairy cows supplemented with kales or swedes during the dry period, but rumen fermentation data was not shown in their work.

Finally, the in situ ruminal degradation parameters of kales and swedes generated in this study can be used by researchers and nutritionists in feeding evaluation models to estimate the nutritive value, nutrient supply, and animal performance of livestock fed with winter brassicas.

5. Conclusions

Relative to winter forage brassica crops, our study demonstrated high digestibility for ruminant feeding in times when pasture availability is low. Differences between the two species were observed in terms of chemical composition, gas production, and dry matter degradation parameters, with swedes exhibiting a faster and more extensive degradation due to their greater concentrations of readily fermentable carbohydrates and lower NDF, which resulted in a lower acetate to propionate ratio and greater butyrate concentrations in the rumen. Additionally, differences among varieties within species were observed and must be considered when selecting certain varieties for use. For example, varieties of swedes showed differences regarding their in situ DM soluble fraction and in vitro tVFA and acetate and butyrate relative proportions of tVFA, whereas kale varieties differed in their in situ soluble CP and insoluble but potentially degradable CP fractions. Continuing with studies under in vivo conditions when feeding winter brassicas, especially to lactating dairy cows, is important because less information is available.

Author Contributions: Conceptualization, J.K., O.B., and R.P.; data curation, D.B., J.D., A.B., and J.K.; funding acquisition, J.K., O.B., and R.P.; investigation, A.B., D.B., J.K., and J.D.; methodology, J.K., O.B., and R.P.; project administration, J.K.; resources, J.K., O.B., and R.P.; supervision, J.K.; validation, J.K., O.B., and R.P.; writing—original draft, D.B., J.D., and J.K.; writing—review and editing, A.B., O.B., and R.P.

References

1. Keogh, B.; French, P.; Murphy, J.J.; Mee, J.F.; McGrath, T.; Storey, T.; Grant, J.; Mulligan, F.J. A note on the effect of dietary proportions of kale (*Brassica oleracea*) and grass silage on rumen pH and volatile fatty acid concentrations in dry dairy cows. *Livest. Sci.* **2009**, *126*, 302–305. [CrossRef]
2. Keim, J.P.; López, I.F.; Balocchi, O.A. Sward herbage accumulation and nutritive value as affected by pasture renovation strategy. *Grass Forage Sci.* **2015**, *70*, 283–295. [CrossRef]
3. Westwood, C.T.; Mulcock, H. Nutritional evaluation of five species of forage brassica. *Proc. N. Z. Grassl. Assoc.* **2012**, *74*, 31–38.
4. Sun, X.Z.; Waghorn, G.C.; Hoskin, S.O.; Harrison, S.J.; Muetzel, S.; Pacheco, D. Methane emissions from sheep fed fresh brassicas (*Brassica spp.*) compared to perennial ryegrass (*Lolium perenne*). *Anim. Feed Sci. Technol.* **2012**, *176*, 107–116. [CrossRef]
5. Rugoho, I.; Edwards, G.R. Dry matter intake, body condition score, and grazing behavior of nonlactating, pregnant dairy cows fed kale or grass once versus twice daily during winter. *J. Dairy Sci.* **2018**, *101*, 257–267. [CrossRef] [PubMed]
6. Keogh, B.; French, P.; McGrath, T.; Storey, T.; Mulligan, F.J. Comparison of the performance of dairy cows offered kale, swedes and perennial ryegrass herbage in situ and perennial ryegrass silage fed indoors in late pregnancy during winter in Ireland. *Grass Forage Sci.* **2009**, *64*, 49–56. [CrossRef]
7. Sun, X.Z.; Pacheco, D.; Luo, D.W. Forage brassica: A feed to mitigate enteric methane emissions? *Anim. Prod. Sci.* **2016**, *56*, 451–456. [CrossRef]
8. Hackmann, T.J.; Sampson, J.D.; Spain, J.N. Variability in in situ ruminal degradation parameters causes imprecision in estimated ruminal digestibility. *J. Dairy Sci.* **2010**, *93*, 1074–1085. [CrossRef]
9. Barry, T.N. The feeding value of forage brassica plants for grazing ruminant livestock. *Anim. Feed Sci. Technol.* **2013**, *181*, 15–25. [CrossRef]
10. Dijkstra, J.; Kebreab, E.; Bannink, A.; France, J.; López, S. Application of the gas production technique to feed evaluation systems for ruminants. *Anim. Feed Sci. Technol.* **2005**, *123–124*, 561–578. [CrossRef]
11. France, J.; Dijkstra, J.; Dhanoa, M.S.; Lopez, S.; Bannink, A. Estimating the extent of degradation of ruminant feeds from a description of their gas production profiles observed in vitro: Derivation of models and other mathematical considerations. *Br. J. Nutr.* **2000**, *83*, 143–150. [CrossRef] [PubMed]
12. Ørskov, E.R.; McDonald, I. The estimation of protein degradability in the rumen from incubation measurements weighted according to rate of passage. *J. Agric. Sci.* **1979**, *92*, 499–503. [CrossRef]
13. Theodorou, M.K.; Williams, B.A.; Dhanoa, M.S.; McAllan, A.B.; France, J. A simple gas production method using a pressure transducer to determine the fermentation kinetics of ruminant feeds. *Anim. Feed Sci. Technol.* **1994**, *48*, 185–197. [CrossRef]

14. Niderkorn, V.; Baumont, R.; Le Morvan, A.; Macheboeuf, D. Occurrence of associative effects between grasses and legumes in binary mixtures on in vitro rumen fermentation characteristics. *J. Anim. Sci.* **2011**, *89*, 1138–1145. [CrossRef]
15. Getachew, G.; DePeters, E.J.; Robinson, P.H.; Fadel, J.G. Use of an in vitro rumen gas production technique to evaluate microbial fermentation of ruminant feeds and its impact on fermentation products. *Anim. Feed Sci. Technol.* **2005**, *123–124*, 547–559. [CrossRef]
16. Keim, J.P.; Cabanilla, J.; Balocchi, O.A.; Pulido, R.N.G.; Bertrand, A. In vitro fermentation and in situ rumen degradation kinetics of summer forage brassica plants. *Anim. Prod. Sci.* **2019**, *59*, 1271–1280. [CrossRef]
17. Valderrama, X.; Anrique, R. In situ rumen degradation kinetics of high-protein forage crops in temperate climates. *Chil. J. Agric. Res.* **2011**, *71*, 572–577. [CrossRef]
18. Hvelplund, T.; Wesiberg, M. In situ techniques for the estimation of protein degradability and postrumen availability. In *Forage Evaluation in Ruminant Nutrition*; Givens, D.I., Ed.; CABI Publishing: Wallingford, UK, 2000; pp. 233–258.
19. Tilley, J.M.A.; Terry, R.A. A two-stage technique for the in vitro digestion of forage crops. *Grass Forage Sci.* **1963**, *18*, 104–111. [CrossRef]
20. Van Soest, P.J.; Robertson, J.B.; Lewis, B.A. Methods for Dietary Fiber, Neutral Detergent Fiber, and Nonstarch Polysaccharides in Relation to Animal Nutrition. *J. Dairy Sci.* **1991**, *74*, 3583–3597. [CrossRef]
21. AOAC. *Official Methods of Analysis*, 16th ed.; Association of Official Analytical Chemists: Washington, DC, USA, 1996.
22. Pelletier, S.; Tremblay, G.F.; Belanger, G.; Bertrand, A.; Castonguay, Y.; Pageau, D.; Drapeau, R. Forage Nonstructural Carbohydrates and Nutritive Value as Affected by Time of Cutting and Species. *Agron. J.* **2010**, *102*, 1388–1398. [CrossRef]
23. Hall, M.B.; Hoover, W.H.; Jennings, J.P.; Webster, T.K.M. A method for partitioning neutral detergent-soluble carbohydrates. *J. Sci. Food Agric.* **1999**, *79*, 2079–2086. [CrossRef]
24. Groot, J.C.J.; Cone, J.W.; Williams, B.A.; Debersaques, F.M.A.; Lantinga, E.A. Multiphasic analysis of gas production kinetics for in vitro fermentation of ruminant feeds. *Anim. Feed Sci. Technol.* **1996**, *64*, 77–89. [CrossRef]
25. Rowe, B.A.; Neilsen, J.E. Effects of irrigating forage turnips, *Brassica rapa* var. *rapa* cv. Barkant, during different periods of vegetative growth. 2. Nutritive characteristics of leaves and roots. *Crop Pasture Sci.* **2011**, *62*, 571–580. [CrossRef]
26. Loaiza, P.A.; Balocchi, O.; Bertrand, A. Carbohydrate and crude protein fractions in perennial ryegrass as affected by defoliation frequency and nitrogen application rate. *Grass Forage Sci.* **2017**, *72*, 556–567. [CrossRef]
27. Keim, J.P.; López, I.F.; Berthiaume, R. Nutritive value, in vitro fermentation and methane production of perennial pastures as affected by botanical composition over a growing season in the south of Chile. *Anim. Prod. Sci.* **2014**, *54*, 598–607. [CrossRef]
28. Oba, M. Review: Effects of feeding sugars on productivity of lactating dairy cows. *Can. J. Anim. Sci.* **2011**, *91*, 37–46. [CrossRef]
29. Cassida, K.A.; Turner, K.E.; Foster, J.G.; Hesterman, O.B. Comparison of detergent fiber analysis methods for forages high in pectin. *Anim. Feed Sci. Technol.* **2007**, *135*, 283–295. [CrossRef]
30. Zhao, X.H.; Liu, C.J.; Liu, Y.; Li, C.Y.; Yao, J.H. Effects of replacing dietary starch with neutral detergent–soluble fibre on ruminal fermentation, microbial synthesis and populations of ruminal cellulolytic bacteria using the rumen simulation technique (RUSITEC). *J. Anim. Physiol. Anim. Nutr.* **2013**, *97*, 1161–1169. [CrossRef]
31. Fulkerson, W.J.; Horadagoda, A.; Neal, J.S.; Barchia, I.; Nandra, K.S. Nutritive value of forage species grown in the warm temperate climate of Australia for dairy cows: Herbs and grain crops. *Livest. Sci.* **2008**, *114*, 75–83. [CrossRef]
32. Aufrere, J.; Graviou, D.; Demarquilly, C. Ruminal degradation of protein of cocksfoot and perennial ryegrass as affected by various stages of growth and conservation methods. *Anim. Res.* **2003**, *52*, 245–261. [CrossRef]
33. Znidarsic, T.; Verbic, J.; Babnik, D.; Velikonja-Bolta, S. The effect of supplementing highly wilted grass silage with maize silage, fodder beet or molasses on degradation of the diets and the efficiency of microbial protein synthesis in the rumen of sheep. *Ital. J. Anim. Sci.* **2010**, *9*, 449–459. [CrossRef]
34. Broderick, G.A.; Cochran, R.C. In vitro and In situ Methods for Estimating Digestibility with Reference to Protein Degradability. In *Feeding Systems and Feed Evaluation Models*; Theodorou, M.K., France, J., Eds.; CAB International: Wallingford, UK, 2000; pp. 53–85.

35. Sun, X.Z.; Waghorn, G.C.; Hatier, J.H.B.; Easton, H.S. Genotypic variation in in sacco dry matter degradation kinetics in perennial ryegrass (*Lolium perenne* L.). *Anim. Prod. Sci.* **2012**, *52*, 566–571. [CrossRef]
36. Xu, M.; Rinker, M.; McLeod, K.R.; Harmon, D.L. Yucca schidigera extract decreases in vitro methane production in a variety of forages and diets. *Anim. Feed Sci. Technol.* **2010**, *159*, 18–26. [CrossRef]
37. Rymer, C.; Huntington, J.A.; Williams, B.A.; Givens, D.I. In vitro cumulative gas production techniques: History, methodological considerations and challenges. *Anim. Feed Sci. Technol.* **2005**, *123–124*, 9–30. [CrossRef]
38. Van Gelder, A.H.; Hetta, M.; Rodrigues, M.A.M.; De Boever, J.L.; Den Hartigh, H.; Rymer, C.; Van Oostrum, M.; Van Kaathoven, R.; Cone, J.W. Ranking of in vitro fermentability of 20 feedstuffs with an automated gas production technique: Results of a ring test. *Anim. Feed Sci. Technol.* **2005**, *123–124*, 243–253. [CrossRef]
39. Seifried, N.; Steingass, H.; Schipprack, W.; Rodehutscord, M. Variation in ruminal in situ degradation of crude protein and starch from maize grains compared to in vitro gas production kinetics and physical and chemical characteristics. *Arch. Anim. Nutr.* **2016**, *70*, 333–349. [CrossRef]
40. Keim, J.P.; Alvarado-Gilis, C.; Arias, R.A.; Gandarillas, M.; Cabanilla, J. Evaluation of sources of variation on in vitro fermentation kinetics of feedstuffs in a gas production system. *Anim. Sci. J.* **2017**, *88*, 1547–1555. [CrossRef]
41. Hall, M.B.; Pell, A.N.; Chase, L.E. Characteristics of neutral detergent-soluble fiber fermentation by mixed ruminal microbes. *Anim. Feed Sci. Technol.* **1998**, *70*, 23–39. [CrossRef]
42. Getachew, G.; Robinson, P.H.; DePeters, E.J.; Taylor, S.J. Relationships between chemical composition, dry matter degradation and in vitro gas production of several ruminant feeds. *Anim. Feed Sci. Technol.* **2004**, *111*, 57–71. [CrossRef]
43. Morvay, Y.; Bannink, A.; France, J.; Kebreab, E.; Dijkstra, J. Evaluation of models to predict the stoichiometry of volatile fatty acid profiles in rumen fluid of lactating Holstein cows. *J. Dairy Sci.* **2011**, *94*, 3063–3080. [CrossRef]
44. Seymour, W.M.; Campbell, D.R.; Johnson, Z.B. Relationships between rumen volatile fatty acid concentrations and milk production in dairy cows: A literature study. *Anim. Feed Sci. Technol.* **2005**, *119*, 155–169. [CrossRef]
45. Moss, A.; Jouany, J.P.; Newbold, J. Methane production by ruminants: Its contri-bution to global warming. *Ann. Zootech.* **2000**, *49*, 231–253. [CrossRef]
46. Beauchemin, K.; McAllister, T.; McGinn, S.M. Dietary mitigation of enteric methane from cattle. *CAB Rev.* **2009**, *4*. [CrossRef]
47. Hall, M.B.; Mertens, D.R. A 100-Year Review: Carbohydrates-Characterization, digestion, and utilization. *J. Dairy Sci.* **2017**, *100*, 10078–10093. [CrossRef] [PubMed]

5

In Vitro Methods of Assessing Protein Quality for Poultry

Dervan D.S.L. Bryan * and Henry L. Classen *

Department of Animal and Poultry Science, University of Saskatchewan, Saskatoon, SK S7N5A8, Canada
* Correspondence: dervan.bryan@usask.ca (D.D.S.L.B.); hank.classen@usask.ca (H.L.C.);

Simple Summary: Over the years, broiler chickens have been selected for rapid growth which makes them very efficient at depositing body protein in a short period of time. This is important since the broiler sector is expected to contribute to the growing global demand for poultry meat. In light of this, the quality of proteins fed to poultry is becoming more important. The concept of protein nutrition is based on the sequential process through which proteins are digested, and the amino acids are absorbed and become available for metabolic processes. The nutritional quality of protein ingredients for poultry is based on their amino acid bioavailability. Animal and plant ingredients are the main sources of protein used in poultry diets and they vary in digestibility and amino acid composition. Although in vivo digestibility assays for poultry are available, they are expensive and time consuming to conduct. In vivo digestibility assays are the optimum tools for characterizing protein sources to be used in commercial production, but it is not always practical to conduct these assays in commercial settings. Commercial production, therefore, relies on the use of other assays such as in vitro assays to evaluate the quality of protein sources.

Abstract: Protein quality assessment of feed ingredients for poultry is often achieved using in vitro or in vivo testing. In vivo methods can be expensive and time consuming. Protein quality can also be evaluated using less expensive and time consuming chemical methods, termed in vitro. These techniques are used to improve the user's efficiency when dealing with large sample numbers, and some mimic the physiological and chemical characteristics of the animal digestive system to which the ingredient will be fed. The pepsin digestibility test is the in vitro method of choice for quick evaluation of protein sample during quality control and in most research settings. Even though the pepsin digestibility test uses enzymes to liberate the amino acids from the protein, it does not mimic normal in vivo digestive conditions. The results obtained with this method may be misleading if the samples tested contain fats or carbohydrates which they often do. Multi-enzyme tests have been proposed to overcome the problem encountered when using the pepsin digestibility test. These tests use a combination of enzymes in one or multiple steps customized to simulate the digestive process of the animal. Multi enzyme assays can predict animal digestibility, but any inherent biological properties of the ingredients on the animal digestive tract will be lost.

Keywords: dietary protein; poultry; digestibility assay; in vitro; pH stat method; pepsin digestibility assay

---

1. Introduction

Over the years, broiler chickens have been selected for rapid growth which makes them very efficient at depositing body protein in a short period of time. This is important since the broiler sector is expected to contribute to the growing global demand for poultry meat. In light of this, the quality of proteins fed to poultry is becoming more important. Animal and plant ingredients are the main sources of protein used in poultry diets and they vary in digestibility and amino acid composition [1–3].

The concept of protein nutrition is based on the sequential process through which proteins are digested, and the amino acids are absorbed and become available for metabolic processes. The nutritional quality of protein ingredients for poultry is based on their amino acid bioavailability. Animal proteins are composed of twenty-two amino acids [4]. Ten of the twenty-two amino acids in poultry meat proteins cannot be synthesized in large enough quantity and, therefore, must be provided in the diet for proper growth and metabolic function [5].

Digestibility is used in practice as an estimator of the amino acid bioavailability in poultry diets [6]. Digestible protein is the proportion of protein that is digested and absorbed in the form of amino acids [6]. On the other hand, amino acid bioavailability is the proportion of an amino acid in a form that is suitable for protein synthesis after the protein has been digested and amino acids absorbed [7]. Since the 1990s, most poultry nutrition research used digestibility assays when evaluating protein feed ingredients instead of bioavailability [5], because they do not require the free form of the amino acid during the evaluation [7]. The digestibility coefficient obtained can be used directly by nutritionist during ration formulation [5].

Although in vivo digestibility assays for assessing protein quality for poultry are available, they are expensive and time consuming to conduct. In vivo digestibility assays are the optimum tool for characterizing protein sources to be used in commercial production, but it is not practical to conduct these assays in a commercial setting. Commercial production therefore, relies on the use of other assays such as in vitro assays to evaluate the quality of protein sources. The pros and cons of in vitro and in vivo assays are covered in the subsequent review. It was clear that there is a need for a poultry specific in vitro protein digestibility assay for assessing protein sources commonly fed to poultry. This review presents a critical overview of current in vitro protein digestibility assays relevant to poultry and the application of their methodology in assessing protein quality of ingredients for poultry. The objectives of this review paper were: (1) To provide a comprehensive review of the in vitro methods currently available which has the potential or has been applied in the assessment of protein quality for poultry, and (2) to explore potential methodological factors which might be important in the assessment protein digestion.

2. Methods of Assessing Protein Quality

Traditionally, protein quality is assessed by evaluating the extent to which amino acids are digested and absorbed from the ingredient. Estimation of protein digestibility is normally achieved by feeding the feed ingredient to the intended animal and assessing protein or amino acid digestibility. This technique is termed in vivo. Protein quality can also be evaluated using less expensive and time consuming in vitro chemical methods. These techniques are used to improve the level of precision while mimicking the physiological and chemical characteristics of the digestive system of the animal to which the ingredient will be feed.

To obtain useful information on the digestibility of nutrients without the use of in vivo assays, researchers often employ the use of in vitro assays. In theory, in vitro digestibility assays should closely simulate the digestive process of the intended animal [8]. Depending on the nature of the research, it is expected that an intended in vitro assay should be reproducible, cheaper than available in vivo assays and simple to perform while giving fast results [9]. Methods for evaluating nutrient digestibility in vitro for simple stomach animals have been reviewed by others [8,10]. Only those methods applicable to protein digestion in poultry will be discussed.

2.1. Chemical In Vitro Methods

Evaluating protein quality using chemical method provides less precision than in vivo techniques but can be used as a routine quality control measure. In the chemical engineering literature, it was known as early as the 1930s that an alkali solution could extract up to 95% of the protein from plant meal sources [11]. In the late 1960s, Rinehart was one of the first to employ the protein solubility technique as a measure of protein quality of soybean meal in the poultry industry [12]. While working at Purina Mills Inc., Rinehart evaluated the suitability of protein from soybean meal derived from different processing systems using potassium hydroxide (KOH).

The ability to predict animal performance is one of the most important criteria of any chemical assay [13]. It was not until the 1950s that Lyman et al. [14] established a relationship between bird performance and the solubility of protein feed ingredients used in poultry diets. The study evaluated the correlation between a chick growth assay and the use of a protein solubility technique using sodium hydroxide as the alkali solution. In the solubility technique, one gram of cottonseed meal with four glass beads was placed in an Erlenmeyer flask with 100 mL of 0.02 N sodium hydroxide solution. The flask was agitated continuously at 37 °C for an h, and then the mixture centrifuged for 5 min at 3000× *g*. After centrifuging, the solution was filtered and aliquots evaluated for protein concentration [14].

The solubility index method was not adopted as a routine measure of protein quality in the poultry feed industry until the test was validated. A study was reported in which the protein solubility technique was used to evaluate soybean quality in poultry feed [12]. This study provided the foundation for the evaluation of protein quality using the solubility technique. The researcher [13] revived the technique when they proposed the use of sodium tetraborate at 40 °C as a more sensitive test for detecting changes in protein quality due to overcooking of meals. By the end of the late 1990s, protein solubility using KOH became a routine technique in research evaluating dietary protein [13,15–17]. Researchers used the protein solubility index to evaluate canola meal quality and found that the 0.5% sodium hydroxide assay did not accurately predict canola meal lysine digestibility in broiler chickens [18]. This suggests that the relationship between protein solubility and amino acid digestibility is ingredient specific.

Protein dispersibility index (PDI) is another method used to evaluate the quality of protein ingredients. This technique involves high speed mixing of a protein sample in water, followed by the assessment of solubility [17]. In the literature, PDI may be referred to as water dispersible protein or water-soluble protein [19]. In 1970, the PDI technique was published as two official and tentative methods of the American Oil Chemists Society [19]. Veltmann and coworkers [20] evaluated the quality of soybean meal used in poultry diets employing the PDI method. The PDI method was able to distinguish between normal processed meals and meal heat-treated to escape rumen degradation. In that same study, a chick growth assay showed that there was no difference between the bioavailability of the protein from the two meals [20]. This suggested that the PDI method did not correlate well to the bioavailability of protein from the ingredient tested.

In 1978, the American Oil Chemists Society published a revised PDI method, which was corrected in 1979 as method Ba 10-65. In brief, 20 g of protein is mixed for 10 min at 7800× *g* with 300 mL of water. A portion of the mixture is centrifuged and the nitrogen content of the solid fraction and the original protein sample measured [21]. The percent dispersed protein is calculated as the protein loss from the original sample to the water. Batal and coworkers [17] compared the revived PDI method against the urease index and KOH solubility test. Of the three tests, the PDI method was more effective and more sensitive in detecting the minimum adequate heat processing conditions required for soybean meal fed to chickens.

Since the 1980s, the PDI method has become a routine technique used worldwide by researchers [17,20,22–24] to assess the quality of protein sources used in monogastric animal feeds. While chemical methods provide an overview of the protein quality of feed ingredients, they do not give a good indication of how much of the nutrient will be absorbed by the animal. Protein solubility index and PDI methods are used as measures of ingredient quality in most poultry nutritional research evaluating high protein ingredients. The information gained from the PDI method and protein solubility index does not provide useful information for diet formulation in a commercial setting, but they are often used in quality control programs.

2.2. pH-Stat/Drop Method

As protein samples are hydrolyzed by digestive enzymes, they release protons from the cleaved peptide bonds, which changes the pH of the reaction media [25]. In the early 1970s, Maga, Lorenz, and Onayemi evaluated the extent to which dietary protein undergoes proteolysis. They realized that there was a close relationship with the initial rate of hydrolysis of the proteins from 0 to 10 min and the digestibility of the protein samples. The rates of hydrolysis of the protein samples were evaluated as an indirect measure of the pH of the reaction mixture over time. In their system, the protein samples were incubated with trypsin at 37 °C in a water bath for 10 min while evaluating the pH change. However, this method lacked precision in predicting the bioavailability of protein [26,27].

To improve precision in predicting bioavailability with the Maga et al. [25] method, Vavak modified the above procedure in a master's thesis while working with distiller's grain protein concentrate [26]. During the modification of the procedure, various enzyme combinations were tested in an effort to gain improvement in the correlation coefficients between pH drop and protein digestibility in rats. The trypsin-chymotrypsin combination gave superior correlation coefficients compared to the initial single trypsin proposed by Maga et al. [25]. Hsu and coworkers [27] suggested that the methods presented by Maga et al. [25] and Vavak [26] were too time consuming and complicated for routine quality control.

A faster method was developed, which could be completed in 1 h [27]. In this method, the trypsin-chymotrypsin enzyme combination was replaced with a multi-enzyme mixture composed of trypsin, chymotrypsin and peptidase. The correlation coefficient with the apparent digestibility of protein for rats was 0.9 using this new multi-enzyme system after evaluating 23 food protein sources. The method was also able to detect the effects of trypsin inhibitor, chlorogenic acid and heat processing on the digestibility of the protein tested. The pH drop method was susceptible to the buffering capacity of the protein source since high ash content affected the digestibility results [27]. A researcher used the pH drop method proposed by Hsu et al. [27] modified by Satterlee et al. [28] to evaluate various high protein feed ingredients while correlating the results to true digestibility in caecectomized cockerels [10]. There was a good correlation with lysine digestibility in caecectomized cockerels and the pH drop test across the ingredients tested. The test, however, showed no relationship to lysine digestibility and protein efficiency ratio in various qualities of feather meal and meat meal samples.

To overcome the susceptibility of the pH drop test to the buffering capacity of protein samples, Pederson and Eggum revised the pH drop method proposed by Hsu et al. [27]. During revision, the consumption of alkali was used as an indirect measure of true protein digestibility values in rats. The pH of the reaction was held constant at 8 during titration with alkali over a 10 min period [29]. The correlation coefficient was improved from 0.9 [27] to 0.96 [29] with a residual error of 1.29 after evaluating 30 protein samples. The authors [29] suggested that the effects of ash content on the test results were due to differences in mineral content, which was mostly due to the influence of calcium. The authors proposed the use of two different regression equations to accurately predict the digestibility of protein samples from plant and animal origins. Using literature derived prediction equation for a specific kind of protein source was unreliable when using the pH-stat method [30]. To measure the degree of hydrolysis of protein, the method requires knowledge of the average dissociation of the α-amino groups of the protein sample and the number of peptide bonds present in the territory structure of the main proteins present in the ingredient [29].

Due to the limitations mentioned, the pH-stat test has been used mostly in food science research to predict the digestibility of highly digestible pure protein sources [29–31]. Such pure protein sources typically have data about the average dissociation of the α–amino groups and the number of peptide bonds present. Since the early 1990s, the pH-stat method has been used to evaluate only aquatic animal feed ingredients [32,33]. To address the limitations of the method, casein average dissociation constant and the number of peptide bonds were used as the standards when calculating the degree of hydrolysis [33]. So far, the data generated with the pH-stat method has been consistent with *in vivo* digestibility assays, especially with the use of purified enzymes extracted from the species to which the ingredient has been fed [32,33]. The pH-stat method has become a valuable tool for aquatic nutritional research, but not for terrestrial animals. The good digestibility correlations seen with aquatic species are probably due to the simple nature of their digestive tract and the use of highly digestible protein sources such as fish meal.

2.3. *Closed Enzymatic Methods*

These systems are used to evaluate the digestibility of nutrients with multiple or single enzymes while simulating part or all of the in vivo digestive process [8]. The system is flexible, so the procedure and enzymes used may vary to meet the specific needs of the research objectives. Only those procedures used specifically to evaluate the digestibility of protein samples will be reviewed. The digestibility of protein is tied to the amino acid content and to the specificity of the digestive enzyme used to free them from complex peptides [34].

2.3.1. Pepsin Assay

The pepsin digestibility assay is one of the most widely used assays to evaluate the quality of feed and protein ingredients. Gehrt and coworkers and Sheffner and coworkers were the first group of researchers to employ a single enzymatic method to evaluate the digestibility of protein using pepsin [35,36]. In their procedure, 1 g of protein was incubated with 25 mg of pepsin in 30 mL of 0.1 N sulfuric acid at 37 °C for 24 h, during this time, the samples were stirred intermittently [35]. After incubation, the samples were placed in a boiling water bath for 10 min. Samples were cooled and the pH adjusted to 2 followed by the addition of one volume each of 10% sodium tungstate and 2/3 N sulfuric acid. The mixtures were filtered after standing for 10 min, and then the filtrate adjusted to pH 6.8 and analyzed for amino acids. When the digestibility data were regressed against the biological value of the protein samples for rats, there was a 0.998 correlation [36].

The pepsin digestibility assay was not accepted as a routine protein quality evaluation until 1959. The Association of Official Analytical Chemists (AOAC) adopted a revived version of the method proposed by Gehrt et al. and Sheffner et al. [35,36]. Hydrochloric acid was used instead of sulfuric acid, and all the fat was extracted from the samples using ether before digestion. The sodium tungstate and pH steps were eliminated. In 1972, the procedure was revised to improve the filtration step and the pepsin concentration was defined as 0.2%.

Since the 1959 AOAC publication of the pepsin digestibility method, it has been used extensively to evaluate high protein feed ingredient quality of both plant and animal origin [15,37]. Johnston and coworkers were one of the first group of researchers to use this method to evaluate poultry feed ingredients of animal origin [37]. After evaluating 20 commercial animal by-product meals, they were able to get a 0.91 correlation with the net protein utilization and the protein efficiency ratio for chickens. The pepsin digestibility procedure proposed by Johnson et al. [37], adjusted the pepsin concentration to 0.002% while eliminating the preliminary grinding and defatting steps.

In another study, the same group of researchers evaluated various levels of pepsin in order to find a suitable level for use in the assay during routine evaluation of meat and bone meal samples fed to poultry [38]. Lower levels of pepsin (0.002%) were able to detect differences between the quality of the meat and bone meal samples, which was contrary to that of the AOAC 0.2% pepsin. Parsons and coworkers did a comparative study on the ability of 0.2%, 0.002%, and 0.0002% pepsin to detect differences in quality among 14 meat and bone meal samples [1]. They confirmed the findings of Johnson et al. [38] that the 0.002% pepsin level gave the best correlation with lysine digestibility in chickens.

2.3.2. Pancreatin

Some testing systems involve the use of pancreatin as the only enzyme source to digest protein samples. Riesen and coworkers described a single enzymatic method that used pancreatin to evaluate the quality of soybean meal in poultry [39]. The samples were ground in a power-driven mortar,

100 mg or 300 mg of pancreatin was added to 2 g of the ground samples in 50 mL of 0.2 M disodium phosphate buffer at pH 8.3. One mL of toluene was added to the solution, and the mixture incubated for 5 d or 12 h at 37 °C. At the end of each digestion period, the samples were heated with steam for 15 min to facilitate enzyme deactivation. The pH of the mixture was adjusted with glacial acetic acid to precipitate the undigested proteins. This method was able to detect the difference between overheated and the normal heated meals, but not the difference between the normal and under heated soybean meals.

Ingram and coworkers modified the procedure by adding 1.2 g of pancreatin to 12 g of sample in 300 mL of buffer for 6 h [40]. The pattern of amino acid released from the samples correlated with the growth of chickens fed the same samples of soybean meal [40]. In another study by Anwar [41], the pancreatin in vitro test was used to evaluate the quality of cottonseed meal, groundnut meal, meat meal and fish meal [41]. The method was not reliable for fish meal and groundnut meal but gave fair results for meat meals. The one-step pancreatin method has been used routinely by many food scientists to evaluate the digestibility of various protein foods, but not by poultry nutritionists [42].

Pancreatic digestion is controlled by substrate concentration in vivo. An increase in protein concentration will promote an increase in proteolytic enzyme secretion [8]. In vitro digestibility methods using pancreatin as the only enzyme, source keeps the enzyme concentration constant when evaluating a range of protein sources [41]. However, the method lacks precession when evaluating a variety of protein sources [41]. Other researchers have found no difference between in vivo chicken ileal digestibility and the pepsin or pancreatin assay when ranking feather meal digestibility [43].

2.3.3. Multi-Enzymatic Assays

A multi-enzyme method may use two or more enzymes while simulating one, two or all stages of the digestive process [8]. Multi-enzyme methods are more comparable to in vivo conditions since many enzymes are involved in the digestion of proteins. The digestion of proteins starts in the stomach under the action of pepsin and hydrochloric acid. The partially digested protein enters the small intestine where they are hydrolyzed by trypsin, chymotrypsin, elastase and carboxypeptidases [8].

In 1964, Akeson and Stahmann described a method using pepsin and pancreatin as enzyme sources. The method was developed to evaluate large numbers of food protein samples while reducing the labour load associated with the pepsin digestibility assay [44]. The method involved incubating 100 mg of protein sample with 1.5 mg pepsin in 15 mL 0.1 N hydrochloric acid for 3 h at 37 °C [44]. The reaction was neutralized with 7.5 mL of 0.2 N sodium hydroxide solution and then 4 mg pancreatin dissolved in 7.5 mL phosphate buffer with pH 8 was added. Fifty parts per million merthiolate were added to the mixture, which was incubated at 37 °C for 24 h. Samples of the digestion mixture were precipitated with acid and centrifuged at 1000× *g* for 30 min after which the supernatant was analyzed for amino acids.

In 1973, Saunders and coworkers described a two enzyme system using pepsin and trypsin. The test occurred in a closed system using centrifuge tubes containing 1 g of protein sample suspended in 20 mL of 0.1 N hydrochloric acid and then mixed with 50 mg pepsin dissolved in 1 mL 0.01 N hydrochloric acid [45]. The mixture was incubated at 37 °C while gently shaken for 48 h, centrifuged at 20,000× *g* for 5 min, and the supernatant removed. The solid was suspended in 10 mL water and 10 mL of 0.1 M phosphate buffer with pH 8 and 5 mg of dissolved trypsin. The mixture was incubated at 23 °C for another 12 h then centrifuged and the solids washed with 30 mL of water five times, with centrifuging and removal of the supernatant each time. The solid was filtered through a 1.2 μm Millipore filter, air-dried, and analyzed for amino acids.

Both the pepsin-pancreatin and pepsin-trypsin methods were able to give good correlation between the in vivo digestibility values for various food proteins using rats [44,45]. The pepsin-pancreatin assay is known to give good correlation with amino acid digestibility of 0.84 in cereal gains with true amino acid availability in chickens but was less reliable for soybean meal and corn gluten meal [46]. However, the pepsin-pancreatin test gave an excellent correlation of 0.91 between the in vivo ileal digestibility of protein of 15 feedstuffs in pigs [47]. The test proposed by Saunders et al. [45] has been used to some extent to evaluate protein digestibility in poultry [48–50].

Dialysis cell method is a non-static system in which products of digestion are removed from the substrate as they become available. When simulating in vivo protein digestion with in vitro techniques, the rate of hydrolysis may be compromised by the accumulation of end products in the system [46,51]. The rate of hydrolysis can be improved if the digestion products are removed from the system as digestion occurs [46]. To prevent the inhibition of proteolysis by the end products, dialysis has been proposed to remove digestion products [52,53]. They conducted their experiments in dialysis bags to facilitate the removal of the end products during incubation of the protein source with the enzymes.

In 1982, Gauthier and coworkers et al. adopted the dialysis principle of [52,53] and presented a method in which the dialysis solution was continually replaced as the incubation proceeded [34]. The content of the dialysis bag was stirred constantly during the digestion process. In brief, 400 mg nitrogen (6.25 × %N) of protein was suspended in a beaker with 100 mL of 0.1 N hydrochloric acid. The beaker was shaken and placed in a water bath at 37 °C for 30 min. The pH of the solution was adjusted to 1.9 then 20 mL of solution containing 5 mg pepsin per mL of 0.1N hydrochloric acid added. The mixture was incubated for 30 min, the pH was adjusted to 8 and transferred to a dialysis bag with a 1000 Da molecular weight cut off. The bag was placed in a U-shaped container with inlets from a peristaltic pump and outlets to a beaker. Twenty mL of a solution containing 5 mg pancreatin per mL sodium phosphate buffer was added to the dialysis bag, which was continuously washed with 37 °C sodium phosphate buffer at a flow rate of 212 mL/h. Samples of dialysate were collected at different time intervals and analyzed. The method was able to detect the effects of heat and alkali treatment on protein digestibility in foods [34].

The digestion unit size plus the use of handmade apparatus were limitations for its use in routine protein evaluation [54]. Savoie and Gauthier modified the design presented by Gauthier et al. [34]. The improvements included the use of a magnetic stir bar and the construction of a cell with an inner compartment fitted with a dialysis membrane. The cell was 100 mm long in comparison to the 298 mm original unit. There was free access to the reaction chamber without disruption of the reaction. Each cell was designed to work as a single unit or part of the multi-unit system. The system developed was very flexible and could be used to measure the release of any product from enzymatic hydrolysis [54].

The dialysis cell method has been applied to study protein digestibility across a number of disciplines [55–57]. This method was able to identify differences in the rate of release of amino acids from different sea bream feed samples [57]. The system was flexible to accommodate the use of crude enzyme extract from sea bream as the digestive enzyme. A comparison between the pH-stat and the dialysis cell method showed that the dialysis cell method was able to identify which products were released from the protein as well as the digestion kinetics of the protein samples [55]. The effects of different processing methods on the digestibility of legume proteins were identified with the dialysis cell method [56]. A detailed description of the availability of different amino acids and the rate at which they were released during digestion was obtained from different protein sources [56,58]. The main disadvantages of the dialysis cell method are the complexity and the number of samples which can be digested in a given run. This method uses custom made dialysis cells, peristalsis pumps, and fraction collectors which can be expensive. Savoie and Gauthier recommend that no more than 6 cells should be used simultaneously due to the manual inputs needed. From a practical point of view, an in vitro method must be simple and easy to implement for it to be adopted by poultry nutritionist [54].

3. Factors Influencing Protein Digestion

The digestibility data obtained by in vitro methods vary even within the same method for the same ingredient. This variation may be due to a number of issues associated with in vitro digestibility systems. Enzymes and their concentration seemed to be one of the most important factors influencing in vitro digestion [1,26,59]. The specificity of enzymes and their ratio to the substrate will determine the level of hydrolysis achieved [8,59].

3.1. Enzyme Specificity

Table 1 shows a list of enzymes involved in protein digestion. The first enzyme responsible for the initiation of protein digestion in poultry is pepsin [8]. This enzyme will only cleave the N-terminal of aromatic amino acids like tyrosine, tryptophan and phenylalanine [4] at low pH. Hydrolysis by pepsin results in smaller peptides that enter the duodenum for further hydrolysis by pancreatic protease [8]. As suggested by Assoumani and Nguyen [10], trypsin will only break a lysyl or arginyl peptide bonds to expose lysine or arginine terminal residues at basic pH. Trypsin binds only to the positive side group of arginine and lysine, where the peptide is cleaved at those amino acids [4].

Table 1. Enzyme and their specific bond cleavage preferences.

Enzymes	Bond Cleave	Reference
Pepsin	N-terminal of aromatic amino acids phenylalanine, tryptophan and tyrosine	[4,8]
Trypsin	Lysyl or arginyl peptide bond to expose lysine or arginine	[10]
Chymotrypsin	Aromatic or large hydrophobic amino acid residues such as tyrosine, phenylalanine, tryptophan, leucyl, methionyl, asparaginyl, and glutamyl	[4,8,60]
Elastase	Glycine and alanine of elastin	[4,8]
Carboxypeptidase A	Peptide bond adjacent to the C-terminal end of a polypeptide chain,	[4,8]
Carboxypeptidase B	Basic amino acids from the C-terminal end of polypeptide chains	[8]
Collagenase	Alpha peptides and hydrogen bonds in the superhelix of tropocollagen and collagen	[61]

The ability of enzymes to hydrolyze substrate may depend on the presence of other enzymes. The activation of chymotrypsin is dependent on the presence of trypsin [4]. Chymotrypsin will act on proteins and peptides, but will also hydrolyze esters and amides [60]. Chymotrypsin cleaves peptides over a wider range of sites than trypsin, both aromatic and hydrophobic side chains of amino acids residues [4]. Peptide bonds involving tyrosine, tryptophan, phenylalanine and glutamyl, leucyl, asparaginyl residues are cleaved by chymotrypsin [4,8].

Lysine or arginine are released from small peptides by carboxypeptidase-B, which is specific for C-terminal basic groups [8]. Animal protein meals may contain high levels of collagen due to the nature of the type of rendering material. Digestion of this meal in vitro may need additional collagenase enzymes during the pancreatic digestion stage [61]. Bonds hydrolyzed in protein feed samples are enzyme-specific, so in vitro digestion models should take this into account by using multiple enzymes [8].

3.2. Protein Structure and Forms

The structure of the protein samples and the food matrix in which the samples are presented will influence protein in vitro digestibility [10]. Protein feed ingredients may contain free amino acids, peptides of various lengths, secondary structure proteins (α-helix, β-pleated sheets, β-turns and superhelix), tertiary structure proteins and quaternary structure proteins [4]. Secondary structure proteins such as scleroproteins, which include collagen, elastin, and keratin, are poorly digested in simple stomach animals [62]. Protein sources containing high levels of these proteins will have limited bioavailability. Higher protein structural configuration requires more time and higher enzyme concentration to achieve greater hydrolysis [4,62].

Secondary structure proteins resist digestion due to the nature of their individual structures. Feather meal, for example, contains high levels of keratin [43], which has highly cross-linked disulphide bonds along the pleated sheet configuration [62]. This makes the protein almost insoluble in water and thereby reduces the action of pepsin and subsequent pancreatic actions [43]. Samples of meat and bone meal may contain elastin and collagen after being produced from tendons, ligaments and bone scraps of animals. Elastin and collagen also contain cross-linking in their helix structures which may influence digestion [62].

The matrix in which the protein is presented in the protein source may limit the access of proteolytic enzymes. Plant proteins are often presented in a matrix with cell walls, lipids, and complex sugars, and may also be organized into specialized storage vacuoles [10,62]. The ability of proteolytic enzymes to access those proteins may depend on the ability of other enzymes to free protein from the matrices [8]. The digestion of protein from plant sources in monogastric animals is closely linked to the protein associated with plant cell wall components [63]. Non-starch polysaccharides are known to protect proteins from enzymatic digestion in a variety of plant feed ingredients in poultry [64]. Solubilization of the cell wall components of plant-sourced protein meals with various carbohydrase enzymes were able to improve the availability of the protein to chickens [63,64].

3.3. *Enzyme Activity*

In vitro digestion may be influenced by the activity of the enzymes used while enzyme activity is affected by factors such as pH, temperature, ratio of enzyme to substrate, and incubation time [8]. As proteins are hydrolyzed by enzyme in vitro, the pH of the mixture will be reduced by the release of protons from the cleaved peptide bonds [25]. If the original pH of the reaction moisture is further away from the optimum pH of the enzyme, the rate of hydrolysis will be reduced drastically in a short period of time. In the pH-stat method, pH is held constant in the optimal range for the enzyme via automated alkali titration [29]. To achieve optimal reaction conditions, most in vitro assays select appropriate starting pH for the enzyme used [10]. The pepsin digestibility assay requires an acidic condition [36], while the pancreatin assay requires a basic environment [39].

The temperature may play a regulatory role as it relates to enzyme activity. Like all chemical reactions, temperature increases the amount of kinetic energy and increases the velocity at which molecules collide in an enzymatic reaction [4]. In vitro digestibility assays using protease keep the temperature of their reaction between 37 and 45 °C [8]. Enzymes are proteins and all proteins can be denatured at high temperatures; therefore the optimal temperature for a given enzyme is always close to the body temperature of the organism from which the enzyme was derived [4]. In vitro assays should reflect in vivo conditions so the temperature at which the reaction takes place is often that of the animal's internal temperature [33,47].

The ratio of enzyme to substrate and the incubation time varies across individual in vitro assays [1,8,51,59]. Generally, the incubation time can range from 0.5 to 45 h depending on the kind of in vitro assay [10]. The enzyme to substrate ratio is often a function of the specific activity of the enzyme. The specific activity of an enzyme is often defined as the amount of product produced from a specific substrate over time while maintaining the reaction at a fixed pH and temperature range [4]. Enzymes from different preparations with different specific activities are often used for the same in vitro assay [54,58]. The ratio of pepsin used with 4 mg nitrogen of sample in the dialysis method ranged from 5 to 7 mg/mL pepsin [34,54]. Pepsin concentration used in the pepsin digestibility test ranged from 0.02 to 2.5 g/L and the sample size of the protein may be expressed as g of nitrogen per sample [8]. To avoid confusion in the literature, an in vitro method should define the enzyme to substrate ratio and the specific activity of each enzyme in the assay [65].

3.4. *Anti Nutritive Agents of Test Samples*

Anti-nutritional compounds are often secondary metabolites and structural components of plants that interfere with the metabolic activities of animals when present in feed ingredients [66]. These compounds provide structural support and some metabolites have evolved into defence chemicals to protect plants from insect damage [67]. Some anti-nutritional compounds represent important storage minerals and intermediate molecules used in various pathways by the plant [66]. The main action of these compounds tends to disrupt the digestive process via multiple modes of action.

3.4.1. Sinapine and Tannins

A phenolic compound found in many plant feed ingredients is sinapine, which is a choline ester derived from 3, 5-dimethoxy-4-hydroxyinnamic acid or tannins [68]. Growing plants use sinapine as their main source of sinapic acid and choline [69]. High levels of sinapic acid can react with other compounds to create a colour change and produce a bitter taste in plant feed ingredients [70]. During oxidation, phenolic acids may react with proteins to form indigestible complexes like quinines which bind to the functional group of lysine and methionine [68].

Tannins are another set of water soluble polyphenolic compounds that may be found in protein meals of plant origin [71]. They are normally present in legume seeds, cereal grains, and oilseeds [68,72]. Tannins are generally grouped into hydrolyzable and condensed tannins. Hydrolyzable tannins may have esters of gallic, m-digallic, or hexahydroxydiphenic acids, which are easily hydrolyzed [71]. Condensed tannins resist hydrolysis and are polymers of flavan-2, 4-diol and flavan-3-ol or a mixture of both [72]. Tannins precipitate protein out of solution through the formation of soluble and insoluble complexes [68], and are known to reduce the digestibility of amino acids in poultry [73]. Tannins inhibit the absorption of protein from the digestive tract [72,73]. Low molecular weight tannins may be absorbed from the intestine and cause toxicity through the inhibition of key metabolic pathways [72,73].

3.4.2. Protease Inhibitors

Almost all plant protein sources available for use in animal production contain some type of protease inhibitor [74]. Even commonly consumed foods such as legumes, cereal grains, and tomatoes contain protease inhibitors [72]. Protease inhibitors block the activity of trypsin, chymotrypsin [62], elastase, and carboxypeptidase [75]. Trypsin inhibitor can be found in field pea, peanut, wheat, soybean, rapeseed, lupin, and sunflower seeds [62,75].

Of the plant protein sources used in poultry production, soybean is generally considered to have the highest trypsin inhibitor activity [72]. The inhibitors bind to the active site of the enzyme, thereby reducing their ability to lower the kinetic energy needed during proteolytic cleavage [4]. The two main inhibitors found in soybean are from the Kunitz and Bowman-Birk inhibitor families [62]. Kunitz is about 21.4 kDa with high affinity for trypsin, while Bowman-Birk is about 8 kDa and has high affinity for both trypsin and chymotrypsin [72].

When birds were fed diets containing raw soybean, the granules of the pancreatic acini were totally depleted in 2 h after feeding and the size of the pancreas increased after 8 d [76]. The pancreatic activity of the birds at 16 d was twice the activity before they were given the diet and the birds growth was reduced drastically. Protease inhibitor activities can be reduced through various heat processes, but complete elimination is often not possible in commercial soybean products [15,74].

3.4.3. Phytate

Feed ingredients derived from plants contain some level of phosphorus stored as phytic acid or phytate which are also known as myo-inositol hexaphosphoric acid and myo-inositol hexaphosphate respectively [77]. Phytate is predominantly found in the seeds of plants, which makes animal feed derived from oilseeds and cereal grains a source of phytate [78]. During germination, the inorganic phytate is hydrolyzed by enzymes to produce phosphate which the plant uses for its growth [79]. Phytic acid has strong mineral binding capacity through its six phosphate groups, which actively bind zinc, iron, calcium, and magnesium [79]. Phytate's chelating ability results in complexes with nutrients such as proteins and minerals [80].

The anti-nutritional effects of phytic acid on protein digestion can occur via direct or indirect modes of action. During protein digestion, phytate may bind to metal cofactors needed for the activity of aminopeptidases and carboxypeptidases [4,72]. Phytate may also bind with protein to form complexes in acidic and neutral pH conditions [80], which may inhibit the activities of digestive enzymes [81]. Intestinal phytase activity observed in poultry [82] may depend on magnesium as a cofactor. In such a case intestinal phytase may not be able to hydrolyze a substantial amount of the dietary phytate if sufficient magnesium is not present. However, in practical feeding situation the poultry industry has incorporated exogenous phytase in poultry diets [83]. The exogenous phytase hydrolyzes the ester bond between the inositol ring and phosphate group, thereby releasing phosphorus and reducing the anti-nutritive effects on protein digestibility. This elicits a question of whether exogenous phytase enzymes should be part of an in vitro protein digestibility assay for poultry.

3.4.4. Effects of Ingredient Processing

Proteins used in animal production are often by-products of other processing industries. The nutritional quality of these proteins is a function of the processes used in meal production. Plant-based protein sources generally will contain some form of anti-nutrient and thus require processing to reduce their effects when fed to animals. Protein meals of animal origin are waste products from food processing facilities. As such, the raw materials may contain higher levels of microbial contamination and require additional processing before it is fed to animals.

The major anti-nutritional compounds found in plant-based protein sources can be reduced through some form of heat treatment. Unfortunately, amino acid digestibility in chickens may be compromised if the heat treatment used is excessive [12] or not enough [22]. Autoclaving flaxseed at 120 °C for 20, 40, and 60 min resulted in changes in the α-helix to β-sheet ratio of the protein fraction [84]. Rumen degradable protein is reduced with increased autoclaving time which suggested that the protein resisted digestion as a result of the change in α-helix to β-sheet ratio. This would be true if that same protein was fed to non-ruminants and the effects would be more severe.

During the commercial production of canola meal using the prepress-solvent extraction system, the meal is subjected to toasting during hexane removal [18]. Amino acid digestibility and the content of the meal are reduced after toasting. The elimination of the spurge steam during toasting could alleviate the loss of amino acids [18]. Soybean meal production involves solvent extraction as well. Ideally, the soybean is exposed to 105 °C for half h [85], but if the meal is heated to 121 °C, the concentration and digestibility of amino acids, especially lysine, are reduced [86]. The loss of amino acids during the production of meals from the solvent extraction process may result in poor growth in chickens fed meals processed under such conditions [13,87].

Amino acid loss during heating processing of protein meal may involve Maillard reactions, were a sugar-amine complex is formed from the reaction of sugars and ketones with amino acids, proteins, and peptides in food [88]. Mauron suggested that Maillard reactions involve early, advanced, and final stage reactions. Early Maillard reaction involves a reversible condensation of the carbonyl group of the sugar with the amino group of the amino acid, peptide, or protein to form a hydrolyzable N-substituted glycosylamine and then 1-amino-1-deoxy-2-ketose. At the early stage, food does not have any browning or flavour, but its nutritive value is reduced. During the advanced stage of the reaction, amines are released and are used as catalysts in reactions to form intermediate flavour products such as acetaldehyde and pyruvaldehyde [89]. The final reaction produces a dark brown nitrogen-containing pigment composed of decomposed amino acids, heterocyclic amines, melanoidin polymers and aldol condensation products [88].

The stages of the Maillard reaction requires specific reaction conditions to be successful [88]. Temperature and moisture are the two important parameters which govern each stage of the Maillard reaction [88]. Experimental simulations of Maillard reaction generally take place in solutions and the formation of melanoidin polymers is an exponential function of heating [90]. Reactions of D-xylose and glycine in aqueous solution at 22, 68, and 100 °C produce a temperature-dependent increase in aromaticity or high molecular weight melanoidin polymers [90]. The rate of the Maillard reaction is defined as the function Q_{10} which is the increase in rate for every 10 °C. As the temperature increases from 22 to 100 °C the quantity of high molecular weight melanoidin increases and the low soluble intermediate products of the Maillard reaction decrease [90].

Protein meals of animal origin do not contain the high levels of sugars found in meals of plant origin, so are less likely to undergo Maillard reaction when exposed to heat treatment. The natural soluble carbohydrate concentration of dried animal protein meals range from 0.3 to 1.3% [91], which is far less than what would normally be present in plant-based meals [13,87]. The meals are prone to Maillard reaction if they are exposed to soluble carbohydrate during autoclaving which has been shown to reduce meal digestibility [91].

Large amounts of meat and bone meal are produced by the rendering industry, but the quality of those meals can vary [37]. The variability in the quality of meat and bone meal can limit its use in poultry production [1]. Oxidation and enzymatic denaturing may occur depending on location and source of the raw material used in the rendering process. Polyunsaturated fats are known to react with atmospheric oxygen which results in the production of peroxides and other auto-oxidation products [92]. If the meal is kept in warm conditions, this could increase the formation of peroxides and secondary oxidation products. The application of heat in the presence of oxygen and polyunsaturated fats is known to increase the production of peroxides and secondary oxidation products [92]. This could be a factor during rendering if parameters such a temperature, time, and raw material polyunsaturated fat content are not controlled during meal production.

4. In Vitro Digestibility Systems Validation

One major challenge often encountered when developing in vitro models to evaluate protein digestion is the ability of a single model to effectively assay multiple kinds of feed ingredients. Due to this challenge, multiple quality control assays such as those based on the physicochemical properties of ingredients have been developed to help the feed industry. The in vitro model developed by Bryan et al. [59] evaluated nine different protein sources which are known to have variable digestibility and physicochemical properties. Correlation analysis between the PDI and KOH solubility of the ingredients and in vitro extent of crude protein (CP) digestion were all significant, with correlation coefficients (r) of 0.64 and 0.84, respectively. There was no correlation between the in vitro CP digestibility and the reactive lysine assay. This might be an indication that the reactive lysine assay was not a useful physiochemical candidate assay to compare with the developed in vitro assay. In Figure 1, such correlation with in vivo data is generally used as the last step for validating in vitro systems [93]. It is a common practice to conduct correction analysis in such circumstances but the validation of in vitro digestibility systems requires more analysis.

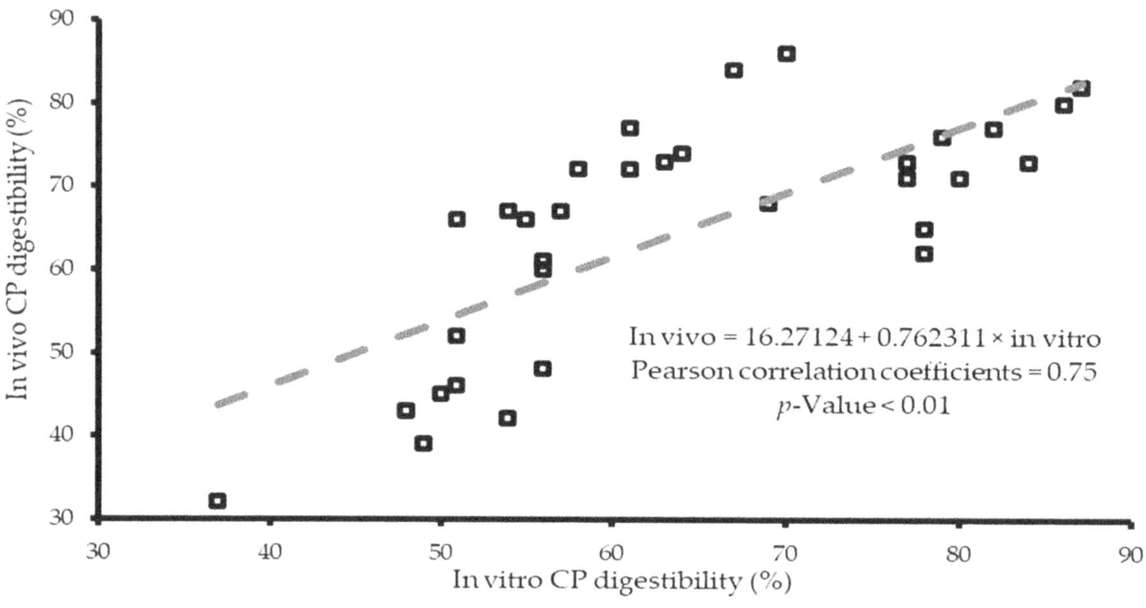

Figure 1. Plot of correlation between in vivo and in vitro crude protein (CP) digestible of nine high protein poultry feed ingredients [93].

In order to know if in vitro CP digestibility data are representative of the in vivo amino acid digestibility, regression and correlation analysis together (Table 2) were performed between the two sets of data generated by Bryan et al. [94,95]. The in vitro CP digestibility was positively correlated with all amino acids except for cysteine (CYS), which had a regression estimate P-value of 0.1. The correlation coefficients ranged from 0.43 to 0.71, except for CYS which was 0.30. The data presented in Table 2 shows the complexity of factors that might affect the interpatient of correlation validation of in vitro systems to in vivo. The in vitro model was developed using soybean meal (SBM) as the model protein source which has both pros and cons. Using SBM might have put the other ingredients at a slight disadvantage since the method optimized SBM digestibility for each stage of digestion and not the other meals. This could have accounted for some of the variation seen in the correlation coefficients of the amino acid with the in vitro CP digestibility. Based on the data presented in Table 2, the in vitro CP digestibility can be used as a predictor of in vivo amino acid digestibility; however, the correlation coefficients varied among amino acids so more samples need to be tested to form stronger prediction equations.

Table 2. Simple linear regression and Pearson correlation of in vitro digestible crude protein (CP) and in vivo standardized ileal amino acids digestibility of the nine meal samples [93].

Item	Regression Coefficients		ANOVA		In Vitro Digestible CP	
	Intercept	**In Vitro Digestible CP**	**R^2**	**MSE**	**Correlation Coefficients**	***P*-Value**
Aspartic acid	3.27	0.83	0.35	252.55	0.59	<0.01
Estimate SE	13.68	0.21	-	-	-	-
Estimate *p*-Value	0.81	<0.01	-	-	-	-
Threonine	27.35	0.55	0.35	115.73	0.59	<0.01
Estimate SE	9.25	0.14	-	-	-	-
Estimate *p*-Value	<0.01	<0.01	-	-	-	-
Serine	38.29	0.43	0.18	168.08	0.43	0.02
Estimate SE	11.15	0.17	-	-	-	-
Estimate *p*-Value	<0.01	0.02	-	-	-	-
Glutamic acid	22.19	0.75	0.50	112.11	0.71	<0.01
Estimate SE	9.11	0.14	-	-	-	-
Estimate *p*-Value	0.02	<0.01	-	-	-	-
Proline	13.72	0.75	0.35	206.60	0.59	<0.01
Estimate SE	12.36	0.19	-	-	-	-
Estimate *p*-Value	0.28	<0.01	-	-	-	-
Glycine	41.58	0.40	0.25	94.69	0.50	<0.01
Estimate SE	8.37	0.13	-	-	-	-
Estimate *p*-Value	<0.01	<0.01	-	-	-	-
Alanine	35.01	0.56	0.39	95.38	0.63	<0.01
Estimate SE	8.40	0.13	-	-	-	-
Estimate *p*-Value	<0.01	0.56	-	-	-	-
Cysteine	26.18	0.41	0.09	326.99	0.30	0.09
Estimate SE	15.55	0.23	-	-	-	-
Estimate *p*-Value	0.10	0.09	-	-	-	-
Valine	42.74	0.40	0.21	113.32	0.46	<0.01
Estimate SE	9.16	0.14	-	-	-	-
Estimate *p*-Value	<0.01	<0.01	-	-	-	-
Methionine	32.256	0.63	0.45	97.07	0.67	<0.01
Estimate SE	8.47	0.13	-	-	-	-
Estimate *p*-Value	<0.01	<0.01	-	-	-	-
Isoleucine	43.74	0.44	0.26	110.68	0.51	<0.01
Estimate SE	9.05	0.14	-	-	-	-
Estimate *p*-Value	<0.01	<0.01	-	-	-	-

Table 2. *Cont.*

Item	Regression Coefficients		ANOVA		In Vitro Digestible CP	
	Intercept	In Vitro Digestible CP	R^2	MSE	Correlation Coefficients	*P*-Value
Leucine	35.38	0.56	0.35	113.76	0.59	<0.01
Estimate SE	9.17	0.14	-	-	-	-
Estimate *p*-Value	<0.01	<0.01	-	-	-	-
Tyrosine	28.97	0.63	0.39	121.17	0.62	<0.01
Estimate SE	9.47	0.14	-	-	-	-
Estimate P-Value	<0.01	<0.01	-	-	-	-
Phenylalanine	39.97	0.5	0.29	120.73	0.54	<0.01
Estimate SE	9.45	0.14	-	-	-	-
Estimate P-Value	<0.01	<0.01	-	-	-	-
Lysine	34.44	0.57	0.50	62.50	0.71	<0.01
Estimate SE	6.80	0.10	-	-	-	-
Estimate *p*-Value	<0.01	<0.01	-	-	-	-
Histidine	12.17	0.84	0.48	150.04	0.70	<0.01
Estimate SE	10.54	0.16	-	-	-	-
Estimate *p*-Value	0.26	<0.01	-	-	-	-
Arginine	33.27	0.63	0.40	119.31	0.63	<0.01
Estimate SE	9.39	0.14	-	-	-	-
Estimate *p*-Value	<0.01	<0.01	-	-	-	-

R^2: R-squared (variance for a dependent variable explained by variables in the regression model); MSE: Means square error; SE: Standard error.

Another approach in the validation step is to add more analysis. A one sample T-Test was performed comparing the difference between the in vitro and in vivo CP digestibility data of Figure 1 to a mean of 0 to see if there were differences between the two methods of assessing CP digestibility. This comparison suggests that there is no difference between in vivo and in vitro CP extent of digestion for the meals evaluated. The Bland Altman plot of the data presented in Figure 2 shows that there was no proportional bias between in vitro and in vivo CP digestibility data for any of the nine meals evaluated and all the data points collected during the assay fell in the 95% confidence limit.

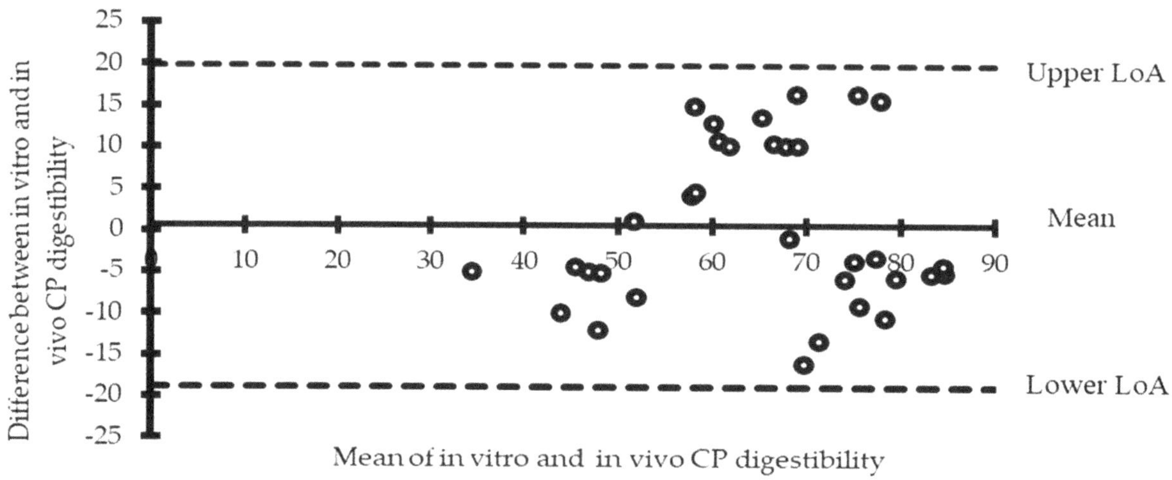

Figure 2. Bland Altman Plot of the difference between in vivo and in vitro crude protein (CP) digestible of nine high protein feed ingredients [93]; LoA: limits of agreement

This indicates that the in vivo and the in vitro CP digestible data were in agreement for the digestibility of nine meals. Based on the correlation, the T-Test, and the Bland Altman plot results, the in vitro assay was able to predict the in vivo CP digestibility of the ingredients. The in vitro assay could, therefore, serve as a tool for assaying CP digestible of meals for broiler chickens.

5. Conclusions

Protein quality assessment of feed ingredients for poultry is often achieved using in vitro or in vivo testing. The disadvantages associated with in vivo methods lead to the commercial acceptance of in vitro methods as the gold standard for assessing protein quality. These techniques are used to improve the user's efficiency when dealing with large numbers of sample and some mimic the physiological and chemical characteristics of the animal digestive system to which the ingredient will be fed. Despite all of the advantages of these in vitro methods, they do not give a true replication of normal in vivo digestive conditions. This is because of the inability of those methods to mimic numerous biological factors involved in in vivo digestion and the complex interaction which exists with various ingredients. Multi enzyme assays can predict animal digestibility of proteins if they are designed properly. However, any inherent biological properties of the ingredients which might impact the animal digestive tract will be lost. Users of in vitro digestibility data should be aware of these disadvantages and take the necessary steps to validate in vitro methods and their data. In any case, in vitro digestibility methods are just estimates of in vivo digestion, which serve as a substitute in situations where in vivo digestion is not possible.

Author Contributions: Conceptualization, D.D.S.L.B. and H.L.C.; writing—original draft preparation, D.D.S.L.B; writing—review and editing, H.L.C.; supervision, H.L.C.; funding acquisition, H.L.C. All authors have read and agreed to the published version of the manuscript.

References

1. Parsons, C.; Castanon, F.; Han, Y. Protein and amino acid quality of meat and bone meal. *Poult. Sci.* **1997**, *76*, 361–368. [CrossRef] [PubMed]
2. Adedokun, S.A.; Adeola, O.; Parsons, C.M.; Lilburn, M.S.; Applegate, T.J. Standardized ileal amino acid digestibility of plant feedstuffs in broiler chickens and turkey poults using a nitrogen-free or casein diet. *Poult.Sci.* **2008**, *87*, 2535–2548. [CrossRef] [PubMed]
3. Kim, E.J.; Utterback, P.L.; Applegate, T.J.; Parsons, C.M. Comparison of amino acid digestibility of feedstuffs determined with the precision-fed cecectomized rooster assay and the standardized ileal amino acid digestibility assay. *Poult. Sci.* **2011**, *90*, 2511–2519. [CrossRef] [PubMed]
4. Bhagavan, N.V.; Bhagavan, N.V. *Medical Biochemistry*; Jones and Bartlett Publishers: Boston, UK, 1992.
5. Ravindran, V.; Bryden, W.L. Amino acid availability in poultry—In vitro and in vivo measurements. *Aust. J. Agric. Res.* **1999**, *50*, 889–908. [CrossRef]
6. Lemme, A.; Ravindran, V.; Bryden, W.L. Ileal digestibility of amino acids in feed ingredients for broilers. *World's Poult. Sci. J.* **2004**, *60*, 423–438. [CrossRef]
7. Batterham, E.S. Availability and utilization of amino acids for growing pigs. *Nutr. Res. Rev.* **1992**, *5*, 1–18. [CrossRef]
8. Boisen, S.; Eggum, B.O. Critical evaluation of in vitro methods for estimating digestibility in simple-stomach animals. *Nutr. Res. Revi.* **1991**, *4*, 141–162. [CrossRef]
9. Clunies, M.; Leeson, S. In vitro estimation of dry matter and crude protein digestibility. *Poult. Sci.* **1984**, *63*, 89–96. [CrossRef]
10. Fuller, M.F. *Vitro Digestion for Pigs and Poultry*; C.A.B. International: Wallingford, UK, 1991; ISBN 0-85198-719-2.
11. Smith, A.K.; Circle, S.J.; Brother, G.H. Peptization of soybean proteins. The effect of neutral salts on the quantity of nitrogenous constituents extracted from oil-free meal. *J. Am. Chem. Soc.* **1938**, *60*, 1316–1320. [CrossRef]
12. Araba, M.; Dale, N.M. Evaluation of protein solubility as an indicator of over processing soybean meal. *Poult. Sci.* **1990**, *69*, 76–83. [CrossRef]
13. Lee, H.; Garlich, J.D. Effect of overcooked soybean meal on chicken performance and amino acid availability. *Poult. Sci.* **1992**, *71*, 499–508. [CrossRef] [PubMed]
14. Lyman, C.M.; Chang, W.Y.; Couch, J.R. Evaluation of protein quality in cottonseed meals by chick growth and by a chemical index method. *J. Nutr.* **1953**, *49*, 679–690. [CrossRef] [PubMed]

15. Parsons, C.M.; Hashimoto, K.; Wedekind, K.J.; Baker, D.H. Soybean protein solubility in potassium hydroxide: An in vitro test of in vivo protein quality. *J. Anim. Sci.* **1991**, *69*, 2918–2924. [CrossRef] [PubMed]
16. Fernandez, S.R.; Zhang, Y.; Parsons, C.M. Determination of protein solubility in oilseed meals using coomassie blue dye binding. *Poult. Sci.* **1993**, *72*, 1925–1930. [CrossRef]
17. Batal, A.; Douglas, M.; Engram, A.; Parsons, C. Protein dispersibility index as an indicator of adequately processed soybean meal. *Poult. Sci.* **2000**, *79*, 1592–1596. [CrossRef]
18. Newkirk, R.W.; Classen, H.L.; Scott, T.A.; Edney, M.J. The digestibility and content of amino acids in toasted and non-toasted canola meals. *Can. J. Anim. Sci.* **2003**, *83*, 131–139. [CrossRef]
19. Johnson, D.W. Functional properties of oilseed proteins. *J. Am. Oil Chem. Soc.* **1970**, *47*, 402–407. [CrossRef]
20. Veltmann, J.R.; Hansen, B.C.; Tanksley, T.D.; Knabe, D.; Linton, A.S. Comparison of the nutritive value of different heat-treated commercial soybean meals: Utilization by chicks in practical type rations. *Poult. Sci.* **1986**, *65*, 1561–1570. [CrossRef]
21. Clarke, E.; Wiseman, J. Effects of variability in trypsin inhibitor content of soya bean meals on true and apparent ileal digestibility of amino acids and pancreas size in broiler chicks. *Anim. Feed Sci. Technol.* **2005**, *121*, 125–138. [CrossRef]
22. De Coca-Sinova, A.; Valencia, D.G.; Jiménez-Moreno, E.; Lázaro, R.; Mateos, G.G. Apparent ileal digestibility of energy, nitrogen, and amino acids of soybean meals of different origin in broilers. *Poult. Sci.* **2008**, *87*, 2613–2623. [CrossRef]
23. Serrano, M.P.; Valencia, D.G.; Méndez, J.; Mateos, G.G. Influence of feed form and source of soybean meal of the diet on growth performance of broilers from 1 to 42 days of age. 1. Floor pen study. *Poult. Sci.* **2012**, *91*, 2838–2844. [CrossRef] [PubMed]
24. Pérez-Calvo, E.; Castrillo, C.; Baucells, M.D.; Guada, J.A. Effect of rendering on protein and fat quality of animal by-products. *J. Anim. Physiol. Anim. Nutr.* **2010**, *94*, e154–e163. [CrossRef] [PubMed]
25. Maga, J.A.; Lorenz, K.; I, O.O. Digestive acceptability of proteins as measured by the initial rate of in vitro proteolysis. *J. Food Sci.* **1973**, *38*, 173–174. [CrossRef]
26. Vavak, D.L.R. A Nutritional Characterization of the Distiller's Grain Protein Concentrates. Ph.D. Thesis, University of Nebraska—Lincoln, Lincoln, NE, USA, 1975.
27. Hsu, H.W.; Vavak, D.L.; Satterlee, L.D.; Miller, G.A. A multienzyme technique for estimating protein digestibility. *J. Food Sci.* **1977**, *42*, 1269–1273. [CrossRef]
28. Satterlee, L.D.; Kendrick, J.G.; Marshall, H.F.; Jewell, D.K.; Ali, R.A.; Heckman, M.M.; Steinke, H.F.; Larson, P.; Phillips, R.D.; Sarwar, G. In vitro assay for predicting protein efficiency ratio as measured by rat bioassay: Collaborative study Milk, chicken, soy protein, cereals, wheat flour, nutritional quality. *J. AOAC Int.* **1982**, *65*, 798–809. [CrossRef]
29. Pedersen, B.; Eggum, B.O. Prediction of protein digestibility by an in vitro enzymatic pH-stat procedure. *Z. Tierphysiol. Tierernähr. Futtermittelkd.* **1983**, *49*, 265–277. [CrossRef] [PubMed]
30. Linder, M.; Rozan, P.; EL Kossori, R.L.; Fanni, J.; Villaume, C.; Mejean, L.; Parmentier, M. Nutritional value of veal bone hydrolysate. *J. Food Sci.* **1997**, *62*, 183–189. [CrossRef]
31. Wang, H.; Faris, R.J.; Wang, T.; Spurlock, M.E.; Gabler, N. Increased in vitro and in vivo digestibility of soy proteins by chemical modification of disulfide bonds. *J. Am. Oil Chem. Soc.* **2009**, *86*, 1093–1099. [CrossRef]
32. Dimes, L.E.; Haard, N.F. Estimation of protein digestibility I. Development of an in vitro method for estimating protein digestibility in salmonids (Salmo gairdneri). *Comp. Biochem. Physiol. A Physiol.* **1994**, *108*, 349–362. [CrossRef]
33. Tibbetts, S.M.; Milley, J.E.; Ross, N.W.; Verreth, J.A.J.; Lall, S.P. In vitro pH-Stat protein hydrolysis of feed ingredients for Atlantic cod, Gadus morhua. 1. Development of the method. *Aquaculture* **2011**, *319*, 398–406. [CrossRef]
34. Gauthier, S.F.; Vachon, C.; Jones, J.D.; Savoie, L. Assessment of protein digestibility by in vitro enzymatic hydrolysis with simultaneous dialysis. *J. Nutr.* **1982**, *112*, 1718–1725. [CrossRef] [PubMed]
35. Gehrt, A.J.; Caldwell, M.J.; Elmslie, W.P. Feed digestibility, chemical method for measuring relative digestibility of animal protein feedstuffs. *J. Agric. Food Chem.* **1955**, *3*, 159–162. [CrossRef]
36. Sheffner, A.L.; Eckfeldt, G.A.; Spector, H. The pepsin-digest-residue (PDR) amino acid index of net protein utilization. *J. Nutr.* **1956**, *60*, 105–120. [CrossRef] [PubMed]
37. Johnson, J.; Coon, C.N. A comparison of six protein quality assays using commercially available protein meals. *Poult. Sci.* **1979**, *58*, 919–927. [CrossRef]

38. Johnson, J.; Coon, C.N. The use of varying levels of pepsin for pepsin digestion studies with animal proteins. *Poult. Sci.* **1979**, *58*, 1271–1273. [CrossRef]
39. Riesen, W.H.; Clandinin, D.R.; Elvehjem, C.A.; Cravens, W.W. Liberation of essential amino acids from raw, properly heated, and overheated soy bean oil meal. *J. Biol. Chem.* **1947**, *167*, 143–150. [PubMed]
40. Ingram, G.R.; Riesen, W.W.; Cravens, W.W.; Elvehjem, C.A. Evaluating soybean oil meal protein for chick growth by enzymatic release of amino acids. *Poult. Sci.* **1949**, *28*, 898–902. [CrossRef]
41. Anwar, A. Evaluation of proteins by in vitro pancreatin digestion. *Poult. Sci.* **1962**, *41*, 1120–1123. [CrossRef]
42. Altangerel, B.; Sengee, Z.; Kramarova, D.; Rop, O.; Hoza, I. The determination of water-soluble vitamins and in vitro digestibility of selected Czech cheeses. *Int. J. Food Sci. Technol.* **2011**, *46*, 1225–1230. [CrossRef]
43. Bielorai, R.; Harduf, Z.; Iosif, B.; Alumot, E. Apparent amino acid absorption from feather meal by chicks. *Brit. J. Nutr.* **1983**, *49*, 395–399. [CrossRef]
44. Akeson, W.R.; Stahmann, M.A. A pepsin pancreatin digest index of protein quality evaluation. *J. Nutr.* **1964**, *83*, 257–261. [CrossRef] [PubMed]
45. Saunders, R.M.; Connor, M.A.; Booth, A.N.; Bickoff, E.M.; Kohler, G.O. Measurement of digestibility of alfalfa protein concentrates by in vivo and in vitro methods. *J. Nutr.* **1973**, *103*, 530–535. [CrossRef] [PubMed]
46. Cave, N.A. Bioavailability of amino acids in plant feedstuffs determined by in vitro digestion, chick growth assay, and true amino acid availability methods. *Poult. Sci.* **1988**, *67*, 78–87. [CrossRef] [PubMed]
47. Boisen, S.; Fernández, J.A. Prediction of the apparent ileal digestibility of protein and amino acids in feedstuffs and feed mixtures for pigs by in vitro analyses. *Anim. Feed Sci. Technol.* **1995**, *51*, 29–43. [CrossRef]
48. Saleh, F.; Ohtsuka, A.; Tanaka, T.; Hayashi, K. Effect of enzymes of microbial origin on in vitro digestibilities of dry matter and crude protein in maize. *J. Poult. Sci.* **2003**, *40*, 274–281. [CrossRef]
49. Saleh, F.; Ohtsuka, A.; Tanaka, T.; Hayashi, K. Carbohydrases are digested by proteases present in enzyme preparations during in vitro digestion. *J. Poult. Sci.* **2004**, *41*, 229–235. [CrossRef]
50. Tahir, M.; Saleh, F.; Ohtsuka, A.; Hayashi, K. An effective combination of carbohydrases that enables reduction of dietary protein in broilers: Importance of hemicellulase. *Poult. Sci.* **2008**, *87*, 713–718. [CrossRef]
51. Robbins, R.C. Effect of ratio of enzymes to substrate on amino acid patterns released from proteins in vitro. *Int. J. Vitam. Nutr. Res.* **1978**, *48*, 44–53.
52. Mauron, J.; Mottu, F.; Bujard, E.; Egli, R.H. The availability of lysine, methionine and tryptophan in condensed milk and milk powder. In vitro digestion studies. *Arch. Biochem. Biophys.* **1955**, *59*, 433–451. [CrossRef]
53. Steinhart, H.; Kirchgessner, M. In vitro digestion apparatus for the enzymatic hydrolysis of proteins. *Arch. Tierernahr.* **1973**, *23*, 449–459. [CrossRef]
54. Savoie, L.; Gauthier, S.F. Dialysis cell for the in vitro measurement of protein digestibility. *J. Food Sci.* **1986**, *51*, 494–498. [CrossRef]
55. Moyano, F.J.; Savoie, L. Comparison of in vitro systems of protein digestion using either mammal or fish proteolytic enzymes. *Comp. Biochem. Physiol. A Mol. Integr. Physiol.* **2001**, *128*, 359–368. [CrossRef]
56. Siddhuraju, P.; Becker, K. Nutritional and antinutritional composition, in vitro amino acid availability, starch digestibility and predicted glycemic index of differentially processed mucuna beans (*Mucuna pruriens* var. utilis): An under-utilised legume. *Food Chem.* **2005**, *91*, 275–286. [CrossRef]
57. Sáenz de Rodrigáñez, M.A.; Gander, B.; Alaiz, M.; Moyano, F.J. Physico-chemical characterization and in vitro digestibility of commercial feeds used in weaning of marine fish. *Aquacult. Nutr.* **2011**, *17*, 429–440. [CrossRef]
58. Savoie, L.; Galibois, I.; Parent, G.; Charbonneau, R. Sequential release of amino acids and peptides during in vitro digestion of casein and rapeseed proteins. *Nutr. Res.* **1988**, *8*, 1319–1326. [CrossRef]
59. Bryan, D.D.S.L.; Abbott, D.A.; Classen, H.L. Development of an in vitro protein digestibility assay mimicking the chicken digestive tract. *Anim. Nutr.* **2018**, *4*, 401–409. [CrossRef]
60. Appel, W. Chymotrypsin: Molecular and catalytic properties. *Clin. Biochem.* **1986**, *19*, 317–322. [CrossRef]
61. Straumfjord, J.V.; Hummel, J.P. Collagen digestion by dog pancreatic juice. *Exp. Biol. Med.* **1957**, *95*, 141–144. [CrossRef]
62. Becker, P.M.; Yu, P. What makes protein indigestible from tissue-related, cellular, and molecular aspects? *Mol. Nutr. Food Res.* **2013**, *57*, 1695–1707. [CrossRef]
63. Theander, O.; Westerlund, E.; Åman, P.; Graham, H. Plant cell walls and monogastric diets. *Anim. Feed Sci. Technol.* **1989**, *23*, 205–225. [CrossRef]

64. Meng, X.; Slominski, B.; Nyachoti, C.; Campbell, L.; Guenter, W. Degradation of cell wall polysaccharides by combinations of carbohydrase enzymes and their effect on nutrient utilization and broiler chicken performance. *Poult. Sci.* **2005**, *84*, 37–47. [CrossRef] [PubMed]
65. Gauthier, S.F.; Vachon, C.; Savoie, L. Enzymatic conditions of an in vitro method to study protein digestion. *J. Food Sci.* **1986**, *51*, 960–964. [CrossRef]
66. Bones, A.M.; Rossiter, J.T. The myrosinase-glucosinolate system, its organisation and biochemistry. *Physiol. Plant.* **1996**, *97*, 194–208. [CrossRef]
67. Chen, S.; Andreasson, E. Update on glucosinolate metabolism and transport. *Plant Physiol. Biochem.* **2001**, *39*, 743–758. [CrossRef]
68. Shahidi, F.; Naczk, M. An overview of the phenolics of canola and rapeseed: Chemical, sensory and nutritional significance. *J. Am. Oil Chem. Soc.* **1992**, *69*, 917–924. [CrossRef]
69. Campbell, L.D.; Smith, T.K. Responses of growing chickens to high dietary contents of rapeseed meal. *Brit. Poult. Sci.* **1979**, *20*, 231–237. [CrossRef]
70. Kozlowska, H.; Naczk, M.; Shahidi, F.; Zadernowski, R. Phenolic acids and tannins in rapeseed and canola. In *Canola and Rapeseed*; Shahidi, F., Ed.; Springer: Boston, MA, USA, 1990; pp. 193–210, ISBN 978-1-4613-6744-4.
71. Mangan, J.L. Nutritional effects of tannins in animal feeds. *Nutr. Res. Rev.* **1988**, *1*, 209–231. [CrossRef]
72. Sarwar Gilani, G.; Wu Xiao, C.; Cockell, K.A. Impact of antinutritional factors in food proteins on the digestibility of protein and the bioavailability of amino acids and on protein quality. *Brit. J. Nutr.* **2012**, *108*, S315–S332. [CrossRef]
73. Elkin, R.G.; Freed, M.B.; Hamaker, B.R.; Zhang, Y.; Parsons, C.M. Condensed tannins are only partially responsible for variations in nutrient digestibilities of sorghum grain cultivars. *J. Agric. Food Chem.* **1996**, *44*, 848–853. [CrossRef]
74. Francis, G.; Makkar, H.P.S.; Becker, K. Antinutritional factors present in plant-derived alternate fish feed ingredients and their effects in fish. *Aquaculture* **2001**, *199*, 197–227. [CrossRef]
75. Friedman, M.; Brandon, D.L. Nutritional and health benefits of soy proteins. *J. Agric. Food Chem.* **2001**, *49*, 1069–1086. [CrossRef] [PubMed]
76. Applegarth, A.; Furuta, F.; Lepkovsky, S. Response of the Chicken Pancreas to Raw Soybeans: Morphologic Responses, Gross and Microscopic, of the Pancreases of Chickens on Raw and Heated Soybean Diets. *Poult. Sci.* **1964**, *43*, 733–739. [CrossRef]
77. Nelson, T.S.; Ferrara, L.W.; Storer, N.L. Phytate phosphorus content of feed ingredients derived from plants. *Poult. Sci.* **1968**, *47*, 1372–1374. [CrossRef] [PubMed]
78. O'Dell, B.L.; De Boland, A.R.; Koirtyohann, S.R. Distribution of phytate and nutritionally important elements among the morphological components of cereal grains. *J. Agric. Food Chem.* **1972**, *20*, 718–723. [CrossRef]
79. Urbano, G.; López-Jurado, M.; Aranda, P.; Vidal-Valverde, C.; Tenorio, E.; Porres, J. The role of phytic acid in legumes: Antinutrient or beneficial function? *J. Physiol. Biochem.* **2000**, *56*, 283–294. [CrossRef]
80. Selle, P.H.; Ravindran, V.; Caldwell, A.; Bryden, W.L. Phytate and phytase: Consequences for protein utilisation. *Nutr. Res. Rev.* **2000**, *13*, 255–278. [CrossRef]
81. Li, Z.; Alli, I.; Kermasha, S. In-vitro α-amylase inhibitor activity-phytate relationships in proteins from Phaseolus beans. *Food Res. Int.* **1993**, *26*, 195–201. [CrossRef]
82. Maenz, D.; Classen, H. Phytase activity in the small intestinal brush border membrane of the chicken. *Poult. Sci.* **1998**, *77*, 557–563. [CrossRef]
83. Adeola, O.; Cowieson, A.J. Board-invited review: Opportunities and challenges in using exogenous enzymes to improve nonruminant animal production. *J. Anim. Sci.* **2011**, *89*, 3189–3218. [CrossRef]
84. Doiron, K.; Yu, P.; McKinnon, J.J.; Christensen, D.A. Heat-induced protein structure and subfractions in relation to protein degradation kinetics and intestinal availability in dairy cattle. *J. Dairy Sci.* **2009**, *92*, 3319–3330. [CrossRef]
85. Ljøkjel, K.; Harstad, O.M.; Skrede, A. Effect of heat treatment of soybean meal and fish meal on amino acid digestibility in mink and dairy cows. *Anim. Feed Sci. Technol.* **2000**, *84*, 83–95. [CrossRef]
86. Parsons, C.M.; Hashimoto, K.; Wedekind, K.J.; Han, Y.; Baker, D.H. Effect of overprocessing on availability of amino acids and energy in soybean meal. *Poult. Sci.* **1992**, *71*, 133–140. [CrossRef]
87. Newkirk, R.; Classen, H. The effects of toasting canola meal on body weight, feed conversion efficiency, and mortality in broiler chickens. *Poult. Sci.* **2002**, *81*, 815–825. [CrossRef] [PubMed]
88. Mauron, J. The Maillard reaction in food; a critical review from the nutritional standpoint. *Prog. Food Nutr. Sci.* **1981**, *5*, 5–35. [PubMed]

89. Hodge, J.E. Dehydrated foods, chemistry of browning reactions in model systems. *J. Agri. Food Chem.* **1953**, *1*, 928–943. [CrossRef]
90. Benzing-Purdie, L.M.; Ripmeester, J.A.; Ratcliffe, C.I. Effects of temperature on maillard reaction products. *J. Agric. Food Chem.* **1985**, *33*, 31–33. [CrossRef]
91. Schroeder, L.J.; Iacobellis, M.; Smith, A.H. Influence of heat on the digestibility of meat proteins. *J. Nutr.* **1961**, *73*, 143–150. [CrossRef]
92. Labuza, T.P.; Ragnarsson, J.O. Kinetic History Effect on Lipid Oxidation of Methyl Linoleate in a Model System. *J. Food Sci.* **1985**, *50*, 145–147. [CrossRef]
93. Bryan, D.D.S.L. Characterization of Protein Sources and Their Effects on Broiler Performance, Digestive Tract Morphology and Caecal Fermentation Metabolites. Ph.D. Thesis, University of Saskatchewan, Saskatoon, SK, Canada, 2018.
94. Bryan, D.D.S.L.; Abbott, D.A.; Van Kessel, A.G.; Classen, H.L. In vivo digestion characteristics of protein sources fed to broilers. *Poult. Sci.* **2019**, *98*, 3313–3325. [CrossRef]
95. Bryan, D.D.S.L.; Abbott, D.A.; Classen, H.L. Digestion kinetics of protein sources determined using an in vitro chicken model. *Anim. Feed Sci. Technol.* **2019**, *248*, 106–113. [CrossRef]

6

Fermentation Pattern of Several Carbohydrate Sources Incubated in an In Vitro Semicontinuous System with Inocula from Ruminants Given Either Forage or Concentrate-Based Diets

Zahia Amanzougarene, Susana Yuste and Manuel Fondevila *

Departamento de Producción Animal y Ciencia de los Alimentos, Instituto Agroalimentario de Aragón (IA2), Universidad de Zaragoza-CITA, M. Servet 177, 50013 Zaragoza, Spain; zahiaagro@yahoo.fr (Z.A.); susana_monre@hotmail.com (S.Y.)

* Correspondence: mfonde@unizar.es;

Simple Summary: A sudden change from a milk/forage diet to a high concentrate diet in young ruminants increases the rate and extent of rumen microbial fermentation, leading to digestive problems, such as acidosis. The magnitude of this effect depends on the nature of the ingredients. Six carbohydrate sources were tested: three cereal grains (barley, maize and brown sorghum), as high starch sources of different availability, and three byproducts (sugarbeet pulp, citrus pulp and wheat bran), as sources of either insoluble or soluble fibre. An in vitro semicontinuous incubation system was used to compare the fermentation pattern of substrates incubated with inocula-simulating concentrate or forage diets, under the pH and liquid outflow rate conditions of intensive feeding systems. The magnitude of microbial fermentation was higher with the concentrate than the forage inoculum, and the drop in pH in the first part of incubation was more profound. Among the substrates, citrus pulp had a greater acidification potential and was fermented at a higher extent, followed by wheat bran and barley. In conclusion, the acidification capacity of substrates plays an important role in the environmental conditions, depending on the type of diet given to the ruminant. This in vitro system allows us to compare the substrates under conditions simulating high-concentrate feeding.

Abstract: The fermentation pattern of several carbohydrate sources and their interaction with the nature of microbial inoculum was studied. Barley (B), maize (M), sorghum, (S), sugarbeet pulp (BP), citrus pulp (CP) and wheat bran (WB) were tested in an in vitro semicontinuous system maintaining poorly buffered conditions from 0 to 6 h, and being gradually buffered to 6.5 from 8 to 24 h to simulate the rumen pH pattern. Rumen fluid inoculum was obtained from lambs fed with either concentrate and barley straw (CI) or alfalfa hay (FI). The extent of fermentation was higher with CI than FI throughout the incubation ($p < 0.05$). Among the substrates, S, BP and M maintained the highest pH ($p < 0.05$), whereas CP recorded the lowest pH with both inocula. Similarly, CP recorded the highest gas volume throughout the incubation, followed by WB and B, and S recorded the lowest volume ($p < 0.05$). On average, the total volatile fatty acid (VFA), as well as lactic acid concentration, was higher with CP than in the other substrates ($p < 0.05$). The microbial structure was more affected by the animal donor of inoculum than by the substrate. The in vitro semicontinuous system allows for the study of the rumen environment acidification and substrate microbial fermentation under intensive feeding conditions.

Keywords: cereals; fibrous byproducts; gas volume; pH; volatile fatty acids; in vitro fermentation

1. Introduction

Reaching a high productive performance in the fattening of young ruminants requires high-energy diets that promote a high rate and extent of rumen microbial fermentation, with acidosis as a frequent consequence [1]. In practice, ruminants reared at pasture are often abruptly introduced to intensive feeding systems without being previously adapted to high-concentrate diets, promoting variable responses in the rate and extent of fermentation [2]. Cereals are commonly considered as ingredients of concentrate diets for ruminants. Their energy value depends on starch availability, which differs according to their chemical structure, protein matrix or, in some cases, the presence of phenolic compounds [3,4]. Fibrous byproducts with either insoluble (cellulose, hemicelluloses) or soluble (mostly pectin) polysaccharides and variable proportions of either starch or sugars [5,6] are also included among the carbohydrate sources currently used. Fitting substrate characteristics to the fermentative ability of rumen microbiota while environmental conditions are maintained at an optimal range is a key factor for maximising the efficiency of energy utilisation, and the risk of physiological impairment is also reduced. The characteristics of the specific rumen microbial community promoted by a certain diet also affect substrate utilisation [7], as the activity of the bacterial species able to fermenting starch or fibrous polysaccharides depends on environmental characteristics [8,9].

The comparison of these energy sources and their effects in the rumen under in vivo conditions is laborious and expensive, and is often biased by the feeding pattern and hardly controlled fermentation conditions [10]. On the other hand, in vitro studies are cheaper and faster and allow for a good insight into rumen fermentation processes [11]. However, most of these in vitro methods are designed to mimic the environment promoted by high forage diets, including the use of inoculum from forage-fed animals [12], and it is not easy to adapt the main physiological conditions, such as pH and rate of passage to conditions promoted by high-concentrate diets [13]. Amanzougarene and Fondevila [14] succeeded in maintaining a low incubation pH in an in vitro closed-batch system by reducing the bicarbonate concentration in the incubation solution, allowing to compare the fermentation of different carbohydrate sources under conditions simulating high-concentrate feeding [15,16]. However, this is not the real physiological situation in vivo, as pH changes across a wide range throughout the day [17] and, besides, rumen outflow rate cannot be assessed in this system. In this regard, the semicontinuous incubation system [18] modified by Prates et al. [19], applying the procedure proposed by Amanzougarene and Fondevila [14] for controlling incubation pH, appears to be a useful tool to mimic the rumen pH pattern and liquid outflow rates under in vitro conditions.

Therefore, in a semicontinuous in vitro incubation system, we compared the acidification potential and the rumen microbial fermentation pattern of several carbohydrate sources of variable composition when a different rumen environment is promoted by either high-forage or high-concentrate diets, aiming to minimise, where possible, the risk of acidosis during feeding transition from a fibrous to a high-concentrate diet.

2. Materials and Methods

2.1. Substrates and Inocula

Six carbohydrate sources were chosen as substrates: three cereal grains (barley var. Gustav (B), maize var. Dekalb 6667Y (M), and a brown sorghum of unknown variety (S)) and three by-product feeds (sugarbeet pulp (BP), citrus pulp (CP) and wheat bran (WB)). All substrates were ground in a hammer mill (Retsch Gmbh/SK1/417449, Haan, Germany) through a 1 mm sieve. The chemical compositions of the substrates are given in Table 1.

Table 1. Chemical composition (g/kg DM) of feeds used as incubation substrates.

Component	B	M	S	BP	CP	WB
OM	978	986	979	953	940	944
CP	105	75	113	107	59	161
EE	24	34	11	5	14	31
Starch	672	706	647	-	-	245
aNDFom	173	91	97	437	207	499
ADF	56	25	60	272	192	145
ADL	18	2	5	75	21	37
NDSF	4	77	110	457	423	155
Sugars	1.6	13	1.3	9	243	31
TP	-	-	2.6	-	-	-
TT	-	-	1.3	-	-	-

Barley (B); maize (M); sorghum (S); sugar beet pulp (BP); citrus pulp (CP); wheat bran (WB). Dry matter (DM); organic matter (OM); crude protein (CP); ether extract (EE); neutral detergent fibre (aNDFom); acid detergent fibre (ADF); acid detergent lignin (ADL); neutral detergent soluble fibre (NDSF). Total phenolics (TP); total tannins (TT).

Rumen fluid was obtained from six lambs housed in the facilities of the Servicio de Apoyo a la Experimentación Animal of the Universidad de Zaragoza. The animal care and procedures for extraction of rumen inoculum were approved by the Ethics Committee for Animal Experimentation. The care and management of animals agreed with the Spanish Policy for Animal Protection RD 53/2013, which complies with EU Directive 2010/63 on the protection of animals used for experimental and other scientific purposes. The lambs were weaned at 49 ± 8 days (average weight 13.6 ± 0.78 kg) and, thereafter, three lambs (1, 2 and 3) were fed ad libitum with a concentrate mixture (composed of barley, maize, wheat, and soybean meal) and barley straw (88:12 concentrate to straw ratio) for 35 days, and then slaughtered (average weight 20.6 ± 1.85 kg) to obtain concentrated inoculum (CI). The other three lambs (4, 5 and 6) were fed ad libitum with alfalfa hay and slaughtered after 45 days (average weight 16.5 ± 0.33 kg) to obtain forage inoculum (FI). The rumen contents of each animal were individually filtered through a cheesecloth and dispensed in 16 mL aliquots into 110 × 16 mm tubes, which were immediately frozen in liquid nitrogen and preserved at −80 °C until use [19]. Immediately before incubation, the rumen inoculum was thawed in a water bath at 39 °C (about 2 min).

2.2. Experimental Conditions

The in vitro semicontinuous system of Fondevila and Pérez-Espés [18], modified by Prates et al. [19], was used. The substrate samples (800 mg) were dispensed into 4 × 4 cm nylon bags (45 μm pore size) that were sealed and introduced into duplicated bottles (116 mL total volume). The bottles were filled under CO_2 flux with 80 ml of incubation solution, including 16 mL (0.20 of total volume) thawed rumen inoculum, without resazurin and microminerals [20], and were incubated in a water bath at 39 °C for 24 h in three incubation series, each corresponding to a different donor animal, for each type of inoculum. The buffer solution was modified to include 0.006 M bicarbonate ion in order to get a poorly buffered medium [14].

The pressure produced on each bottle was measured every 2 (from 0 to 12 h incubation) or 4 h (from 12 to 24 h) with a HD8804 manometer provided with a TP804 pressure gauge (DELTA OHM, Caselle di Selvazzano, Italy). The readings, corrected for the atmospheric pressure, were converted to volume (ml) using a pre-established linear regression recorded in this type of bottle (n = 48, R^2 = 0.993), and were expressed per unit of incubated organic matter (OM). Along the incubation, an aliquot volume of the medium was extracted immediately after each gas measurement and replaced anaerobically by the same volume of incubation solution (without microbial inoculum) to simulate an approximate liquid turnover rate of 0.08/h. In order to simulate daily rumen pH fluctuations, from 0 to 6 h, the incubation solution was poorly buffered, as explained above, to allow the incubation pH to drop as fermentation proceeded, whereas, from 8 h onwards, the replacing incubation solution was made up with 0.058 M bicarbonate ion to allow the pH to increase to around 6.5.

The incubation pH was recorded on every extraction. In addition, the medium was sampled at 6 and 10 h for determination of volatile fatty acids concentration (VFA; 2 mL on a 0.5 mL solution of 0.5 M phosphoric acid with 1 mg 4-methyl-valeric acid as the internal standard) and at 6 h for the determination of the lactic acid concentration (2 mL). The samples were stored at -20 °C until analysis. Moreover, another sample (6 mL) was also taken at 8 h and immediately frozen (−80 °C) for the determination of microbial biodiversity by terminal restriction fragment length polymorphism (tRFLP). At the end of incubation, the substrate bags were removed from the bottles, rinsed and dried at 60 °C for 48 h for the determination of dry matter disappearance (DMd).

2.3. Chemical and Microbiological Analyses

The dry matter (DM) and OM content in the substrates and the incubation residues were analysed following the AOAC [21] procedures (methods ref. 934.01 and 942.05). The substrates were also analysed for crude protein (CP) and ether extract (EE) (ref. 976.05 and 2003.05) [21], and their concentration of neutral detergent fibre (aNDFom) was analysed as described by Mertens [22] in an Ankom 200 Fibre Analyser (Ankom Technology, New York, NY, USA), using α–amylase and sodium sulphite, with results being expressed exclusive of residual ashes. The acid detergent fibre (ADF) (ref. 973.18) and acid detergent lignin (ADL) were determined as described by AOAC [21] and Robertson and Van Soest [23], respectively. Neutral detergent soluble fibre (NDSF) was estimated following Hall et al. [24], discounting the aNDFom and the ethanol insoluble EE, CP and starch fractions from the insoluble OM. The total starch content in B, M, S and WB substrates was determined enzymatically from samples ground to 0.5 mm using a commercial kit (Total Starch Assay Kit K-TSTA 07/11, Megazyme, Bray, Ireland). The total phenolic (TP) content in S was analysed following the colourimetric method of Makkar et al. [25] using the Folin–Ciocalteau reagent and with tannic acid (MERCK Chemicals, Madrid, Spain) as the reference standard. The total tannins (TT) were estimated as the difference between TP before and after treatment with polyvinyl polypyrrolidone.

The frozen samples of medium incubation were thawed and centrifuged at 13,000 *g* for 15 minutes at 4 °C in order to analyse their lactic acid and VFA. The VFA were determined by gas chromatography on an Agilent 6890 apparatus equipped with a flame detector and a capillary column (HP-FFAP Polyethylene glycol TPA, 30 m × 530 µm id). The lactic acid concentration was determined by the colourimetric method proposed by Barker and Summerson [26]. For the microbial diversity analysis, frozen microbial samples were freeze-dried, thoroughly mixed and disrupted (Mini-Bead Beater, Biospec Products, Bartlesville, OK, USA). The DNA was extracted using the Qiagen QIAmp DNA Stool Mini Kit (Qiagen Ltd., West Sussex, UK) following the manufacturer's recommendations, except that the samples were initially heated at 95 °C for 5 min to maximise the lysis of the bacterial cells. The concentration of extracted DNA was tested in a Nanodrop ND-1000 (Nano-Drop Technologies, Inc., Wilmington, DE, USA). PCR was performed using a 16S rRNA bacteria-specific primer (cyanine-labelled forward 27F, 5′-AGA GTT TGA TCC TGG CTCAG-3′ and unlabelled reverse 1389R, 5′-AGG GGG GGT GTG TAG AAG-3′; [27]) using a DNAEngine® Gradient Cycler (Bio-Rad, Spain). The PCR product was purified using a Purelink PCR purification kit (ref. K3100-01; Invitrogen) and diluted to 10 µL. The DNA concentration of each amplified and purified sample was obtained by spectrophotometry (Nanodrop® ND-1000 spectrophotometer) to enable a standardised quantity of 50 ng DNA to be used per restriction enzyme digest reaction. The digestion of samples was carried out using HhaI, HaeIII and MspI (Promega, Spain), following the manufacturer recommendations, except for HhaI, where the recommended addition of bovine serum albumin was omitted. The restriction digests were purified by ethanol precipitation [28] in 35 µL sample loading solution buffer, including a 600 bp size standard (Beckman Coulter Inc., Fullerton, CA, USA) before being applied to a 3500 × L Genetic Analyzer (Applied Biosystems). Once the size and height of every peak was obtained, 1% of the second highest peak was used as the criteria for the lower threshold for peaks, to detect and eliminate smaller, broader peaks that would not be indicative of single true OTUs.

2.4. Calculations And Statistical Analyses

The tRFLP results were analysed from a matrix generated for each data list obtained, and results were presented in the form of relative abundance. The three matrices resulting from each series and enzyme were concatenated and analysed with R statistical software (https://cran.r-project.org/bin/windows/base/, version 3.5.0, R Foundation for Statistical Computing, Vienna, Austria). FactoMineR, Factoextra, MixOmics, Vegan, MASS, and Ggplot2 packages were used to carry out the analysis of hierarchical classification on the principal components for obtaining the cluster dendrogram.

The results were analysed statistically by ANOVA using the Statistix 10 package [29]. On each sampling time, the effect of the incubation series (equivalent to the donor animal; interaction inoculum x incubation series, random effect), the type of inoculum, the type of substrate, and the interaction of both factors on pH, gas production, total VFA and lactic acid concentration, and VFA profile were studied as factors. The treatment differences among the means with $p < 0.05$ and $0.05 < p < 0.10$ were accepted as representing statistically significant differences and a trend to the differences, respectively. When significant, the differences were contrasted by the Tukey *t*-test. Simple and multiple linear regressions were established to study the relationships among the different parameters studied.

3. Results

3.1. Pattern of Incubation pH

The mean inoculum pH at the start of the incubation series was 6.45 ± 0.15 and 6.87 ± 0.02 for CI and FI, respectively ($n = 3$). The average minimum pH was recorded at 6 h incubation (5.96) for CI, and at 8 h (6.22) for FI. Thereafter, the pH increased to reach its maximum (6.64 for both inocula) at 24 and 20 h for CI and FI. The pH differences in the incubation medium among inocula ($p < 0.05$) were ± 0.3 units from 2 to 6 h, decreasing gradually to ± 0.1 at 12 h. A significant interaction inoculum × substrate ($p < 0.05$) observed on pH at 4, 8, 10, 12, 16 and 20 h and a tendency ($p = 0.052$) at 2 h incubation indicates the different behaviour of the substrates depending on the inoculum. Therefore, a comparison of the pH pattern among the incubated substrates is presented in Figure 1 separately for each inoculum. With CI (Figure 1a), the lowest incubation pH from 2 to 12 h was recorded with CP ($p < 0.05$), reaching its minimum at 6 h (5.60), although recovered thereafter to 6.63 at 24 h incubation. In ascending order, WB and B reached their minimum pH at 8 h (5.89 and 5.97, respectively), whereas BP, M, and S maintained a higher medium pH from 4 to 8 h ($p < 0.05$). The differences among M, S, BP, and B disappeared from 10 to 16 h ($p > 0.05$), and no differences were detected among the substrates ($p > 0.05$) at the end of incubation. When the substrates were incubated using FI (Figure 1b), CP recorded the lowest pH from 4 to 10 h incubation ($P < 0.05$), and its minimum value was 5.90, whereas S, M and BP maintained the highest medium pH during this period (6.30 to 6.46), and B and WB were grouped at intermediate values ($p < 0.05$). At 16 and 20 h incubation only, B recorded a lower value (6.44 and 6.57; $p < 0.05$) and, again, no differences were detected among the substrates at the end of incubation ($p > 0.05$).

3.2. Pattern of in Vitro Gas Production

The volume of gas produced with the CI inoculum was higher than that obtained with FI at all incubation times ($p < 0.05$). Because of the interaction inoculum x substrate at 4 h and from 8 to 24 h ($p < 0.05$), for an easier understanding, the gas production is presented separately for CI and FI (Figure 2). The major difference among substrate fermentative behaviour between the inocula is manifested in the magnitude of differences among them. Thus, with CI (Figure 2a), CP recorded the highest gas volume from 4 h onwards, at 12 h being on average 0.42 times higher than the other substrates, while also recorded differences at 2 h with BP and S ($p < 0.05$). The gas volume with WB was higher than BP and S from 4 h onwards, and higher than M from 6 h and B from 8 to 20 h ($p < 0.05$). Differences were also recorded between B and S from 8 to 16 h and at 24 h ($p < 0.05$). A similar pattern was observed with FI (Figure 2b), but the magnitude of differences was lower. Thus, CP was higher than B, M, BP and S from 6 to 24 h ($p < 0.05$), with differences at 12 h reaching 0.59 of their average, but did not differ from

WB, which was higher than BP and S in that period and also higher than M from 6 to 10 h ($p < 0.05$). Differences between B and M respect to S were also detected from 16 and 20 h onwards, respectively ($p < 0.05$).

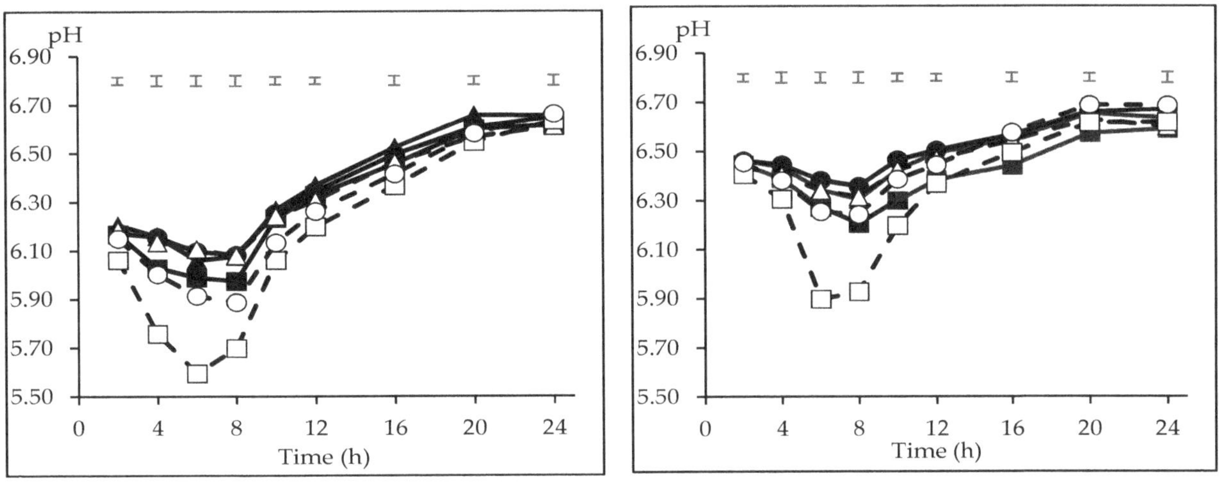

(a) (b)

Figure 1. Pattern of medium pH of carbohydrate substrates (barley (B) □, maize (M) ▲, sorghum (S) •; solid lines, citrus pulp (CP) ■, sugar beet pulp (BP) △, wheat bran (WB) ○; dashed lines) incubated with inoculum from concentrate (CI, Figure 1a) or forage (FI, Figure 1b) diets. The initial pH was 6.45 (Figure 1a) and 6.87 (Figure 1b). The upper bars show the standard error of the means (n = 3).

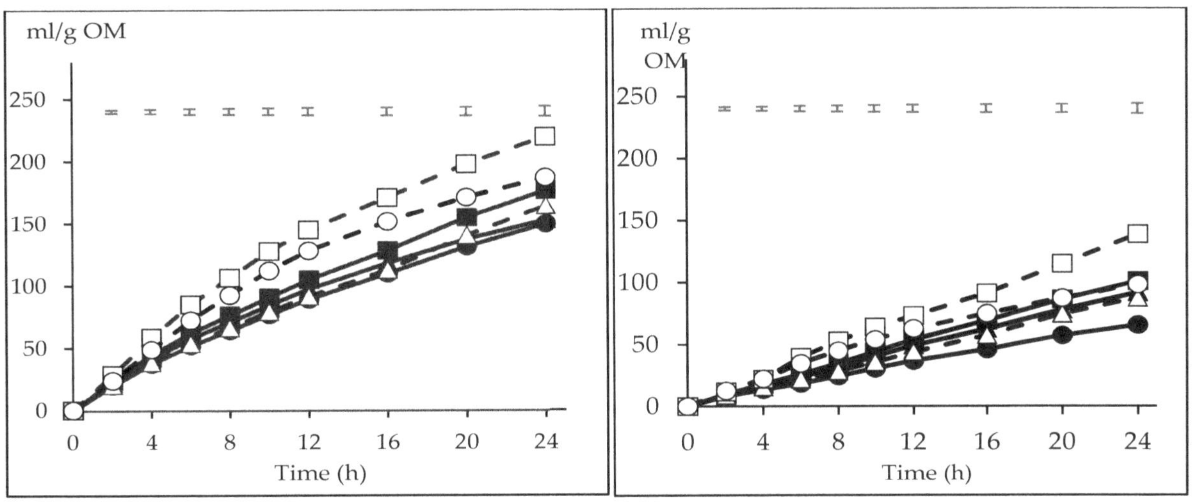

(a) (b)

Figure 2. Pattern of gas production from the carbohydrate substrates (barley (B) ■, maize (M) ▲, sorghum (S) •; solid lines, citrus pulp (CP) □, sugar beet pulp (BP) △, wheat bran (WB) ○; dashed lines) incubated with inoculum from concentrate (Figure 2a) or forage (Figure 2b) diets. The upper bars show the standard error of the means (n = 3).

3.3. *Dry matter Disappearance (DMd)*

Inoculum differences in DMd after 24 h of incubation were not detected, although CI was numerically higher than FI (proportions of 0.382 *vs.* 0.339 from the substrate weight; $p > 0.05$). The substrates ranked according to the proportion of DMd as follows: CP, 0.502 > B, 0.449 > WB, 0.360, M, 0.343 > BP, 0.265, S, 0.243 ($p < 0.001$; SEM = 0.0120). The interaction inoculum x substrate was not significant (p = 0.21), indicating that the substrates behaved similarly with both inocula.

3.4. Volatile Fatty Acids and Lactic Acid Production

Tables 2 and 3 show that CI promoted a higher ($p < 0.05$) concentration of total VFA than FI at both 6 (23.2 *vs.* 9.8 mM) and 10 h (22.2 *vs.* 9.3 mM). The molar proportions of acetate, propionate, and butyrate did not manifest the differences between inocula ($p > 0.05$), whereas with CI, valerate was higher and branched-chain volatile fatty acids (BCVFA, sum of isobutyrate and isovalerate) lower than with FI at both incubation times ($p < 0.05$).

Table 2. Average of total volatile fatty acids concentration (VFA, mM) and molar VFA proportions (mmol/mmol), together with lactate concentration (mM) recorded at 6 h of the different carbohydrate sources incubated as substrates with concentrate (CI) or forage (FI) inoculum.

Substrates	VFA	Acetate	Propionate	Butyrate	Valerate	BCVFA	Lactic acid
			with CI				
B	22.34 ^{ab}	0.570	0.240	0.153	0.022	0.014	3.83 ^{b}
M	20.18 ^{bc}	0.578	0.235	0.149	0.023	0.016	2.35 ^{c}
S	21.72 ^{abc}	0.588	0.245	0.131	0.022	0.015	1.93 ^{c}
BP	21.66 ^{abc}	0.593	0.235	0.135	0.021	0.016	0.90 ^{c}
CP	26.55 ^{a}	0.595	0.238	0.135	0.020	0.012	8.70 ^{a}
WB	26.94 ^{a}	0.590	0.245	0.132	0.020	0.013	2.95 ^{c}
			With FI				
B	10.31 ^{xyz}	0.633	0.226	0.102	0.009	0.030 ^{y}	3.05 ^{y}
M	8.83 ^{yz}	0.632	0.229	0.095	0.010	0.034 ^{xy}	2.69 ^{y}
S	8.31 ^{z}	0.642	0.216	0.094	0.009	0.039 ^{x}	0.84 ^{y}
BP	9.52 ^{yx}	0.665	0.207	0.088	0.008	0.032 ^{y}	1.64 ^{y}
CP	12.12 ^{yz}	0.686	0.209	0.075	0.008	0.023 ^{z}	8.67 ^{x}
WB	9.83 ^{yz}	0.653	0.225	0.083	0.009	0.030 ^{y}	2.97 ^{y}
SEM	1.065	0.0156	0.0076	0.0085	0.0010	0.0009	0.588
			p-Value				
Inoculum	0.002	0.077	NS	NS	0.005	<0.001	NS
Substrate	<0.001	NS	NS	NS	NS	<0.001	<0.001
Inoc. × Subs.	NS	NS	NS	NS	NS	<0.001	NS

Means within a column with different superscripts for CI (^{a,b,c}) or FI (^{x,y,z}) differ ($p < 0.05$). Standard error of the means (SEM). Branched-chain volatile fatty acids (BCVFA) (sum of isobutyrate + isovalerate). NS: $p > 0.10$.

Table 3. Average of total volatile fatty acids concentration (VFA, mM) and molar VFA proportions (mmoL/mmoL), recorded at 10 h of the different carbohydrate sources incubated as substrates with concentrate (CI) or forage (FI) inoculum.

Substrates	VFA	Acetate	Propionate	Butyrate	Valerate	BCVFA
			With CI			
B	21.16 ^{b}	0.561	0.225	0.172	0.028	0.014
M	19.42 ^{b}	0.548	0.227	0.178	0.031	0.017
S	19.87 ^{b}	0.557	0.249	0.152	0.026	0.017
BP	20.25 ^{b}	0.604	0.229	0.130	0.022	0.016
CP	27.32 ^{a}	0.553	0.240	0.164	0.031	0.012
WB	25.10 ^{ab}	0.537	0.261	0.158	0.029	0.015
			With FI			
B	10.46	0.597	0.256 ^{xy}	0.117	0.009	0.022
M	8.51	0.620	0.236 ^{xy}	0.109	0.009	0.026
S	8.48	0.620	0.222 ^{y}	0.120	0.009	0.029
BP	8.34	0.611	0.227 ^{y}	0.126	0.009	0.027
CP	10.47	0.579	0.267 ^{x}	0.124	0.010	0.019
WB	9.76	0.575	0.274 ^{x}	0.117	0.010	0.025
SEM	1.137	0.0233	0.0078	0.0169	0.0019	0.0024
			p-Value			
Inoculum	0.011	NS	NS	NS	0.008	0.008
Substrate	0.001	NS	<0.001	NS	NS	0.046
Inoc. × Subs.	0.058	NS	0.017	NS	NS	NS

Means within a column with different superscripts for CI (^{a,b,c}) or FI (^{x,y,z}) differ ($p < 0.05$). SEM: standard error of the means. Branched-chain volatile fatty acids (BCVFA) (sum of isobutyrate + isovalerate).

Among the substrates, at 6 h (Table 2) CP recorded a higher total VFA concentration than BP, M and S (average values of 19.3, 15.6, 14.5 and 15.0 mM, respectively), whereas WB (15.9 mM) was also higher than M and S ($p < 0.05$). Differences in the molar VFA profile were only recorded for BCVFA, with the highest proportions in S and M and the lowest with CP ($p < 0.05$); however, the interaction inoculum x substrate ($p < 0.001$) indicates that differences in BCVFA proportion were only observed with FI. Regarding the concentration of lactic acid at 6 h among the substrates, CP recorded the highest concentration and BP and S the lowest (8.7 *vs.* 1.3 and 1.4 mM; $p < 0.05$). Similar trends were observed at 10 h in total VFA concentration (Table 3), with CP rendering a higher concentration than BP, M and S, but tending to be significant only with CI interaction inoculum x substrate, ($p = 0.058$). The interaction inoculum x substrate in the proportion of propionate ($p = 0.017$) indicates that values recorded with WB and CP were higher than those with BP and S, but differences were only manifested with FI, whereas no differences ($p > 0.05$) among the substrates were recorded on acetate, butyrate and valerate proportions. The highest proportion of BCVFA was promoted by S and the lowest by CP ($p < 0.05$).

3.5. *Bacterial Biodiversity*

Bacterial biodiversity after 8 h of incubation was markedly affected by the source of rumen inoculum. Thus, the substrates incubated with rumen inoculum from lambs fed the high-concentrate diet clustered together, except for WB in the first incubation run, as well as substrates that were incubated with FI (Figure 3). Bacterial biodiversity was also markedly affected by the incubation series—that is, the donor animal—for both inocula.

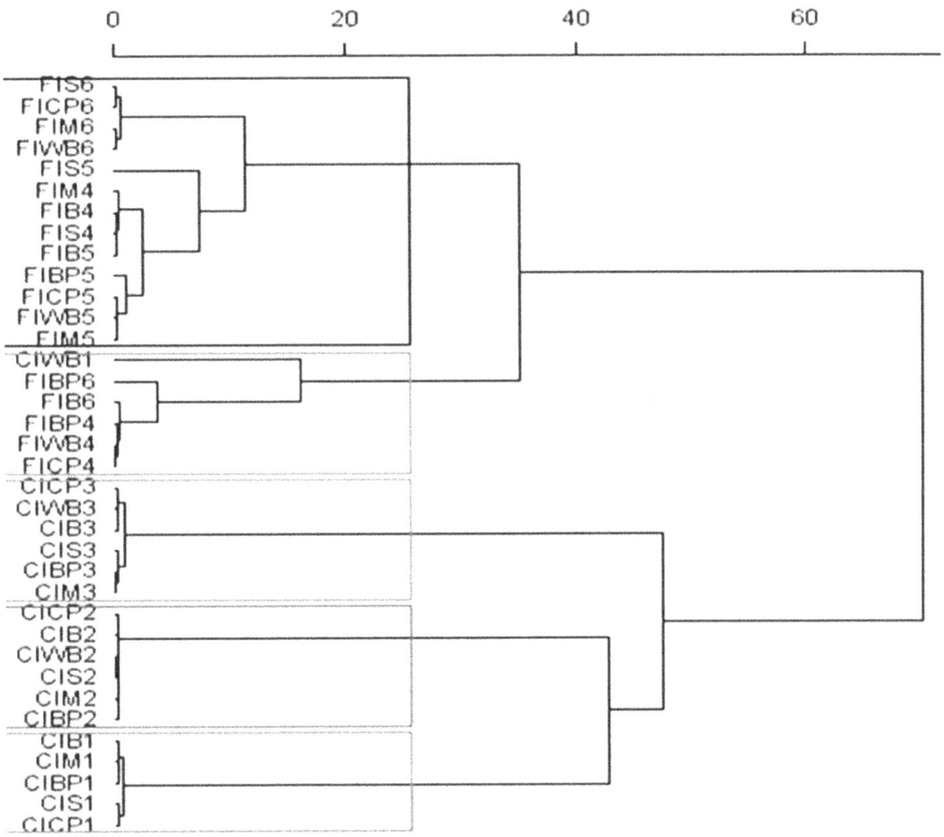

Figure 3. Dendrogram of bacteria diversity from terminal restriction fragment length polymorphism (tRFLP) data generated by enzyme digestion (HhaI, MspI, and HaeIII) for the carbohydrate substrates (B, M, S, BP, CP, and WB) incubated for 8 h with inoculum from concentrate (CI) or forage (FI) diets. The scale bar shows the Euclidean distances, "Ward method".

4. Discussion

Conventional in vitro closed batch systems are adapted for the study of microbial fermentation under conditions mimicking high forage diets, which is not applicable to evaluate diets given in intensive ruminant-fattening systems. An in vitro semicontinuous incubation system [18,19], adapted to control of the pH by modifying the bicarbonate ion concentration [14] allows us to approach the ruminal fermentation pattern of the different carbohydrate sources to the rumen physiological conditions that occur in intensive feeding systems, either during a transition process to high-concentrate diets (i.e., when rumen conditions are still modulated by a forage diet) or when animals are adapted to such feeding conditions (as promoted by a concentrate diet). The pH pattern obtained along the in vitro incubation with CI and FI, reaching a minimum value at 6–8 h after substrate availability and then progressively increasing to final pH values of around 6.4–6.5, fitted well with the circadian evolution of rumen pH observed with practical forage or concentrate feeding of ruminants [30]. Thus, we can assume it allows the different substrates to express their acidification potential at the time that their fermentation is compared under more realistic conditions.

4.1. Effect of the Inoculum Source on the in Vitro Fermentation Pattern

The source of rumen fluid has an important role in the pattern of in vitro fermentation [15,31,32], with an inoculum promoted by a concentrate diet having a higher fermentative potential than another from a forage diet. In our experiment, the lower buffering of the incubation medium during the first 6–8 hours allowed for a clear expression of the acidification potential of the incubated substrates, which was expressed at a higher extent with CI than FI (average pH along the 24 h incubation period ranging from 6.45 to 5.96 *vs.* 6.87 to 6.22) as pH dropped to values close to those considered as a threshold for microbial activity [33], whereas a higher pH was maintained with FI. Despite this, substrates incubated with CI rendered almost two-fold gas volume more than with FI, irrespective of the chemical nature (starch- or fibre-rich) of those substrates. Despite the more pronounced drop of pH with CI, the incubation environment promoted by a concentrate diet given to the donor animals was more favourable for fermentation of non-fibrous carbohydrates than that induced by a fibrous diet [20,34], probably because of the lack of adaptation of microbiota to ferment starch and sugar substrates with a forage inoculum [15,35] and the inherent buffering capacity of forage legumes such as alfalfa. However, assuming that a part of the gas produced comes from the activity of bicarbonate ion in the buffering of fermentation acids produced, such differences in gas production could be partly associated with the lower pH promoted by CI inoculum, although the contribution of this indirect gas is hard to quantify [14]. In the case of the byproducts, characterised by their richness in rapidly fermentable fibre, microbiota might easily counterbalance the lack of adaptation for their degradation [36,37]. In contrast, the low pH occurring during the initial part of incubation may affect, at a higher extent, the activity of the bacterial species adapted to fibre degradation, causing a lower magnitude of fermentation of structural polysaccharides like cellulose and hemicelluloses [12,38,39].

Contrary to what might be expected, the results of the gas production were not supported by those of DMd. This parameter was especially low compared to the extent of the rumen degradation of starch-rich sources (around 0.70–0.80 [40]) or fibrous sources (ranging from 0.40 to 0.70 [41]). This is difficult to explain, but we have also observed this low response in previous in vitro experiments [42], partly associated with a low pH [13]. Calsamiglia et al. [43] justified similar results by the differences between rumen and in vitro microbial ecosystems, partly because the dilution of inoculum in the latter reduces the extent of the degradation. In contrast to DMd, the concentration of total VFA followed a similar trend than that of gas production, being higher for CI at both sampling times, as has been observed by others [15,32,43]. Calsamiglia et al. [43] did not observe any inoculum effect on the acetate and butyrate proportions, and propionate proportion was higher with the concentrate inoculum, as observed in our study at 10 h incubation. However, differences in the proportion of BCVFA, which resulted from fermentation of protein and branched-chain amino acids [44] were higher with FI between 6 and 10 h, probably because of the fermentation of protein from the alfalfa hay fed to the donor lambs.

The effect of the inoculum source was also observed in microbial diversity, suppporting the recent findings reported by Tapio et al. [45] and Nagata et al. [46] who showed the difference in the rumen microbial population when bulls were fed with forage or concentrate diets.

4.2. *Effect of Different Substrates on the In Vitro Fermentation Pattern*

Despite the marked differences in the magnitude of fermentation between CI and FI, the fermentation pattern among the substrates was almost the same between both inocula. The results of the measured parameters showed a strong correlation between gas production and the other parameters (pH, VFA and lactic acid concentrations) at 6 h ($n = 36$; adjusted $R^2 = 0.90$; $p < 0.001$). Similarly, at 10 h incubation, the volume of gas produced was strongly correlated with incubation pH and VFA ($n = 36$; adjusted $R^2 = 0.84$; $p < 0.001$). These results confirm that the gas production and, equally, the concentration of total VFA and lactic acid, are the main factors indicating the acidification potential of the incubated substrates [31,47,48]. Citrus pulp had a higher acidification capacity than the other substrates, which is associated with a higher magnitude of fermentation that is manifested in high gas production, as well as VFA and lactic acid concentration. Despite the high concentration of lactic acid with CP at 6 h (Table 2), it did not achieve the range considered as a risk of acidosis in vivo [30] and, in fact, did not promote the values of incubation pH below 5.5 that are considered as a threshold for the onset of subacute acidosis [49]. These results were in agreement with those found by Amanzougarene et al. [42] in a batch culture with a minimum buffer concentration, and could be associated with its richness in soluble sugars [50,51], estimated as 0.24 g/kg DM (Table 1), which are fermented at a very fast rate. Although CP also has a high proportion of soluble fibre (0.42, Table 1), this response cannot be directly associated with the fast fermentation of pectin [37,52], since BP includes a similar NDSF proportion (0.46) and it was fermented at a slower rate and magnitude. In fact, Strobel and Russel [53] reported that at a pH of 6.00, the extent of pectin fermentation was reduced with respect to a higher pH. The lower fermentation rate of BP and, thus, its lower acidification potential can also be related to its high NDF content, which does not ensure its maximum fermentation in the 24 h incubation period [54]. Considering the aforementioned characteristics of BP composition, mainly its high NDF and NDSF proportions as well as its low sugar content, its lower concentration of lactic acid produced with respect to the other incubated substrates could be expected. Others [52,53] have also stated that the yield of lactic acid production from pectins fermentation is very low.

The extent of fermentation of WB and B was lower than that of CP, but higher than those of the remaining substrates, probably linked to the high proportion of rapidly fermentable starch in these substrates, compared with those of M and S. Nocek and Tamminga, [55] indicated that 0.80 to 0.90 of barley or wheat starch is digested in the rumen, compared to only 0.55 to 0.70 of that of corn and sorghum. In addition, WB have a considerable amount of NDSF and highly fermentable NDF (Table 1). The structure of the starch endosperm of maize and sorghum, together with their different proportions of amylose [4], as well as the protein matrix in the endosperm in these cereal species [56] and the presence of phenolic compounds in the brown sorghum [16] explain why the fermentation of starch of barley and wheat bran by ruminal bacteria was higher [4,57,58]. Consequently, the differences in the starch characteristics and fermentation rate promote the response in a medium pH [15]. When incubating several grains in a well buffered medium, Lanzas et al. [59] observed a higher fractional rate of 48 h gas production with barley than maize and sorghum varieties (on average, 0.24, 0.15 and 0.06/h). Opatpatanakit et al. [60], modified the incubation pH similarly to the present work, and also observed the highest gas production with barley, intermediate with maize and lowest with sorghum (on average, 222, 138 and 104 mL/g DM, respectively), under pH values at 7 h incubation ranging from 5.7 to 6.1 for barley, 6.5 to 6.9 for maize and 6.5 to 6.8 for sorghum.

From our findings, it can be indicated that citrus pulp and, to a lower extent, wheat bran had an acidic capacity of an even higher magnitude than cereal sources, including barley. Despite the differences on the magnitude and extent of fermentation between the different incubated substrates, the results of microbial diversity with both inocula showed the major effect of the donor animal on this parameter, partly because of aspects related to in vitro methodology, such as the short period of incubation. These results were in accordance with those reported by Taxis et al. [61] and Söllinger et al. [62], explaining the differences in microbial diversity from one animal to another. However, within each series (donor animal), our results did not demonstrate differences between the substrates.

5. Conclusions

Under the fermentation conditions of high-concentrate feeding, some sources of highly fermentable fibre, such as citrus pulp and, to a lower extent, wheat bran may create a more acidic environment than cereals. Among these, barley promotes a lower pH than maize or sorghum, as this grain is associated with a higher rate and extent of fermentation. The rumen environment promoted by high forage/fibre diets is not adapted for non-fibrous carbohydrates, and fermentation of soluble fibre is not differentially enhanced, producing a lower extent of substrate fermentation than concentrate diets. Therefore, the choosing of ingredients is important when ruminants are changed from a forage to a high-concentrate diet, although this cannot be inferred from this study. In any case, in this experiment, acidification levels did not reach those that may change the fermentation pattern. Care must be taken in substrate comparison in terms of gas production, since the buffering of the medium under low pH conditions may overestimate the fermentation differences by increasing the indirect gas production. The in vitro semicontinuous system, adapted to a variable medium pH, has proven to be useful for the study of rumen microbial fermentation under intensive feeding conditions.

Author Contributions: Conceptualization, M.F.; methodology, Z.A. and M.F.; software, Z.A. and M.F.; validation, Z.A., S.Y. and M.F.; formal analysis, Z.A. and M.F.; investigation, Z.A., M.F. and S.Y.; resources, M.F.; data curation, M.F.; writing—original draft preparation, Z.A.; writing—review and editing, M.F.; visualization, Z.A.; supervision, M.F.; project administration, M.F.; funding acquisition, M.F. All authors have read and approved the final manuscript.

Acknowledgments: Sincere appreciation to the lab team for their support in the laboratory work.

References

1. Bevans, D.W.; Beauchemin, K.A.; Schwartzkopf-Genswein, K.S.; McKinnon, J.J.; McAllister, T.A. Effect of rapid or gradual grain adaptation on subacute acidosis and feed intake by feedlot cattle. *J. Anim. Sci.* **2005**, *83*, 1116–1132. [CrossRef] [PubMed]
2. Fernando, S.C.; Purvis, H.T.; Najar, F.Z.; Sukharnikov, L.O.; Krehbiel, C.R.; Nagaraja, T.G.; Roe, B.A.; de Silva, U. Rumen microbial population dynamics during adaptation to a high-grain diet. *Appl. Environ. Microbiol.* **2010**, *76*, 7482–7490. [CrossRef] [PubMed]
3. O'Brien, L. Genotype and environment effects on feed grain quality. *Aust. J. Agric. Res.* **1999**, *50*, 703–719. [CrossRef]
4. Offner, A.; Bach, A.; Sauvant, D. Quantitative review of *in situ* starch degradation in the rumen. *Anim. Feed Sci. Technol.* **2003**, *106*, 81–93. [CrossRef]
5. De Peeters, E.J.; Fadel, J.G.; Arosemena, A. Digestion kinetics of neutral detergent fiber and chemical composition within some selected by-product feedstuffs. *Anim. Feed Sci. Technol.* **1997**, *67*, 127–140. [CrossRef]
6. Maes, C.; Delcour, J.A. Alkaline hydrogen peroxide extraction of wheat bran non-starch polysaccharides. *J. Cereal Sci.* **2001**, *34*, 29–35. [CrossRef]
7. Mould, F.L.; Kliem, K.E.; Morgan, R.; Mauricio, R.M. *In vitro* microbial inoculum: A review of its function and properties. *Anim. Feed Sci. Technol.* **2005**, *123–124*, 31–50. [CrossRef]
8. McAllister, T.A.; Cheng, K.J.; Rode, L.M.; Forsberg, C.W. Digestion of barley, maize, and wheat by selected species of ruminal bacteria. *Appl. Environ. Microbiol.* **1990**, *56*, 3146–3153. [CrossRef]

9. Klieve, A.V.; Hennessy, D.; Ouwerkerk, D.; Forster, R.J.; Mackie, R.I.; Attwood, G.T. Establishing populations of Megasphaera elsdenii YE34 and Butyrivibrio fibrisolvens YE44 in the rumen of cattle fed high grain diets. *J. Appl. Microbiol.* **2003**, *95*, 621–630. [CrossRef]
10. Dijkstra, J.; Kebreab, E.; Bannink, A.; France, J.; López, S. Application of the gas production technique to feed evaluation systems for ruminants. *Anim. Feed Sci. Technol.* **2005**, *123–124*, 561–578. [CrossRef]
11. Raab, L.; Cafantaris, B.; Jilg, T.; Menke, K.H. Rumen protein degradation and biosynthesis. A new method for determination of protein degradation in rumen fluid *in vitro*. *Br. J. Nutr.* **1983**, *50*, 569–582. [CrossRef]
12. Sari, M.; Ferret, A.; Calsamiglia, S. Effect of pH on *in vitro* microbial fermentation and nutrient flow in diets containing barley straw or non-forage fiber sources. *Anim. Feed Sci. Technol.* **2015**, *200*, 17–24. [CrossRef]
13. Bertipaglia, L.M.A.; Fondevila, M.; van Laar, H.; Castrillo, C. Effect of pelleting and pellet size of a concentrate for intensively reared beef cattle on *in vitro* fermentation by two different approaches. *Anim. Feed Sci. Technol.* **2010**, *159*, 88–95. [CrossRef]
14. Amanzougarene, Z.; Fondevila, M. Fitting of pH conditions for the study of concentrate feeds fermentation by the *in vitro* gas production technique. *Anim. Prod. Sci.* **2018**, *58*, 1751–1757. [CrossRef]
15. Amanzougarene, Z.; Yuste, S.; Castrillo, C.; Fondevila, M. *In vitro* acidification potential and fermentation pattern of different cereal grains incubated with inoculum from animals given forage or concentrate-based diets. *Anim. Prod. Sci.* **2018**, *58*, 2300–2307. [CrossRef]
16. Amanzougarene, Z.; Yuste, S.; de Vega, A.; Fondevila, M. Differences in nutritional characteristics of three varieties of sorghum grain determine their *in vitro* rumen fermentation. *Span. J. Agric. Res.* **2018**, *16*. [CrossRef]
17. Krause, K.M.; Combs, D.K. Effects of forage particle size, forage source and grain fermentability on performance and ruminal pH in mid lactation cows. *J. Dairy Sci.* **2003**, *86*, 1382–1397. [CrossRef]
18. Fondevila, M.; Pérez-Espés, B. A new in vitro system to study the effect of liquid phase turnover and pH on microbial fermentation of concentrate diets for ruminants. *Anim. Feed Sci. Technol.* **2008**, *144*, 196–211. [CrossRef]
19. Prates, A.; de Oliveira, J.A.; Abecia, L.; Fondevila, M. Effects of preservation procedures of rumen inoculum on *in vitro* microbial diversity and fermentation. *Anim. Feed Sci. Technol.* **2010**, *155*, 186–193. [CrossRef]
20. Mould, F.L.; Morgan, R.; Kliem, K.E.; Krystallidou, E. A review and simplification of the *in vitro* incubation medium. *Anim. Feed Sci. Technol.* **2005**, *123–124*, 155–172. [CrossRef]
21. AOAC. *Official Methods of Analysis*, 18th ed.; Horwitz, W., Latimer, G.W., Eds.; Association of Official Analytical Chemists: Gaithersburg, MD, USA, 2005.
22. Mertens, D.R. Gravimetric determination of amylase-treated neutral detergent fiber in feeds with refluxing in beakers or crucibles: Collaborative study. *J. AOAC Int.* **2002**, *85*, 1217–1240. [PubMed]
23. Robertson, J.B.; Van Soest, P.J. The detergent system of analysis and its application to human foods. In *The Analysis of Dietary Fiber in Foods*; James, W.P.T., Theander, O., Eds.; Marcel Dekker: New York, NY, USA, 1981; pp. 123–158.
24. Hall, M.B.; Lewis, B.A.; Van Soest, P.J.; Chase, L.E. A simple method for estimation of neutral detergent-soluble fibre. *J. Sci.Food Agric.* **1997**, *74*, 441–449. [CrossRef]
25. Makkar, H.P.S.; Blummel, M.; Borowy, N.K.; Becker, K. Gravimetric determination of tannins and their correlations with chemical and protein precipitation methods. *J. Sci. Food Agric.* **1993**, *61*, 161–165. [CrossRef]
26. Barker, S.B.; Summerson, W.H. The colorimetric determination of lactic acid in biological material. *J. Biol. Chem.* **1941**, *138*, 535–554.
27. Hongoh, Y.; Yuzawa, H.; Ohkuma, M.; Kudo, T. Evaluation of primers and PCR conditions for the analysis of 16S rRNA genes from a natural environment. *FEMS Microbiol. Lett.* **2003**, *221*, 299–304. [CrossRef]
28. De la Fuente, G.; Belanche, A.; Girwood, S.E.; Pinloche, E.; Wilkinson, T.; Newbold, C.J. Pros and cons of ion-torrent next generation sequencing versus terminal restriction fragment length polymorphism T-RFLP for studying the rumen bacterial community. *PLoS ONE* **2014**, *9*, e101435. [CrossRef]
29. Analytical Software. *Statistix 10 for Windows*; Analytical Software: Tallahasee, FL, USA, 2010.
30. Nagaraja, T.G.; Titgemeyer, E.C. Ruminal acidosis in beef cattle: The current microbiological and nutritional outlook. *J. Dairy Sci.* **2007**, *90*, E17–E38. [CrossRef]
31. Broudiscou, L.P.; Offner, A.; Sauvant, D. Effects of inoculum source, pH, redox potential and headspace di-hidrogen on rumen *in vitro* fermentation yields. *Animal* **2014**, *8*, 931–937. [CrossRef]

32. Kim, S.H.; Mamuad, L.L.; Kim, E.J.; Sung, H.G.; Bae, G.S.; Cho, K.K.; Lee, C.; Lee, S.S. Effect of different concentrate diet levels on rumen fluid inoculum used for determination of in vitro rumen fermentation, methane concentration, and methanogen abundance and diversity. *Ital. J. Anim. Sci.* **2018**, *17*, 359–367. [CrossRef]
33. Hiltner, P.; Dehority, B.A. Effect of soluble carbohydrates on digestion of cellulose by pure cultures of rumen bacteria. *Appl. Environ. Microbiol.* **1983**, *46*, 642–648. [CrossRef]
34. Menke, K.H.; Steingass, H. Estimation of the energy feed value obtained from chemical analysis and in vitro gas production using rumen fluid. *Anim. Res. Dev.* **1988**, *28*, 7–55.
35. Nagadi, S.; Herrero, M.; Jessop, N.S. The influence of diet of the donor animal on the initial bacterial concentration of ruminal fluid and in vitro gas production degradability parameters. *Anim. Feed Sci. Technol.* **2000**, *87*, 231–239. [CrossRef]
36. Hatfield, R.D.; Weimer, P.J. Degradation characteristics of isolated and in situ cell wall lucerne pectic polysaccharides by mixed ruminal microbes. *J. Sci. Food Agric.* **1995**, *69*, 185–196. [CrossRef]
37. Barrios-Urdaneta, A.; Fondevila, M.; Castrillo, C. Effect of supplementation with different proportions of barley grain or citrus pulp on the digestive utilization of ammonia-treated straw by sheep. *Anim. Sci.* **2003**, *76*, 309–317. [CrossRef]
38. Mould, F.L.; Ørskov, E.R. Manipulation of rumen fluid pH and its influence on cellulolysis *in sacco*, dry matter degradation and the ruminal microflora of sheep offered either hay or concentrate. *Anim. Feed Sci. Technol.* **1983**, *10*, 1–14. [CrossRef]
39. Grant, R.H.; Mertens, D.R. Influence of buffer pH and raw corn starch addition on in vitro fiber digestion kinetics. *J. Dairy Sci.* **1992**, *75*, 2762–2768. [CrossRef]
40. Cerneau, P.; Michalet-Doreau, B. In situ starch degradation of different feeds in the rumen. *Reprod. Nutr. Dev.* **1991**, *31*, 65–72. [CrossRef]
41. Demarquilly, C.; Andrieu, J. Les fourrages. In *Alimentation des Bovins Ovins et Caprins*; INRA: Paris, France, 1988; Volume 3, pp. 57–74.
42. Amanzougarene, Z.; Yuste, S.; de Vega, A.; Fondevila, M. *In vitro* fermentation pattern and acidification, potential of different sources of carbohydrates by ruminants giving high concentrate diets. *Span. J. Agric. Res.* **2017**, *15*, e0602. [CrossRef]
43. Calsamiglia, S.; Cardozo, P.W.; Ferret, A.; Bach, A. Changes in rumen microbial fermentation are due to a combined effect of type of diet and pH. *J. Anim. Sci.* **2008**, *86*, 702–711. [CrossRef]
44. Saro, C.; Ranilla, M.J.; Tejido, M.L.; Carro, M.D. Influence of forage type in the diet of sheep on rumen microbiota and fermentation characteristics. *Livest. Sci.* **2014**, *160*, 52–59. [CrossRef]
45. Tapio, I.; Fischer, D.; Blasco, L.; Tapio, M.; Wallace, R.J.; Bayat, A.R.; Ventto, L.; Kahala, M.; Negussie, E.; Shingfield, K.J.; et al. Taxon abundance, diversity, co-occurrence and network analysis of the ruminal microbiota in response to dietary changes in dairy cows. *PLoS ONE* **2017**, *12*, e0180260. [CrossRef] [PubMed]
46. Nagata, R.; Kim, Y.H.; Ohkubo, A.; Kushibiki, S.; Ichijo, T.; Sato, S. Effects of repeated subacute ruminal acidosis challenges on the adaptation of the rumen bacterial community in Holstein bulls. *J. Dairy Sci.* **2018**, *101*, 4424–4436. [CrossRef] [PubMed]
47. Russell, J.B.; Hino, T. Regulation of lactate production in streptococcus bovis—A spiraling effect that contributes to rumen acidosis. *J. Dairy Sci.* **1985**, *68*, 1712–1721. [CrossRef]
48. Sauvant, D.; Giger-Reverdin, S.; Meschy, F. Le contrôle de l'acidose ruminale latente. *INRA Prod. Anim.* **2006**, *19*, 69–78.
49. Krause, K.M.; Oetzel, G.R. Understanding and preventing subacute ruminal acidosis in dairy herds: A review. *Anim. Feed Sci. Technol.* **2006**, *126*, 215–236. [CrossRef]
50. Hall, M.B.; Pell, A.N.; Chase, L.E. Characteristics of neutral detergent-soluble fiber fermentation by mixed ruminal microbes. *Anim. Feed Sci. Technol.* **1998**, *70*, 23–39. [CrossRef]
51. Ariza, P.; Bach, A.; Stern, M.D.; Hall, M.B. Effects of carbohydrates from citrus pulp and hominy feed on microbial fermentation in continuous culture. *J. Anim. Sci.* **2001**, *79*, 2713–2718. [CrossRef]
52. Bampidis, V.A.; Robinson, P.H. Citrus by-products as ruminant feeds: A review. *Anim. Feed Sci. Technol.* **2006**, *128*, 175–217. [CrossRef]
53. Strobel, H.J.; Russel, J.B. Effects of pH and energy spilling on bacterial protein synthesis by carbohydrate-limited cultures of mixed rumen bacteria. *J. Dairy Sci.* **1986**, *69*, 2941–2947. [CrossRef]

54. Sauvant, D.; Dorleans, M.; Delacour, C.; Bertrand, D.; Giger, S. La modélisation des cinétiques de dégradation des constituants pariétaux et cellulaires des aliments dans le rumen. *Reprod. Nutr. Dev.* **1986**, *26*, 303–304. [CrossRef]
55. Nocek, J.E.; Tamminga, S. Site of digestion of starch in the gastrointestinal tract of dairy cows and its effect on milk yield and composition. *J. Dairy Sci.* **1991**, *74*, 3598–3629. [CrossRef]
56. McAllister, T.A.; Philippe, R.C.; Rode, L.M.; Cheng, K.J. Effect of the protein matrix on the digestion of cereal grains by ruminal microorganisms. *J. Anim. Sci.* **1993**, *71*, 205–212. [CrossRef] [PubMed]
57. Overton, T.R.; Cameron, M.R.; Elliott, J.P.; Clark, J.H.; Nelson, D.R. Ruminal fermentation and passage of nutrients to the duodenum of lactating cows fed mixtures of corn and barley. *J. Dairy Sci.* **1995**, *78*, 1981–1998. [CrossRef]
58. Firkins, J.L.; Eastridge, M.L.; St-Pierre, N.R.; Noftsger, S.M. Effects of grain variability and processing on starch utilization by lactating dairy cattle. *J. Anim. Sci.* **2001**, *79*, E218–E238. [CrossRef]
59. Lanzas, C.; Fox, D.G.; Pell, A.N. Digestion kinetics of dried cereal grains. *Anim. Feed Sci. Technol.* **2007**, *136*, 265–280. [CrossRef]
60. Opatpatanakit, Y.; Kellaway, R.C.; Lean, I.J.; Annison, G.; Kirby, A. Microbial fermentation of cereal grains in vitro. *Aust. J. Agric. Res.* **1994**, *45*, 1247–1263. [CrossRef]
61. Taxis, T.M.; Wolff, S.; Gregg, S.J.; Minton, N.O.; Zhang, C.; Dai, J.; Schnabel, R.D.; Taylor, J.F.; Kerley, M.S.; Pires, J.C.; et al. The players may change but the game remains: Network analyses of ruminal microbiomes suggest taxonomic differences mask functional similarity. *Nucleic Acids Res.* **2015**, *43*, 9600–9612. [CrossRef]
62. Söllinger, A.; Tveit, A.T.; Poulsen, M.; Noel, S.J.; Bengtsson, M.; Bernhardt, J.; Frydendahl Hellwing, A.L.; Lund, P.; Riedel, K.; Schleper, C.; et al. Holistic assessment of rumen microbiome dynamics through quantitative metatranscriptomics reveals multifunctional redundancy during key steps of anaerobic feed degradation. *mSystems* **2018**, *3*. [CrossRef]

7

Effect of Dietary Insoluble and Soluble Fibre on Growth Performance, Digestibility and Nitrogen, Energy and Mineral Retention Efficiency in Growing Rabbits

Carlos Farías-Kovac [1], Nuria Nicodemus [1], Rebeca Delgado [1], César Ocasio-Vega [1], Tamia Noboa [1], Ramadan Allam-Sayed Abdelrasoul [2], Rosa Carabaño [1] and Javier García [1,*]

[1] Departamento de Producción Agraria, Escuela Técnica Superior de Ingeniería Agronómica, Agroalimentaria y de Biosistemas, Universidad Politécnica de Madrid, C/Senda del Rey 18, 28040 Madrid, Spain; scarlos_03@hotmail.com (C.F.-K.); nuria.nicodemus@upm.es (N.N.); rebeca_dm85@yahoo.es (R.D.); cesar.ocasio1@gmail.com (C.O.-V.); taminobo31@hotmail.com (T.N.); rosa.carabano@upm.es (R.C.)

[2] Poultry Department, Faculty of Agriculture, Fayoum University, Fayoum 63514, Egypt; ras05@fayoum.edu.eg

* Correspondence: javier.garcia@upm.es;

Simple Summary: Rabbits, like other herbivores, require a minimal level of insoluble fibre in the diet to warrant an adequate digestive function. Epizootic rabbit enteropathy (ERE) is the main digestive trouble in growing rabbits that increases the use of antibiotics. The increase of soluble fibre (once met insoluble fibre requirements) limits the incidence of ERE and improves nitrogen and energy balance, while high levels of insoluble fibre seem to favour ERE. This study evaluated whether the increase of soluble and insoluble fibre above the current requirements of insoluble fibre had a positive impact on mortality, growth performance, diet digestibility, and energy, nitrogen, and mineral balance. Treatments had no effect on mortality, which was low (1%). The increase of insoluble fibre reduced the dietary digestible energy, while soluble fibre only increased it when combined with a low insoluble fibre level. The group fed with the lowest insoluble and soluble fibre levels showed the best energy and mineral balance, while the increase of insoluble or soluble fibre did not improve any growth trait. We conclude that in healthy rabbits, and once the minimal insoluble fibre requirements are met, no increase of insoluble or soluble fibre is recommended.

Abstract: Dietary soluble fibre limits the incidence of epizootic rabbit enteropathy (ERE) and improves the energy and nitrogen balance in low-insoluble fibre diets, while high-insoluble fibre diets seem to favour ERE. This study assessed whether the positive effects of soluble fibre are influenced by the level of insoluble fibre. Four diets (2 × 2 factorial arrangement) were used with two levels of insoluble fibre (314 vs. 393 g/kg DM) and soluble fibre (87 vs. 128 g/kg DM), resulting in four diets with increasing total dietary fibre levels. Growth performance and chemical composition (body and carcass) (28–62 days of age), faecal digestibility (54–57 days of age), and jejunal morphometry functionality (39 days of age) were determined. Mortality was low (<1%) and treatments did not influence it. Insoluble and soluble fibre tended to reduce the growth rate ($p \leq 0.109$), body protein, and fat accretion ($p = 0.049$ to 0.120), but only insoluble fibre impaired feed efficiency ($p < 0.001$). The efficiency of digestible energy used for growth was impaired with the increase of total dietary fibre ($p = 0.027$), while that of nitrogen remained majorly unaffected. In conclusion, in healthy rabbits, the increase of either insoluble or soluble fibre had no benefit.

Keywords: insoluble fibre; soluble fibre; feed efficiency; whole body and carcass chemical composition; energy nitrogen and mineral balance; fibre digestibility; mucosa morphology; energy nitrogen and mineral retention efficiency; rabbit

---

1. Introduction

The inclusion of a moderated level of soluble fibre (~12%; SF), usually derived from sugar beet pulp (SBP) inclusion, enhanced gut functionality [1,2] and modified intestinal microbiota [3–5], which was related to a decrease of the mortality rate in rabbits affected by epizootic rabbit enteropathy (ERE) [6–8]. These positive effects associated with SBP were accounted not only for its SF but also for its high content of easily fermentable insoluble fibre [9].

The inclusion of around 15–25% of SBP is required to meet the recommended SF level. However, the energy and nutrient retention efficiency in growing rabbits seem to be directly affected by the level of fermentable fibre. When SBP was included above 15% in the diet replacing barley, it reduced the energy and nitrogen retention efficiency [10]. Nevertheless, a 30% SBP level of inclusion, in substitution of barley and alfalfa, did not affect these traits in low-fibrous diets [11]. On the opposite, energy and nitrogen retention efficiency increased when a lower level of SBP was used (18%) in low-fibre diets (31% neutral detergent fibre, NDF, on DM basis and free of ash and protein) [8], probably due to the improvement of the rabbit health status and the reduction of urine nitrogen losses. In this context, the increase of the insoluble fibre level (NDF) impaired the nutritive value of SBP [12], which might be related with the role of NDF in modulating the microbial activity and rate of passage [13,14], and finally might affect the dietary energy and nitrogen efficiency. The use of high-insoluble fibrous diets for fattening rabbits is not strange at the field level, even when they might increase the incidence of ERE [15,16].

The aim of this work was to clarify whether the positive effect of the moderated inclusion of SF is influenced or not by the NDF level on growth performance, diet digestibility, gut mucosa morphometry and functionality, and energy, protein and mineral retention efficiency of growing rabbits.

2. Materials and Methods

2.1. Animals and Housing

A total of 264 crossbred hybrid rabbits (New Zealand White × Californian, V × R line from UPV, Valencia, Spain) weaned at 28 days of age were used. They came from 71 multiparous rabbit does from a farm periodically affected by ERE. Rabbits after weaning received the same diets as those offered to their mothers (described below). Rabbits were individually caged in flat deck cages (610 × 250 × 330 mm), except during the faecal digestibility trial in which they were kept in wire metabolism cages (405 × 510 × 320 mm) that allowed the separation of faeces and urine. Rabbits had *ad libitum* access to feed and water with no antibiotic supplemented throughout the experimental period and their health was checked daily. The temperature ranged between 18 and 23 °C. All the experimental procedures used were in compliance with the Spanish guidelines for care and use of animals in research [17], and authorized by the Dirección General de Agricultura y Ganadería from the Community of Madrid (PROEX 328/15).

2.2. Diets

Four diets in a 2 × 2 factorial arrangement were used with two levels of insoluble fibre quantified as NDF (314 vs. 393 g/kg DM, free of ash and protein) and two levels of SF quantified by the difference of total dietary fibre and NDF (87 vs. 128 g/kg DM) (Table 1). A control diet was formulated to meet most nutrient requirements for growing rabbits [18], with an NDF level (344 g NDF free of ash/kg DM) similar to that proposed for post-weaned rabbits [15] (342 g NDF free of ash/kg DM), and with a level of SF lower than the one recommended for growing rabbits under ERE risk [7] (LIF-LSF diet). The increase of NDF was obtained by replacing barley and wheat from the LIF-LSF diet for wheat straw, defatted grape seed meal, dehydrated alfalfa, and lard, the latter to minimize the reduction of digestible energy (DE) content (HIF-LSF diet). The increase of SF was obtained by replacing wheat bran, sunflower meal, gluten feed, wheat straw, and grape seed meal from the LIF-LSF diet for SBP (LIF-HSF diet). A fourth diet high in both NDF and SF was formulated (HIF-HSF diet).

Table 1. Ingredients and chemical composition of the experimental diets.

Item [1]	Low-Insoluble Fibre		High-Insoluble Fibre	
	Low SF	High SF	Low SF	High SF
	LIF-LSF	LIF-HSF	HIF-LSF	HIF-HSF
Ingredient, g/kg as fed				
Barley	150.0	141.0	70.0	70.0
Wheat	150.0	141.0	70.0	70.0
Dehydrated alfalfa	101.8	101.8	154.0	154.0
Cereal straw	84.0	63.0	169.3	146.6
Defatted grape seed meal	36.0	27.0	72.6	62.9
Wheat bran	176.5	60.0	133.8	10.0
Sugar beet pulp	0.0	170.0	0.0	170.0
Gluten feed	93.5	30.5	71.3	5.0
Sunflower meal	110.0	171.0	130.0	185.0
Soybean meal	60.0	60.0	60.0	60.0
Lard	0.0	0.0	35.0	35.0
Cane molasses	10.0	10.0	10.0	10.0
L-Lysine 50	3.0	2.4	3.2	2.6
Alimet	0.5	0.5	0.9	0.9
L-Threonine	1.0	0.6	1.2	0.8
L-Tryptophan	0.2	0.2	0.2	0.2
Sodium chloride	2.5	2.5	2.5	2.5
Calcium carbonate	16.0	8.0	10.0	4.0
Calcium phosphate	3.5	9.0	4.5	9.0
Mineral/vitamin premix [1]	1.5	1.5	1.5	1.5
Chemical composition, g/kg DM				
DM	903	905	908	906
Ash	83.9	89.1	90.9	96.3
Total dietary fibre [2]	398	444	484	519
Neutral detergent fibre (NDF [2])	311	317	396	390
Acid detergent fibre [3]	162	179	230	247
Acid detergent lignin [3]	36.6	38.3	66.4	65.2
Soluble fibre [4]	86.7	127.5	88.1	129.0
Crude protein (CP)	197	198	187	183
CP-Total dietary fibre	61.1	67.9	69.2	72.3
CP-NDF	33.8	37.4	33.8	38.0
Starch	212	181	117	86.3
Ether extract	31.6	35.9	79.3	65.7
Gross energy, MJ/kg DM	18.0	17.7	18.9	18.3

[1] Mineral and vitamin composition (per kg of complete diet): 20.0 mg of Mn; 60 mg of Zn; 12 mg of Cu; 1.01 mg of I; 0.30 mg of Co; 42 mg of Fe; 0.12 mg of Se; 9750 UI of vitamin A; 1200 UI of vitamin D_3, 42 UI of vitamin E, 1.5 mg of vitamin K_3; 1.5 mg of vitamin B_1; 3.75 mg of vitamin B_2; 1.5 mg of vitamin B_6; 0.018 mg of vitamin B_{12}; 12 g of Pantothenic acid; 40 g of Nicotinic acid; 0.75 mg of Folic acid; 0.075 mg of Biotin. [2] Corrected for ash and protein. [3] Corrected for ash. [4] Quantified as total dietary fibre NDF (both corrected for ash and protein).

2.3. Growth Performance and Body Chemical Composition Trial

In total, 224 rabbits (56/diet) weighing 510 ± 74 g were weaned at 28 days of age, and growth rate, feed intake, feed efficiency, and mortality was recorded until 62 days of age. The whole body and carcass chemical composition and energy content was estimated in vivo using the bioelectrical impedance analysis (BIA) technique in 39 rabbits/diet (randomly selected from this group), weighing 511 ± 77 g.

Measurements of resistance and reactance were measured in rabbits with a body composition analyser (Model Quantum II, RJL Systems, Detroit, MI, USA) at 28 and 62 days of age [19,20]. Multiple regression equations were used to estimate water, protein, ash, fat, and energy proportions both in the whole body and in the carcass according to these authors. Twenty-four rabbits were discarded due to an excess of feed waste (5, 2, 9, and 8 rabbits from the LIF-LSF, LIF-HSF, HIF-LSF, and HIF-HSF groups, respectively) and two rabbits died from the LIF-HSF group.

2.4. Faecal Digestibility Trial

A group of 40 rabbits (10/diet) weighing 1905 ± 108 g (belonging to the growth performance group) were caged individually in metabolism cages to determine the apparent faecal digestibility of gross energy, protein, and fibre fractions according to Perez et al. [21]. Feed intake and total faecal output were recorded for each rabbit during four consecutive days (from 54 to 57 days of age). Faeces were collected daily and were stored at −20 °C, dried at 80 °C for 48 h, and ground with a 1-mm screen for analysis.

2.5. Calculations of Energy and Nitrogen Efficiency

Estimated values for the total whole-body nitrogen and energy content were used to obtain the nitrogen and energy retention in the whole body (NR whole body and ER whole body, respectively) at 28 and 62 days of age. Values were expressed per kg $BW^{0.75}$ and day (where the BW was calculated as the average of the final and the initial body weight). Estimated values for the carcass nitrogen, and energy content were used to calculate the nitrogen and energy retention in the carcass (NR carcass and ER carcass, respectively) at 28 and 62 days of age. Values were also expressed per kg $BW^{0.75}$ and day. Moreover, nitrogen and gross energy intake (Ni and GEi, respectively) and digestible N and DE intake (DNi and DEi, respectively) were recorded to calculate the overall N and GE whole-body retention efficiency as: NR whole body/Ni, NR whole body/DNi, ER whole body/GEi and ER whole body/DEi. Besides, the overall N and GE carcass retention efficiency was calculated as: NR carcass/DNi and ER carcass/DEi.

Total N and GE excretion as skin and viscera, faeces, or heat production and urine were calculated as follows:

$$\text{N lost as skin and viscera (g/kg BW}^{0.75}\text{ and day)} = \text{(g NR in the whole body} - \text{g NR in the carcass)/kg BW}^{0.75}\text{ and day.} \quad (1)$$

$$\text{N excreted as faeces (g/kg BW}^{0.75}\text{ and day)} = (\text{Ni} - \text{DNi})\text{/kg BW}^{0.75}\text{ and day.} \quad (2)$$

$$\text{N excreted as urine (g/kg BW}^{0.75}\text{ and day)} = (\text{DNi} - \text{NR in the whole body})\text{/kg BW}^{0.75}\text{ and day.} \quad (3)$$

$$\text{Energy lost as skin and viscera (MJ/kg BW}^{0.75}\text{ and day)} = \text{(MJ ER in the whole body} - \text{MJ ER in the carcass)/kg BW}^{0.75}\text{ and day.} \quad (4)$$

$$\text{Energy excreted as faeces (MJ/kg BW}^{0.75}\text{ and day)} = (\text{GEi} - \text{DEi})\text{/kg BW}^{0.75}\text{ and day.} \quad (5)$$

$$\text{Energy excreted as urine and heat production (MJ/kg BW}^{0.75}\text{ and day)} = (\text{DEi} - \text{RE in the whole body})\text{/kg BW}^{0.75}\text{ and day.} \quad (6)$$

Energy retained as protein and fat was calculated taking into account that energy deposited as protein and fat equals 23.15 and 35.65 kJ/g [22]. Energy retained as fat was also calculated as the difference between the total retained energy and the retained energy as protein. Digestible nitrogen and DE intake used for production was obtained by the difference between the total DNi and DEi and the nitrogen and energy requirements for maintenance, respectively (0.464 g DN/kg $BW^{0.75}$ day, and 430 kJ DE/kg $BW^{0.75}$ day) [22]. Mineral balance was estimated in the same way as the nitrogenous and energetic ones.

2.6. *Gut Histology and Enzymatic Activity and Immune Function*

Another different group of 40 rabbits (10/treatment) of 28 days of age and 472 ± 84 g BW were caged in groups of 2 rabbits (5 cages/treatment) and slaughtered by head concussion at 39 days of age. A sample from the middle part of the jejunum (3 cm) were collected in 10% buffered neutral formaldehyde solution (pH 7.2 to 7.4) for histological analysis and processed as described [23]. Another 6 cm were excised from the middle part of the jejunum, flushed with saline solution, frozen in dry ice, and immediately stored at −80 °C to determine sucrose-specific activity as described [1]. Villous height, crypt depth, and goblet cell from each villous were measured individually. Seven samples were discarded due to poor staining or conservation (3, 1, and 3 from the LIF-HSF, HIF-LSF, and HIF-HSF groups, respectively).

2.7. *Chemical Analysis*

Procedures of the AOAC [24] were used to determine DM (method 934.01), ash (method 942.05), crude protein (method 968.06), ether extract (920.39), starch (amyloglucosidase-α-amylase method; method 996.11), and total dietary fibre (985.29; TDF). Dietary NDF, acid detergent fibre, and acid detergent lignin were determined sequentially using the filter bag system (Ankom Technology, New York, NY, USA) [24,25]. Dietary NDF was determined using a thermo-stable amylase without any sodium sulphite added and corrected for ash and protein. The dietary SF was calculated, as TDF–NDF (both corrected for ash and protein), and no correction for mucin was done in the faecal samples. Gross energy (GE) was measured by an adiabatic bomb calorimeter (model 356, Parr Instrument Company, Moline, IL, USA).

2.8. *Statistical Analysis*

Data of faecal digestibility, growth performance, sucrose activity, and nitrogen and energy balances were analysed as a completely randomized design with the level of NDF, SF, and their interaction as the main sources of variation by using the mixed procedure of SAS (SAS Inst., Cary, NC, USA). Weaning weight was used as a linear covariate for growth traits, whole body and carcass chemical composition, and nitrogen, energy and mineral balances. Mortality was analysed using a logistic model (GENMOD procedure of SAS using a binomial distribution) considering the same variables, and the results were transformed from the logit scale. All data were presented as least-squares means. When interactions were significant, comparisons among all the treatment means were made using a *t*-test.

3. Results

In the whole experimental period, rabbits fed with HIF diets showed a higher feed intake (by 11%; $p < 0.001$, Table 2) and tended to decrease the growth rate ($p = 0.109$), leading to a lower feed efficiency (by 12%; $p < 0.001$) compared with the LIF group. In contrast, rabbits fed HSF diets reduced their feed intake (by 2%; $p = 0.048$), with no effect on the feed efficiency. No interaction was observed between NDF and SF in growth traits. Treatments did not affect the mortality rate, which was lower than 1%.

Table 2. Effect of dietary level of insoluble and soluble fibre on rabbits' growth performance from 28 to 62 days of age.

Item	Diets [1]				SEM		p-Value		
	Low-Insoluble Fibre		High-Insoluble Fibre						
	Low-Soluble Fibre LIF-LSF	High-Soluble Fibre LIF-HSF	Low-Soluble Fibre HIF-LSF	High-Soluble Fibre HIF-HSF	IF and SF	IF × SF	IF	SF	IF × SF
N	51	52	47	48					
28–42 days of age									
Body weight 28 day, g	496	459	533	540	6.58	9.30	<0.001	0.103	0.018
Growth rate, g/day	52.5	51.1	52.6	52.0	0.66	0.93	0.660	0.266	0.653
Feed intake, g/day	107	107	118	118	1.89	2.68	<0.001	0.907	0.974
Feed efficiency, g/g	0.519	0.489	0.447	0.445	0.010	0.015	<0.001	0.277	0.340
42–62 days of age									
Body weight 42 day,g	1241	1221	1242	1233	9.172	13.01	0.656	0.263	0.649
Growth rate, g/day	50.7	50.3	49.2	47.9	0.50	0.71	0.012	0.234	0.517
Feed intake, g/day	152	148	170	164	1.51	2.15	<0.001	0.009	0.515
Feed efficiency, g/g	0.336	0.345	0.291	0.294	0.004	0.005	<0.001	0.276	0.581
28–62 days of age									
Body weight 62 g/day	2254	2227	2225	2191	13.39	19.01	0.109	0.098	0.867
Growth rate, g/day	51.4	50.6	50.6	49.6	0.39	0.56	0.109	0.099	0.870
Feed intake, g/day	133	131	149	145	1.22	1.73	<0.001	0.048	0.650
Feed efficiency, g/g	0.388	0.391	0.341	0.344	0.003	0.005	<0.001	0.541	0.968

[1] LIF-LSF = Low-insoluble fibre, Low-soluble fibre; LIF-HSF = Low-insoluble fibre, High-soluble fibre HIF-LSF = High-insoluble fibre, Low-soluble fibre; HIF-HSF = High-insoluble fibre, High-soluble fibre. Weight at weaning (28 days) as a covariate.

The digestibility of dry and organic matter decreased with the NDF level (by 25% and 14%, respectively; $p < 0.001$. Table 3) but increased with the level of SF (by 3.5% and 4%, respectively; $p < 0.001$). In contrast, the ash digestibility decreased with both the level of NDF and SF (by 10% and 8%, respectively; $p \leq 0.014$). An interaction between the level of NDF and SF was observed for the digestibility of energy, protein, total dietary fibre, and NDF ($p \leq 0.037$). Rabbits fed HIF diets had a lower GE and protein digestibility than those fed LIF diets (by 11% and 4%; $p < 0.001$), while the increase of SF improved GE and protein digestibility with LIF but not with HIF diets. The increase of NDF at the low SF level (LIF-LSF vs. HIF-LSF) did not modify the digestibility of any fibre fraction, but the increase of SF improved the digestibility of NDF and SF ($p < 0.001$), being higher in LIF than in HIF groups. These values resulted in very similar ratios of dietary digestible protein/DE among groups, although those of the HSF groups were lower ($p < 0.001$), and that of the HIF-HSF group tended to be the lowest ($p < 0.070$).

At 28 days of age, rabbits from HIF groups showed a higher whole-body fat, and a lower whole-body protein and mineral proportions than those from LIF groups (22.4% vs. 23.4% for fat, 59.1% vs. 58.5% for protein, and 11.7% vs. 11.5% for minerals, all of them on a DM basis; $p < 0.001$. Table S1), with no differences in the carcass protein concentration (Table S2). However, an interaction NDF × SF was found for the fat and mineral proportions, being higher for the LIF-HSF and HIF-LSF groups ($p \leq 0.062$. Table S2). At 62 days of age, there were no differences in the whole-body fat proportion, but the whole-body protein concentration increased in the HIF and HSF groups (in both by 1%; $p \leq 0.032$). Additionally, rabbits fed HSF diets increased in the carcass the protein and decreased the fat proportions (by 0.8 and 3%, respectively; $p \leq 0.028$), and a similar trend was observed for the HIF groups ($p \leq 0.067$), while the HIF diets tended to increase the carcass mineral proportion ($p = 0.055$). The daily whole-body protein and fat accretion tended to reduce with the increase of NDF and SF ($p = 0.049$ to 0.120, Table S3), with no effect on mineral accretion. The composition of the body weight gain had lower fat in the HIF than in the LIF groups ($p = 0.050$), and a lower protein content ($p = 0.034$). The daily carcass fat accretion and the fat proportion in the carcass weight gain decreased in the HSF compared to the LSF groups (by 5% and 2%; $p \leq 0.039$, Table S4), while the daily carcass protein and mineral accretion and the protein and mineral proportions in the carcass decreased in the HIF compared to the LIF groups (by 4% and 2% for protein and by 3 and 1% form minerals; $p \leq 0.060$).

The digestible nitrogen intake was higher for the HIF-LSF group compared with the other three groups (2.41 vs. 2.26 g DNi/kg $BW^{0.75}$ d; $p = 0.003$, Table 4). The nitrogen retained in the whole body and in the carcass decreased by 2–3% when NDF increased ($p \leq 0.040$), with no effect of SF. These results led to the lowest DN retention efficiency in the whole body and in the carcass in the HIF-LSF group compared to the other three groups (0.427 vs. 0.460, and 0.284 vs. 0.310; $p \leq 0.011$), as well as for the retention efficiency of the estimated DN used for growth (0.506 vs. 0.553, and 0.336 vs. 0.372; $p \leq 0.011$). Faecal nitrogen losses were lower in the LIF groups, and HSF diets reduced them when combined with the LIF diet but increased with the HIF diet ($p < 0.001$). They were inversely proportional to the digestible crude protein content of the diets. Urinary nitrogen losses were the highest for the HIF-LSF group, showing similar values to the other three groups (1.38 vs. 1.23 g urinary/kg $BW^{0.75}$ d; $p = 0.002$).

Table 3. Effect of dietary level of insoluble and soluble fibre on faecal apparent digestibility of dietary components in rabbits from 54 to 57 days of age.

Item	Diets [1]				SEM		p-Value		
	Low-Insoluble Fibre		High-Insoluble Fibre						
	Low-Soluble Fibre LIF-LSF	High-Soluble Fibre LIF-HSF	Low-Soluble Fibre HIF-LSF	High-Soluble Fibre HIF-HSF	IF and SF	IF × SF	IF	SF	IF × SF
N	9	8	10	9					
Initial body weight, g	1961	1856	1939	1904	23.26	32.88	0.691	0.042	0.300
Feed intake, g DM/day	135	128	136	142	3.28	4.63	0.113	0.803	0.180
Faecal apparent digestibility, %									
Dry matter	63.9	66.8	55.7	57.0	0.40	0.57	<0.001	<0.001	0.111
Organic matter	65.1	68.5	56.6	58.1	0.44	0.62	<0.001	<0.001	0.147
Ash	50.5	48.3	47.5	41.7	1.08	1.53	0.004	0.014	0.237
Gross energy	63.1 [b]	66.7 [a]	56.2 [c]	57.2 [c]	0.41	0.58	<0.001	<0.001	0.037
Crude protein, CP	75.7 [ab]	76.8 [a]	74.5 [b]	72.1 [c]	0.49	0.70	<0.001	0.377	0.018
Total dietary fibre	35.4 [c]	50.1 [a]	33.5 [c]	40.9 [b]	0.59	0.83	<0.001	<0.001	<0.001
Neutral detergent fibre	28.1 [c]	40.2 [a]	28.0 [c]	31.6 [b]	0.69	0.97	<0.001	<0.001	<0.001
Soluble fibre	61.9	74.7	58.0	69.0	1.51	2.14	0.034	<0.001	0.637
Digestible energy, MJ/kg DM	11.38 [b]	11.78 [a]	10.64 [c]	10.46 [c]	0.075	0.106	<0.001	0.287	0.010
Digestible CP, g/kg DM	14.91 [a]	15.23 [a]	13.89 [b]	13.17 [c]	0.095	0.134	<0.001	0.147	<0.001
Ratio digestible [CP/energy], g/MJ	13.11	12.93	13.105	12.59	0.052	0.074	0.012	<0.001	0.069

[1] LIF-LSF = Low-insoluble fibre, Low-soluble fibre; LIF-HSF = Low-insoluble fibre, High-soluble fibre HIF-LSF = High-insoluble fibre, Low-soluble fibre; HIF-HSF = High-insoluble fibre, High-soluble fibre. [a–c] Fattening mean values in the same row with a different superscript differ $p < 0.050$.

Table 4. Effect of dietary level of insoluble and soluble fibre on nitrogen (N) balance from 28 to 62 days of age.

Item	Diets [1]				SEM		p-Value		
	Low-Insoluble Fibre		High-Insoluble Fibre						
	Low-Soluble Fibre LIF-LSF	High-Soluble Fibre LIF-HSF	Low-Soluble Fibre HIF-LSF	High-Soluble Fibre HIF-HSF	IF and SF	IF × SF	IF	SF	IF × SF
N	34	36	32	33					
kg $BW^{0.75}$	1.27	1.25	1.25	1.24	0.01	0.01	0.101	0.086	0.389
[2] Nitrogen intake, g/kg $BW^{0.75}$ d									
Ni,	2.98	2.97	3.23	3.10	0.03	0.04	<0.001	0.111	0.157
DNi	2.25 ^{b}	2.28 ^{b}	2.41 ^{a}	2.24 ^{b}	0.02	0.03	0.138	0.030	0.003
[3] Nitrogen retained, g/kg $BW^{0.75}$ d									
NR whole body	1.05	1.03	1.02	1.02	0.05	0.01	0.040	0.166	0.510
NR carcass	0.711	0.696	0.682	0.677	0.007	0.010	0.017	0.275	0.570
[4] Nitrogen efficiency									
NR whole body/Ni	0.353	0.351	0.318	0.330	0.004	0.006	<0.001	0.383	0.186
NR whole body/DNi	0.467 ^{a}	0.456 ^{a}	0.427 ^{b}	0.458 ^{a}	0.005	0.007	0.017	0.161	0.005
NR whole body/DNp	0.561 ^{a}	0.549 ^{a}	0.506 ^{b}	0.550 ^{a}	0.007	0.010	0.015	0.111	0.049
NR carcass/DNi	0.316 ^{a}	0.308 ^{a}	0.284 ^{b}	0.305 ^{a}	0.004	0.006	0.004	0.283	0.011
NR carcass/DNp	0.381 ^{a}	0.370 ^{a}	0.336 ^{b}	0.366 ^{a}	0.005	0.008	0.004	0.205	0.008
[5] Nitrogen losses, g/kg $BW^{0.75}$ d									
Skin and viscera	0.337	0.335	0.343	0.342	0.006	0.008	0.469	0.803	0.988
Faeces	0.725 ^{c}	0.691 ^{d}	0.824 ^{b}	0.865 ^{a}	0.008	0.011	<0.001	0.74	<0.001
Urine	1.21 ^{b}	1.25 ^{b}	1.38 ^{a}	1.22 ^{b}	0.02	0.03	0.057	0.082	0.002

[1] LIF-LSF = Low-insoluble fibre, Low-soluble fibre; LIF-HSF = Low-insoluble fibre, High-soluble fibre HIF-LSF = High-insoluble fibre, Low-soluble fibre; HIF-HSF = High-insoluble fibre, High-soluble fibre. [2] Ni (g Ni/kg $BW^{0.75}$ day): Nitrogen intake. DNi (g DNi/kg $BW^{0.75}$ day): Digestible N intake. [3] NR whole body (g/kg $BW^{0.75}$ day): retained N in the whole body. NR carcass (g/kg $BW^{0.75}$ day): retained N in carcass. [4] DNp (g DNp/kg $BW^{0.75}$ day): Intake of digestible N used for growth, obtained by the difference between DNi and the DN used for maintenance (0.464 g DNm/kg $BW^{0.75}$ day). Skin and viscera (g N/kg $BW^{0.75}$ day): (g N retained in the whole body − g retained in carcass)/kg $BW^{0.75}$ day. Faeces (g/kg $BW^{0.75}$ day): (Total N intake − DNi)/kg $BW^{0.75}$ day. [5] N lost as skin and viscera (g/kg $BW^{0.75}$ day) = (g NR in the whole body − g NR carcass)/kg $BW^{0.75}$ day. N excreted as faeces (g/kg $BW^{0.75}$ day) = (Ni − DNi)/kg $BW^{0.75}$ day. Urine (g/kg $BW^{0.75}$ day): (DNi − NR whole body)/kg $BW^{0.75}$ day. Weight at weaning (28 days) as a covariate. [a–d] Fattening mean values in the same row with a different superscript differ $p < 0.050$.

The DE intake was also the highest for the HIF-LSF group compared with the other three groups (1.15 vs. 1.10 MJ DE/kg $BW^{0.75}$ d; $p = 0.027$, Table 5). The energy retained in the whole body and in the carcass tended to decrease when NDF and SF increased ($p \leq 0.084$). Consequently, the LIF-LSF group had the highest DE retention efficiency in the whole body and in the carcass compared with the other three groups (0.330 vs. 0.302, and 0.204 vs. 0.185, $p \leq 0.029$). These differences were even higher for the retention efficiency of estimated DE used for growth in the whole body and in the carcass, decreasing with the increase of total dietary fibre, although showing the lowest values in rabbits fed the HIF-LSF diet ($p \leq 0.027$). Faecal energy losses increased with NDF and decreased with SF ($p < 0.001$), and were inversely proportional to the DE of the diets. The energetic losses as urine and heat production were minimal for the LIF-LSF group and maximal for the HIF-LSF group, showing intermediate values for the other two groups ($p = 0.014$).

The digestible mineral intake was the highest for the HIF-LSF group compared with the other three groups (4.59 vs. 4.01 g MIi/kg $BW^{0.75}$ day; $p < 0.001$, Table 6). The minerals retained in the whole body were not affected by treatments, but those retained in the carcass decreased by 2% when NDF increased (1.021 vs. 0.997 g MI/kg $BW^{0.75}$ day; $p < 0.001$). The LIF-LSF group showed the highest efficiency of retention of digestible minerals in the carcass, which decreased with both NDF and SF, although the HIF-LSF group showed the lowest value ($p < 0.001$). Faecal mineral losses increased with both the level of NDF and SF (by 11% and 8%, respectively; $p < 0.001$). Mineral urine losses evolved inversely to the carcass retention efficiency ($p < 0.001$).

The increase of NDF had no influence on the villous height in the LSF groups but tended to decrease it in LIF-HSF and increase it in the HIF-HSF groups ($p = 0.071$, Table 7). Treatments did not affect the crypt depth, the ratio villous height/crypt depth, and the number of goblet cells per villi in the jejunal mucosa. The increase of SF impaired the sucrose specific activity (by 21%; $p = 0.034$), while the increase of NDF tended to reduce it ($p = 0.059$).

Table 5. Effect of dietary level of insoluble and soluble fibre on energy (E) balance from 28 to 62 days of age.

Item	Diets [1]				SEM		p-Value		
	Low-Insoluble Fibre		High-Insoluble Fibre						
	Low-Soluble Fibre LIF-LSF	High-Soluble Fibre LIF-HSF	Low-Soluble Fibre HIF-LSF	High-Soluble Fibre HIF-HSF	IF and SF	IF × SF	IF	SF	IF × SF
N	34	36	32	33					
$BW^{0.75}$ d	1.27	1.25	1.25	1.24	0.01	0.01	0.101	0.086	0.389
[2] Energy intake, MJ/kg $BW^{0.75}$ d									
GEi	1.70	1.65	2.05	1.94	0.02	0.03	<0.001	0.002	0.24
DEi	1.08 ^{b}	1.10 ^{b}	1.15 ^{a}	1.11 ^{ab}	0.01	0.02	0.016	0.619	0.027
[3] Energy retained, kJ/kg $BW^{0.75}$ d									
ER whole body	354	339	338	333	4.04	5.73	0.068	0.067	0.405
ER carcass	219	209	206	203	2.83	4.01	0.030	0.084	0.348
[4] Energy efficiency [1]									
ER whole body/GEi	0.208	0.207	0.165	0.173	0.003	0.004	<0.001	0.415	0.247
ER whole body/DEi	0.330 ^{a}	0.311 ^{b}	0.294 ^{b}	0.302 ^{b}	0.004	0.006	<0.001	0.350	0.029
ER whole body/DEp	0.677 ^{a}	0.631 ^{ab}	0.557 ^{c}	0.598 ^{bc}	0.014	0.020	<0.001	0.896	0.027
ER carcass/DEi	0.204 ^{a}	0.191 ^{b}	0.179 ^{b}	0.184 ^{b}	0.003	0.004	<0.001	0.323	0.024
ER carcass/DEp	0.419 ^{a}	0.387 ^{a}	0.339 ^{b}	0.365 ^{ab}	0.009	0.012	<0.001	0.831	0.017
[5] Energy losses, MJ/kg $BW^{0.75}$ d									
Skin and viscera,	0.135	0.132	0.131	0.130	0.003	0.004	0.647	0.331	0.794
Faeces	0.628	0.551	0.896	0.828	0.007	0.010	<0.001	<0.001	0.624
Urine + heat production	0.721 ^{c}	0.765 ^{bc}	0.814 ^{a}	0.777 ^{ab}	0.012	0.017	0.004	0.837	0.014

[1] LIF-LSF = Low-insoluble fibre, Low-soluble fibre; LIF-HSF = Low-insoluble fibre, High-soluble fibre HIF-LSF = High-insoluble fibre, Low-soluble fibre; HIF-HSF = High-insoluble fibre High-soluble fibre. [2] GEi: Gross Energy intake (MJ/kg $BW^{0.75}$ day). DEi: Digestible Energy intake (MJ/kg $BW^{0.75}$ day). [3] ER whole body (kJ/kg $BW^{0.75}$ day): GE retained in the whole body ER carcass (kJ/kg $BW^{0.75}$ day): GE retained in carcass. [4] DEp (MJ/kg $BW^{0.75}$ day): Intake of DE used for growth, obtained by the difference between DEi and the DE used for maintenance (430 kJ DE/kg $BW^{0.75}$ day). [5] Skin and viscera (MJ/kg $BW^{0.75}$ day): (MJ GE retained in the whole body − MJ GE retained in carcass)/kg $BW^{0.75}$ day. Faeces (MJ/kg $BW^{0.75}$ and day) (GEi − DEi)/kg $BW^{0.75}$ day. Urine + heat production (MJ/kg $BW^{0.75}$ d): (DEi − GE retained in carcass − GE lost in skin and viscera)/kg $BW^{0.75}$ day. Weight at weaning (28 days) as a covariate. $^{a-c}$ Fattening mean values in the same column with a different superscript differ $p < 0.050$.

Table 6. Effect of dietary level of insoluble and soluble fibre on mineral (MI) balance from 28 to 62 days of age.

Item	Diets [1]				SEM		p-Value		
	Low-Insoluble Fibre		High-Insoluble Fibre						
	Low-Soluble Fibre LIF-LSF	High-Soluble Fibre LIF-HSF	Low-Soluble Fibre HIF-LSF	High-Soluble Fibre HIF-HSF	IF and SF	IF × SF	IF	SF	IF × SF
N	34	36	32	33					
[2] Mineral intake, g/kg $BW^{0.75}$ d									
MIi,	7.92	8.34	9.66	9.66	0.09	0.13	<0.001	0.097	0.095
DMIi	3.97 ^{b}	4.04 ^{b}	4.59 ^{a}	4.02 ^{b}	0.04	0.06	<0.001	<0.001	<0.001
[3] Mineral retained, g/kg $BW^{0.75}$ d									
MIR whole body	1.25	1.22	1.23	1.23	0.01	0.01	0.337	0.223	0.068
MIR carcass	1.027	1.016	0.998	0.996	0.005	0.007	<0.001	0.315	0.531
Minerals efficiency									
MIR whole body/MIi	0.159 ^{a}	0.148 ^{b}	0.127 ^{c}	0.128 ^{c}	0.002	0.002	<0.001	0.032	0.009
MIR whole body/DMIi	0.318 ^{a}	0.306 ^{a}	0.268 ^{b}	0.308 ^{a}	0.003	0.005	<0.001	0.005	<0.001
MIR in carcass/DMIi	0.261 ^{a}	0.254 ^{ab}	0.218 ^{c}	0.249 ^{b}	0.003	0.004	<0.001	0.036	<0.001
[4] Mineral losses, g/kg $BW^{0.75}$ d									
Skin and viscera,	0.227	0.207	0.227	0.236	0.007	0.011	0.192	0.594	0.160
Faeces	3.95	4.30	5.07	5.63	0.05	0.07	<0.001	<0.001	0.110
Urine	2.71 ^{b}	2.81 ^{b}	3.36 ^{a}	2.79 ^{b}	0.04	0.06	<0.001	<0.001	<0.001

[1] LIF-LSF = Low-insoluble fibre, Low-soluble fibre; LIF-HSF = Low-insoluble fibre, High-soluble fibre HIF-LSF = High-insoluble fibre, Low-soluble fibre; HIF-HSF = High-insoluble fibre, High-soluble fibre. [2] MIi (g MIi/kg $BW^{0.75}$ day): Mineral intake. DMIi (g DMIi/kg $BW^{0.75}$ day): Digestible mineral intake. [3] MIR whole body (g/kg $BW^{0.75}$ day): Mineral retained in the whole body. MIR carcass (g/kg $BW^{0.75}$ day): Mineral retained in carcass. [4] Skin and viscera (g Min/kg $BW^{0.75}$ day): (g MIR in the whole body − g MIR in carcass)/kg $BW^{0.75}$ day. Faeces (g/kg $BW^{0.75}$ day): (MIi − DMIi)/kg $BW^{0.75}$ d. Faeces (g/kg $BW^{0.75}$ day) = (MINi − DMIi)/kg $BW^{0.75}$ day. Urine (g/kg $BW^{0.75}$ day): (DMIi − MIR whole body)/kg $BW^{0.75}$ day. Weight at weaning (28 day) as a covariate. $^{a–c}$ Fattening mean values in the same row with a different superscript differ $p < 0.050$.

Table 7. Effect of dietary level of insoluble and soluble fibre on mucosa histology and enzyme activity of 39-day-old rabbits.

Item	Diets [1]				SEM		p-Value		
	Low-Insoluble Fibre		High Insoluble Fibre						
	Low-Soluble Fibre LIF-LSF	High-Soluble Fibre LIF-HSF	Low-Soluble Fibre HIF-LSF	High-Soluble Fibre HIF-HSF	IF and SF	IF × SF	IF	SF	IF × SF
N	10	7	9	7					
Villous height, µm	380	340	383	422	14.8	20.8	0.052	0.995	0.071
Crypt depth, µm	144	156	147	160	5.45	7.69	0.667	0.115	0.989
Ratio villous height/crypt depth	2.71	2.20	2.67	2.70	0.15	0.22	0.293	0.290	0.232
Goblet cells, n°/villous	11.1	9.86	10.3	11.8	1.06	1.49	0.707	0.923	0.350
Mucose protein, mg/g of tissue	61.5	46.2	57.6	52.2	4.81	6.80	0.878	0.136	0.470
Sucrose activity, µmol of glucose/mg of protein	273	212	218	174	17.0	24.0	0.059	0.034	0.703

[1] LIF-LSF = Low-insoluble fibre, Low-soluble fibre; LIF-HSF = Low-insoluble fibre, High-soluble fibre HIF-LSF = High-insoluble fibre, Low-soluble fibre; HIF-HSF = High-insoluble fibre, High-soluble fibre.

4. Discussion

Digestive disorders are one of the main causes of mortality in growing rabbits, with antibiotic treatment being the usual way to control them [26]. One of the nutritional strategies to limit ERE is the inclusion of moderate levels of SF [3,6,27,28], while the increase of NDF might increase its incidence [15,16]. However, in this study, a very low mortality rate was reported, avoiding any conclusion about treatments, and suggesting a stronger influence of other environmental factors than the diet. Even the higher dietary protein levels than those expected did not trigger an ERE outbreak, although a high protein level is recognized as an important risk factor [27,29–31]. Under these good sanitary conditions, there was no positive effect of the level of SF on the mucosa morphology (ratio villous height/crypth depth and number of goblet cells) or functionality (sucrose-specific activity), which contrast with the positive effect reported when there is an ERE outbreak [1,5]. The trend observed for the interaction in the villous height suggests a negative influence of SF in the LIF groups but positive in the HIF groups, an observation that will require further confirmation. The negative influence of NDF on the sucrose-specific activity, similarly to that of SF, might be partially associated to the reduction of starch intake, as they were positively correlated ($n = 4$; $r = 0.90$; $p = 0.099$), rather than to mucosa damage.

The fibre digestibility depends on microbial activity, fermentation time, and digestion rate, which are all affected by the chemical composition of dietary fibre (mainly by the degree of lignification of NDF and the SF content) and particle size [32]. The digestibility of NDF of the LIF-LSF diet was low and close to other low insoluble-soluble fibre diets with medium-high DE concentrations [3,8,27,28,33]. The increase of NDF when SF was low (HIF-LSF vs. LIF-LSF) did not modify NDF digestibility, although the degree of lignification and feed intake increased, indicating that it probably accelerated the rate of passage [34,35]. This lack of effect might be associated to the fast digestion rate of the degradable fraction of NDF, which can be degraded/solubilized up to 50% before the caecum [9]. Comparatively, it was surprising that soluble fibre digestibility decreased from the LIF-LSF to the HIF-LSF group, which might account for the potential reduction in the mean retention time and the changes in the ingredient proportions. However, a higher incidence of these factors on the NDF than on the SF fraction was expected. The increase of SF when NDF is low (LIF-HSF vs. LIF-LSF) improved both NDF and SF digestibility in agreement with previous studies [7,9,23]. It is explained by the faster digestion rate of SF, similar to that observed in vitro for SBP when compared with straw [36]. The increase of NDF when SF is high (HIF-HSF vs. LIF-HSF) improved also the NDF and SF digestibility as expected [37,38] but the NDF to a lesser extent. It might be linked to the faster rate of passage associated to the HIF groups, which might limit the fermentation of SBP fibre and accordingly its nutritive value. This effect was already observed [12], when a higher DE value for SBP using a high energy basal diet than a low one was reported.

The increase of NDF led to a higher feed intake, and impaired feed efficiency as expected. On the opposite, the increase of SF did not modify feed efficiency, although it tended to reduce feed intake and the growth rate. The replacement of starch mainly by SF and the absence of ERE might be behind this lack of effect on feed efficiency as observed previously [26], although other authors found a positive effect [8,37,39]. These contradictory results are not explained by differences in the acid detergent fibre content [7,40], and do not seem associated to the health status. However, a potential improvement of the nitrogenous retention efficiency, and a reduction of nitrogen losses might be behind the positive effect of SF on feed efficiency [8]. In contrast, the inclusion of SF in the substitution of NDF, and minor proportions of starch and fat, usually improves the feed efficiency [3,38,41]. The magnitude of the feed intake reduction with the inclusion of SF widely differs among studies, with no great influence on DEi in some of them, but in others, where the level of SBP inclusion exceeded 15–30%, compromised DEi (and even DNi), impairing the feed conversion ratio and dressing out percentage [6,8,10,42,43]. It was related to the accumulation of digesta in the caecum derived from a combination of factors like the level and type of NDF, and effective particle size [44].

In this study, rabbits tended to grow slower either with the increase of NDF or SF or both, even when they had a similar or even higher DE and DN intake. These results agree with the trend to reduce the whole-body protein and fat accretion with the increase of total dietary fibre, with the reduction being more pronounced for fat than for protein. It led to a progressive reduction of the nitrogen and energy retained in the whole body and in the carcass, even when diets were formulated to meet all nutrient requirements, which contrasts with the lack of an effect of SF on these traits observed in a previous study [8]. These results are explained by the trend to impair the retention efficiency of the DE used for growth (DEp) both in the whole body and in the carcass when total dietary fibre increased. The reduction of the energetic efficiency is parallel to the increase of energetic losses as urine and heat production from the LIF-LSF to the HIF-HSF group (excluding the HIF-LSF group), most probably related to the heat increment associated to the use of nutrients for growth and that associated with the increase of the fermentative activity, as the N losses in urine were similar among these groups. The increase of total dietary fibre (from the LIF-LSF to the HIF-HSF group) implied a higher intake of digestible total dietary fibre (by 65%) and a lower intake of starch (by 55%), accompanied in the HIF groups by an increase of fat intake (by 140%). The higher metabolic efficiency of glucose compared to volatile fatty acids, and the potential need in the HIF groups to obtain glucose from amino acids seemed to be behind the energetic efficiency impairment observed with the increase of total dietary fibre. In relation to fat, other authors found no influence of the increase of fat intake on the overall DE or DE efficiency for growth [45], suggesting a minor influence on the energetic efficiency results of this study. The impairment of the efficiency of the retention of DE (used for growth) was not related with the efficiency of retention of DN (used for growth), which was similar among groups (excluding the HIF-LSF group, which will deserve a separate comment). These results contrast with the lack of an effect of SF on the retained energy and nitrogen observed previously [8], which might be explained by the higher starch and lower fermentable fibre intake obtained in their study [8,23]. Besides, the same authors observed a positive effect of the increase of SF on nitrogen metabolism due to the reduction of urinary nitrogen losses, probably derived from a shift of urinary nitrogen excretion to faecal excretion. This effect was not observed in this study probably due to the excess of dietary digestible protein, despite the increase of the fermentable fibre intake with the level of total dietary fibre. Rabbits fed the HIF-LSF diet showed the worst efficiencies of DN and of DE used for growth, which were lower than those expected according to its NDF and SF levels. It is mainly explained by the high DN intake, derived from the combination of the highest feed intake and a high dietary DN/DE ratio, that led to the highest urinary nitrogen (and energetic) losses. The latter are associated with higher heat production derived from urea synthesis, which helps to explain the trend to increase energetic losses as urine and heat production in rabbits fed HIF-LSF compared with those of the HIF-HSF group ($p = 0.112$). In fact, when experiments with wide differences in dietary digestible protein content are combined, a close relationship between the DN intake and urinary nitrogen losses is found (Figure 1).

The influence of dietary fibre on mineral balance in rabbits has been scarcely studied. The mineral retention in the carcass was impaired with the increase of NDF, although the digestible mineral intake was similar (HIF-HSF) or even higher (HIF-LSF group). It led to a reduction on the efficiency of retention of digested minerals with the increase of total dietary fibre (except for the HIF-LSF group). This result agreed with the reduction of the serum mineral concentrations (per unit of mineral ingested) in pregnant sows when dietary fibre increased [46]. However, the increase of dietary fibre did not influence all minerals in the same way. In fact, the apparent absorption of calcium, phosphorous, and magnesium was not influenced by the fibre level, while it impaired that of sodium and potassium in pigs [47]. In rabbits, the type of fibre (alfalfa, olive pulp, grape pulp) also influenced the apparent absorption of different minerals, but no effect was reported on the plasma levels of most elements [48].

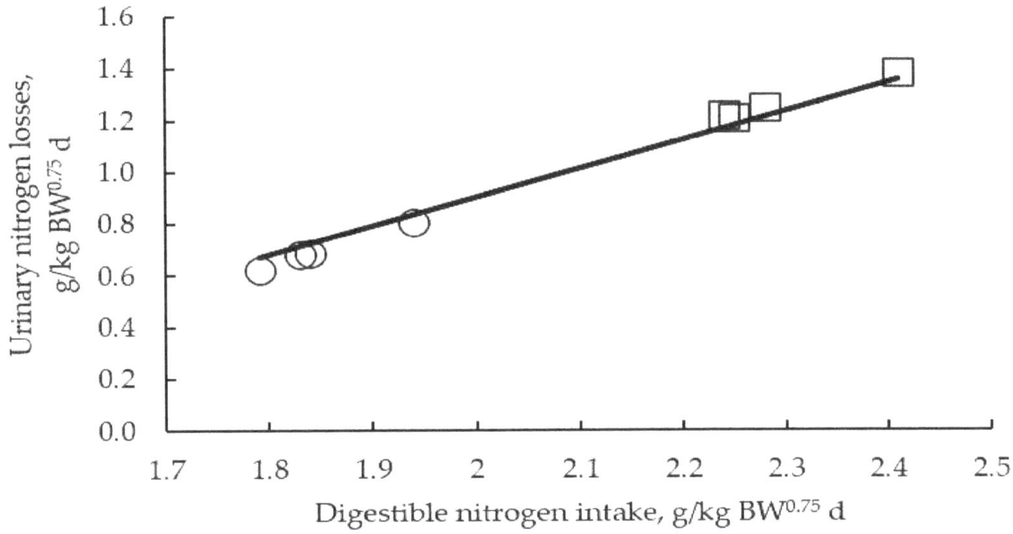

Figure 1. Relationship between digestible nitrogen intake and urinary nitrogen losses during fattening in rabbits (Urinary nitrogen losses = −1.31(±0.13) + 1.11(±0.061) digestible nitrogen intake; N = 8; $p < 0.001$. Original observations plotted with the mean regression line across studies: ○ Delgado et al., 2018. □ Current study).

5. Conclusions

The increase of either NDF or SF in diets for healthy rabbits had no positive effect in terms of growth traits, retention efficiency of digestible nitrogen energy and minerals, and mucosa morphology, once the NDF requirements for growing rabbits are met. Further research is required to identify what is the optimal level of NDF to combine with SF under an enteropathy outbreak.

Supplementary Materials:
Table S1: Effect of dietary level of insoluble and soluble fibre on the whole body chemical composition and energy content of weaning rabbits at 28 and 62 days of age. Table S2: Effect of dietary level of insoluble and soluble fibre on carcass chemical composition and energy content of weaning rabbits at 28 and 62 days of age. Table S3: Effect of dietary level of insoluble and soluble fibre on daily gain of whole body components and their proportions in the daily gain from 28 to 62 days of age. Table S4: Effect of dietary level of insoluble fibre and soluble fibre on daily gain of carcass components and their proportions in the carcass daily gain from 28 to 62 days.

Author Contributions: J.G., R.C. and N.N. contributed to concept and design of the experiment. C.F.-K., N.N., R.D., C.O.-V., T.N., R.A.-S.A. and J.G. performed the animal trial including the faecal and histological sampling. C.F.-K. conducted the chemical and histological determinations (helped by R.D., C.O.-V., T.N. and R.A.-S.A.), statistical analysis, and prepared the tables. All the authors contributed to the interpretation of the results. C.F.-K. and J.G. wrote the manuscript with the collaboration of N.N. and R.C. All authors have read and agreed to the published version of the manuscript.

References

1. Gómez-Conde, M.S.; García, J.; Chamorro, S.; Eiras, P.; Rebollar, P.G.; Pérez de Rozas, A.; Badiola, I.; de Blas, C.; Carabaño, R. Neutral detergent-soluble fiber improves gut barrier function in twenty-five-day-old weaned rabbits1. *J. Anim. Sci.* **2007**, *85*, 3313–3321. [CrossRef] [PubMed]
2. El abed, N.; Delgado, R.; Abad-Guamán, R.; Romero, C.; Villamide, M.J.; Menoyo, D.; Carabaño, R.; García, J. Soluble and insoluble fibre from sugar beet pulp enhance intestinal mucosa morphology in young rabbits. In *Book of Abstracts of the 62nd Annual Meeting of the European Association for Animal Production*; EAAP Book of Abstracts; Wageningen Academic Publishers: Stavanger, Norway, 2011; Volume 17, p. 159, ISBN 978-90-8686-731-8.

3. Gómez-Conde, M.S.; de Rozas, A.P.; Badiola, I.; Pérez-Alba, L.; de Blas, C.; Carabaño, R.; García, J. Effect of neutral detergent soluble fibre on digestion, intestinal microbiota and performance in twenty five day old weaned rabbits. *Livest. Sci.* **2009**, *125*, 192–198. [CrossRef]
4. Rodríguez-Romero, N.; Abecia, L.; Fondevila, M. Bacterial profile from caecal contents and soft faeces in growing rabbits given diets differing in soluble and insoluble fibre levels. *Anaerobe* **2012**, *18*, 602–607. [CrossRef]
5. El abed, N.; Badiola, I.; Pérez de Rozas, A.; González, J.; Menoyo, D.; Carabaño, R.; García, J. Effect of soluble and insoluble fractions of sugar beet pulp on ileal and caecal microbiota of rabbits after weaning. *World Rabbit Sci.* **2013**, *21*, 207–208. [CrossRef]
6. Martínez-Vallespín, B.; Martínez-Paredes, E.; Ródenas, L.; Cervera, C.; Pascual, J.J.; Blas, E. Combined feeding of rabbit female and young: Partial replacement of starch with acid detergent fibre or/and neutral detergent soluble fibre at two protein levels. *Livest. Sci.* **2011**, *141*, 155–165. [CrossRef]
7. Trocino, A.; García Alonso, J.; Carabaño, R.; Xiccato, G. A meta-analysis on the role of soluble fibre in diets for growing rabbits. *World Rabbit Sci.* **2013**, *21*. [CrossRef]
8. Delgado, R.; Nicodemus, N.; Abad-Guamán, R.; Sastre, J.; Menoyo, D.; Carabaño, R.; García, J. Effect of dietary soluble fibre and n-6/n-3 fatty acid ratio on growth performance and nitrogen and energy retention efficiency in growing rabbits. *Anim. Feed Sci. Technol.* **2018**, *239*, 44–54. [CrossRef]
9. Abad-Guamán, R. Identification of the Method to Quantify Soluble Fibre and the Effect of the Source of Fibre on the Ileal and Faecal Digestibility of Soluble and Insoluble Fibre in Rabbits. Ph.D. Thesis, E.T.S.I. Agrónomos (Universidad Politécnica de Madrid), Madrid, Spain, 2015.
10. Garcia, G.; Galvez, J.F.; de Blas, J.C. Effect of substitution of sugarbeet pulp for barley in diets for finishing rabbits on growth performance and on energy and nitrogen efficiency. *J. Anim. Sci.* **1993**, *71*, 1823–1830. [CrossRef]
11. Carabaño, R.; Motta-Ferreira, W.; de Blas, J.C.; Fraga, M.J. Substitution of sugarbeet pulp for alfalfa hay in diets for growing rabbits. *Anim. Feed Sci. Technol.* **1997**, *65*, 249–256. [CrossRef]
12. de Blas, C.; Villamide, M.J. Nutritive value of beet and citrus pulps for rabbits. *Anim. Feed Sci. Technol.* **1990**, *31*, 239–246. [CrossRef]
13. García, J.; Carabaño, R.; de Blas, J.C. Effect of fiber source on cell wall digestibility and rate of passage in rabbits. *J. Anim. Sci.* **1999**, *77*, 898. [CrossRef] [PubMed]
14. Gidenne, T. Effets d'une réduction de la teneur en fibres alimentaires sur le transit digestif du lapin. Comparaison et validation de modèles d'ajustement des cinétiques d'excrétion fécale des marqueurs. *Reprod. Nutr. Dev.* **1994**, *34*, 295–306. [CrossRef] [PubMed]
15. Gutiérrez, I.; Espinosa, A.; García, J.; Carabaño, R.; De Blas, J.C. Effect of levels of starch, fiber, and lactose on digestion and growth performance of early-weaned rabbits. *J. Anim. Sci.* **2002**, *80*, 1029–1037. [CrossRef] [PubMed]
16. Tazzoli, M.; Trocino, A.; Birolo, M.; Radaelli, G.; Xiccato, G. Optimizing feed efficiency and nitrogen excretion in growing rabbits by increasing dietary energy with high-starch, high-soluble fibre, low-insoluble fibre supply at low protein levels. *Livest. Sci.* **2015**, *172*, 59–68. [CrossRef]
17. BOE.es—Documento BOE-A-2013-1337. Available online: https://www.boe.es/diario_boe/txt.php?id=BOE-A-2013-1337 (accessed on 21 May 2020).
18. de Blas, C.; Mateos, G.G. Feed formulation. In *Nutrition of the Rabbit*, 3rd ed.; CABI Publishing CAB International: Wallingford, UK, 2020; pp. 243–253, ISBN 978-1-78924-127-3.
19. Saiz, A.; García-Ruiz, A.I.; Fernández, A.; Nicodemus, N. Application of Bioelectrical impedance analysis (BIA) to study the carcass chemical composition of rabbits from 35 to 63 d of age. *World Rabbit Sci.* **2013**, *21*, 208–209. [CrossRef]
20. Saiz, A.; García-Ruiz, A.I.; Fuentes-Pila, J.; Nicodemus, N. Application of bioelectrical impedance analysis to assess rabbit's body composition from 25 to 77 days of age1. *J. Anim. Sci.* **2017**, *95*, 2782–2793. [CrossRef]
21. Pérez, J.M.; Cervera, C.; Falcao E Cunha, L.; Maertens, L.; Villamide, M.J.; Xiccato, G. European ring-test on in vivo determination of digestibility in rabbits: Reproducibility of a reference method in comparison with domestic laboratory procedures. *World Rabbit Sci.* **1995**, *3*, 171–178. [CrossRef]
22. Xiccato, G.; Trocino, A. Energy and protein metabolism and requirements. In *Nutrition of the Rabbit*, 3rd ed.; CABI Publishing CAB International: Wallingford, UK, 2020; pp. 89–125, ISBN 978-1-78924-127-3.

23. Delgado, R.; Menoyo, D.; Abad-Guamán, R.; Nicodemus, N.; Carabaño, R.; García, J. Effect of dietary soluble fibre level and n-6/n-3 fatty acid ratio on digestion and health in growing rabbits. *Anim. Feed Sci. Technol.* **2019**, *255*, 114222. [CrossRef]
24. Horwitz, W.; AOAC International (Eds.) *Official Methods of Analysis of AOAC International*, 18th ed.; Current Through rev. 1, 2006; AOAC International: Gaithersburg, MD, USA, 2006; ISBN 978-0-935584-77-6.
25. Mertens, D.R. Gravimetric determination of amylase-treated neutral detergent fiber in feeds with refluxing in beakers or crucibles: Collaborative study. *J. AOAC Int.* **2002**, *85*, 1217–1240.
26. Solans, L.; Arnal, L.J.; Sanz, C.; Benito, A.; Chacón, G.; Alzuguren, O.; Fernández, B.A. Rabbit Enteropathies on commercial farms in the iberian peninsula: Etiological agents identified in 2018_2019. *Animals* **2019**, *9*, 1142. [CrossRef]
27. Xiccato, G.; Trocino, A.; Majolini, D.; Fragkiadakis, M.; Tazzoli, M. Effect of decreasing dietary protein level and replacing starch with soluble fibre on digestive physiology and performance of growing rabbits. *Animal* **2011**, *5*, 1179–1187. [CrossRef] [PubMed]
28. Grueso, I.; De Blas, J.C.; Cachaldora, P.; Mendez, J.; Losada, B.; García-Rebollar, P. Combined effects of supplementation of diets with hops and of a substitution of starch with soluble fiber on feed efficiency and prevention of digestive disorders in rabbits. *Anim. Feed Sci. Technol.* **2013**, *180*, 92–100. [CrossRef]
29. Chamorro, S.; Gómez-Conde, M.S.; Pérez de Rozas, A.M.; Badiola, I.; Carabaño, R.; De Blas, J.C. Effect on digestion and performance of dietary protein content and of increased substitution of lucerne hay with soya-bean protein concentrate in starter diets for young rabbits. *Animal* **2007**, *1*, 651. [CrossRef] [PubMed]
30. Gidenne, T.; Kerdiles, V.; Jehl, N.; Arveux, P.; Eckenfelder, B.; Briens, C.; Stephan, S.; Fortune, H.; Montessuy, S.; Muraz, G. Protein replacement by digestible fibre in the diet of growing rabbits: 2-Impact on performances, digestive health and nitrogen output. *Anim. Feed Sci. Technol.* **2013**, *183*, 142–150. [CrossRef]
31. Gutiérrez, I.; Espinosa, A.; García, J.; Carabaño, R.; De Blas, C. Effect of protein source on digestion and growth performance of early-weaned rabbits. *Anim. Res.* **2003**, *52*, 461–471. [CrossRef]
32. Gidenne, T.; Carabaño, R.; Abad-Guamán, R.; García, J.; de Blas, C. Fibre digestion. In *Nutrition of the Rabbit*, 3rd ed.; CABI Publishing CAB International: Wallingford, UK, 2020; pp. 69–88, ISBN 978-1-78924-127-3.
33. Gidenne, T.; Perez, J.-M. Replacement of digestible fibre by starch in the dietof the growing rabbit. I. Effects on digestion, rate of passage and retention of nutrients. *Ann. Zootech.* **2000**, *49*, 357–368. [CrossRef]
34. Nicodemus, N.; Carabaño, R.; García, J.; Méndez, J.; de Blas, C. Performance response of lactating and growing rabbits to dietary lignin content. *Anim. Feed Sci. Technol.* **1999**, *80*, 43–54. [CrossRef]
35. Gidenne, T.; Arveux, P.; Madec, O. The effect of the quality of dietary lignocellulose on digestion, zootechnical performance and health of the growing rabbit. *Anim. Sci.* **2001**, *73*, 97–104. [CrossRef]
36. Ocasio-Vega, C.; Abad-Guamán, R.; Delgado, R.; Carabaño, R.; Carro, M.D.; García, J. Effect of cellobiose supplementation and dietary soluble fibre content on *in vitro* caecal fermentation of carbohydrate-rich substrates in rabbits. *Arch. Anim. Nutr.* **2018**, *72*, 221–238. [CrossRef]
37. Caîsin, L.; Martínez-Paredes, E.; Ródenas, L.; Moya, V.J.; Pascual, J.J.; Cervera, C.; Blas, E.; Pascual, M. Effect of increasing lignin in isoenergetic diets at two soluble fibre levels on digestion, performance and carcass quality of growing rabbits. *Anim. Feed Sci. Technol.* **2020**, *262*, 114396. [CrossRef]
38. Trocino, A.; Fragkiadakis, M.; Majolini, D.; Carabaño, R.; Xiccato, G. Effect of the increase of dietary starch and soluble fibre on digestive efficiency and growth performance of meat rabbits. *Anim. Feed Sci. Technol.* **2011**, *165*, 265–277. [CrossRef]
39. Trocino, A.; Fragkiadakis, M.; Majolini, D.; Tazzoli, M.; Radaelli, G.; Xiccato, G. Soluble fibre, starch and protein level in diets for growing rabbits: Effects on digestive efficiency and productive traits. *Anim. Feed Sci. Technol.* **2013**, *180*, 73–82. [CrossRef]
40. Villamide, M.J.; Carabaño, R.; Maertens, L.; Pascual, J.; Gidenne, T.; Falcao-E-Cunha, L.; Xiccato, G. Prediction of the nutritional value of European compound feeds for rabbits by chemical components and in vitro analysis. *Anim. Feed Sci. Technol.* **2009**, *150*, 283–294. [CrossRef]
41. Trocino, A.; Fragkiadakis, M.; Radaelli, G.; Xiccato, G. Effect of dietary soluble fibre level and protein source on growth, digestion, caecal activity and health of fattening rabbits. *World Rabbit Sci.* **2010**, *18*, 199–210. [CrossRef]
42. Garcia, G.; Galvez, J.F.; De Blas, J.C. Substitution of barley grain by sugar-beet pulp in diets for finishing rabbits. 1. Effect on energy and nitrogen balance. *J. Appl. Rabbit Res.* **1992**, *15*, 1008.
43. Garcia, G.; Galvez, J.F.; De Blas, J.C. Substitution of barley grain by sugar-beet pulp in diets for finishing rabbits. 2. Effect on growth performance. *J. Appl. Rabbit Res.* **1992**, *15*, 1017.

44. García, J.; Gidenne, T.; Gidenne, T.; Falcao-e-Cunha, L.; de Blas, C. Identification of the main factors that influence caecal fermentation traits in growing rabbits. *Anim. Res.* **2002**, *51*, 165–173. [CrossRef]
45. Fernández, C.; Fraga, M.J. Effect of fat inclusion in diets for rabbits on the efficiency of digestible energy and protein utilization. *World Rabbit Sci.* **2010**, *4*. [CrossRef]
46. Girard, C.L.; Robert, S.; Matte, J.J.; Farmer, C.; Martineau, G.P. Influence of high fibre diets given to gestating sows on serum concentrations of micronutrients. *Livest. Prod. Sci.* **1995**, *43*, 15–26. [CrossRef]
47. Stanogias, G.; Pearce, G.R.; Alifakiotis, T.; Michaelidis, J.; Pappa-Michaelidis, B. Effects of dietary concentration and source of fibre on the apparent absorption of minerals by pigs. *Anim. Feed Sci. Technol.* **1994**, *47*, 287–295. [CrossRef]
48. Tortuero, F.; Rioperez, J.; Cosin, C.; Barrera, J.; Rodriguez, M.L. Effects of dietary fiber sources on volatile fatty acid production, intestinal microflora and mineral balance in rabbits. *Anim. Feed Sci. Technol.* **1994**, *48*, 1–14. [CrossRef]

8

Combining Orchardgrass and Alfalfa: Effects of Forage Ratios on In Vitro Rumen Degradation and Fermentation Characteristics of Silage Compared with Hay

Zhulin Xue [1], Nan Liu [1], Yanlu Wang [2], Hongjian Yang [2], Yuqi Wei [1], Philipe Moriel [3], Elizabeth Palmer [3] and Yingjun Zhang [1,*]

1 Key Laboratory of Grasslands Management and Utilization, Ministry of Agriculture and Rural Affairs, College of Grassland Science and Technology, China Agricultural University, Beijing 100193, China; xzl2007.ok@163.com (Z.X.); liunan@cau.edu.cn (N.L.); weiyuqi91@126.com (Y.W.)

2 State Key Laboratory of Animal Nutrition, College of Animal Science and Technology, China Agricultural University, Beijing 100193, China; yanluwang@yeah.net (Y.W.); yang_hongjian@sina.com (H.Y.)

3 IFAS-Range Cattle Research and Education Center, University of Florida, Gainesville, FL 32611, USA; pmoriel@ufl.edu (P.M.); e.palmer@ufl.edu (E.P.)

* Correspondence: zhangyj@cau.edu.cn

Simple Summary: Forages are an essential portion of ruminant rations to maintain rumen function. Exploring how orchardgrass and alfalfa interact in the rumen is necessary to better understand their feed use potential as both hay and silage. This study evaluated in vitro rumen degradation, fermentation characteristics, and methane production responses to different forage ratios of alfalfa and orchardgrass. The results indicate that dry matter and organic matter degradability and methane production were greater for mixed silages compared to mixed hays. A forage ratio of 50:50 for orchardgrass and alfalfa favor the growth of rumen microorganisms without compromising nutrient digestion and rumen fermentation.

Abstract: This study aimed to investigate the effects of different forage ratios of orchardgrass (*Dactylis glomerata*) and alfalfa (*Medicago sativa*) on in vitro rumen degradation and fermentation characteristics. Orchardgrass and alfalfa were harvested separately and prepared as hay and silage mixtures at ratios of 100:0, 75:25, 50:50, 25:75, and 0:100 (w/w on a dry matter basis) and anaerobically incubated for 48 h with rumen fluid obtained from lactating dairy cows. Fermented residues and cultured fluids were used to determine nutrient degradability, fermentation parameters, and associative effect indices. Increasing the proportion of alfalfa in hay and silage mixtures quadratically increased in vitro organic matter disappearance (IVOMD, up +5.14%) and marginally decreased in vitro neutral detergent fiber disappearance (NDFD, down −1.79%). Meanwhile, increasing the proportion of alfalfa accelerated the rumen fermentation process (e.g., gas production) and remarkably enhanced the growth of rumen microbes as indicated by microbial protein production (MCP, 13.4% increase). Increments of rumen degradability and methane production were more pronounced in silage mixtures than hay mixtures. In combination, a forage ratio of 50:50 for orchardgrass and alfalfa is recommended for both hay and silage in order to improve the feed use potential in ruminants.

Keywords: forage quality; gas production; methane; ruminant

1. Introduction

Orchardgrass (*Dactylis glomerata*) is one of the most productive cool-season grasses, fairly drought resistant, and is widely distributed in North America, Europe, and East Asia [1]. Previous studies noted that orchardgrass had greater nutritive value than other forages commonly used as feedstuff for livestock, such as bromegrass, tall fescue, and reed canarygrass [2,3]. However, dietary inclusion of orchardgrass hay and silage in steers and sheep should be around 60–75% to avoid negative effects on rumen fermentation [4,5]. Additionally, polyphenol oxidase activity in orchardgrass was able to decrease protein degradation and lipolysis during aerobic ensiling and rumen fermentation [6]. Alfalfa (*Medicago sativa*) has been shown to be rich in crude protein (CP) as well as rapidly degradable protein in the rumen and high inclusion in the diet usually reduces nitrogen (N) utilization and increases urine N excretion [7,8]. However, it is unclear whether a mixture of orchardgrass and alfalfa has the potential to optimize ruminal digestibility and fermentation characteristics through complementary effects of nutrients in the forages.

Digestive interactions may occur in the rumen when ruminants are fed grass-legume mixtures [9,10] and these interactions are due to the diverse impacts on rumen digestion, fermentation, and metabolism efficiency [11]. Positive or negative interactions indicate that the response of ruminants to grass–legume mixtures is either greater or less than the arithmetical calculation obtained from using the responses for grass or legume alone. It would be helpful to determine the proportion of alfalfa at which maximal benefits can be reached in terms of rumen digestibility and fermentation efficiency in order to more precisely formulate rations [12].

The two most common preservation methods of forages, haymaking or ensiling, result in different nutritional compositions, which coincide with changes in ruminal digestibility and fermentation characteristics. For example, alfalfa preserved as hay has less effective protein degradability and digestibility than alfalfa silage, but replacement of alfalfa hay with alfalfa silage did not change ruminal volatile fatty acids (VFAs) or ammonia N [13,14]. Although some studies have separately evaluated the digestibility and fermentation characteristics of silage or hay mixtures [15,16], digestion parameters of silage in comparison with hay mixtures have not been assessed.

The in vitro digestion and gas production method has been widely used to routinely evaluate the associative effects of basal and supplementary feeds with many combinations and is less expensive and time-consuming than in vivo trials [17]. We hypothesized that mixtures of orchardgrass and alfalfa at suitable ratios would increase rumen degradability and fermentation efficiency, while decreasing methane (CH_4) production compared to alfalfa or orchardgrass alone. Therefore, the objectives of this study were to (1) investigate the associative effects of orchardgrass and alfalfa mixtures on rumen degradation and fermentation characteristics and (2) compare ruminal degradation and fermentation characteristics of hay and silage mixtures.

2. Materials and Methods

2.1. Experimental Site for Forage Planting

The field component of our experiment was conducted in Dongchengfang County (39°37′ N, 115°51′ E) in Zhuozhou City of Hebei Province of China. In September 2014, orchardgrass (cv. Aba) and alfalfa (cv. WL534) were established as monocultures in 2 field plots (21 × 75 m), each with three replicates. Sowing rate was 15 kg/ha and row spacing was 15 cm for alfalfa and orchardgrass monocultures. Monocultures were grown under the same management practices, with weed-free conditions and no insect pests. The sward soil was a sandy loam with organic matter (OM) content of 13.8 g/kg, total N content of 0.850 g/kg, plant-available phosphorus (P) content of 15.42 mg/kg, available potassium (K) content of 80.12 mg/kg and a pH of 7.80. A total of 75 kg/ha N fertilizer was applied annually in the swards.

2.2. Hay and Silage Mixture Preparation

Whole orchardgrass and alfalfa plants were harvested (second cutting) with a reaping hook from the field plots at the jointing and early bloom stages. Ten square areas (1 × 1 m) in each field plot were randomly selected and plants were harvested at a stubble height of 5 cm above ground. The harvested fresh forage was weighed separately and recorded. Representative forage samples (approximately 1 kg) from each square harvest were oven-dried at 65 °C for 48 h to determine the dry matter (DM) yield and chemical composition. Based on the forage DM obtained from four annual harvests, the annual DM yields of orchardgrass and alfalfa averaged 1065 g/m^2 and 1480 g/m^2, respectively. The chemical composition of orchardgrass and alfalfa prior to hay making and ensiling was shown in Table 1. After the forage was harvested, representative samples (approximately 20 kg) of each preservation method were divided equally into two subsamples. The first subsample, of both orchardgrass and alfalfa, was air-dried in the sun for 48 h, oven-dried at 65 °C for 48 h, thoroughly mixed to obtain forage ratios of 100:0, 75:25, 50:50, 25:75, and 0:100 (orchardgrass: alfalfa, w/w), and five representative hay samples of each forage ratio were prepared. All of these hay samples were ground in a mill (FW100; Tianjin Taisite Instrument Co., Ltd., Tianjin, China) to pass through a 1.0 mm sieve and stored in plastic bags for chemical analysis and in vitro batch culturing.

The second subsample, of both orchardgrass and alfalfa, was sun-wilted for 2 h until the DM content increased to 400 g/kg and was chopped into 2.5 cm pieces. A microwave oven (M1-L213B, Midea Group Co., Ltd., Foshan, China) was used for the rapid determination of DM content at regular intervals. The wilted forage samples were mixed at ratios of 100:0, 75:25, 50:50, 25:75, and 0:100 (orchardgrass: alfalfa, w/w) based on DM content and were packed into polyethylene silos (1 L capacity) until the total forage weight reached approximately 750 g. Five silo replicates of each forage ratio were prepared, and all silos were sealed with screw lids and kept in the dark at approximately 25 °C. After 40 days of ensiling, a total of 20 g of each silage mixture was homogenized in a blender with 180 mL of distilled water for 1 min. The samples were filtered through a four-layered cheesecloth to prepare the silage extract for determination of ensiling characteristics such as pH, ammonia N, and organic acids, including lactic acid, acetic acid, propionic acid, and butyric acid. Silage mixtures were then dried in a forced-air oven at 65 °C for 48 h and ground in a mill to pass through a 1.0 mm sieve for subsequent analysis.

Table 1. Chemical composition of orchardgrass and alfalfa prior to hay making and ensiling based on dry matter (g/kg).

Item	Orchardgrass	Alfalfa
Dry matter	260	268
Crude protein	167	214
Neutral detergent fiber	562	484
Acid detergent fiber	424	385
Acid detergent lignin	37.0	72.2
Ash	111	92.0
Ether extract	26.8	24.0
Calcium	7.04	14.1
Phosphorus	3.42	3.36
Potassium	33.6	27.8
Magnesium	3.72	3.16

2.3. Rumen Fluid Collection

All experimental procedures used in this study were conducted in accordance with the Institutional Animal Care Committee of China Agricultural University (CAU20171014-1). Rumen fluid was collected from five rumen-cannulated (Type 2c; Bar Diamond Inc., Parma, ID, USA) lactating Holstein dairy cows with a live weight of 530 ± 31.2 kg. Cows were uniformly fed equal rations twice daily at 7:00 a.m. and 7:00 p.m. and provided free access to water. The ration contained 4.0 kg/day of alfalfa hay, 3.5 kg/day of maize silage and 5.5 kg/day of commercial concentrate with a net energy for lactation of 6.69 MJ/kg DM and a CP content of 160 g/kg. Fresh rumen contents (mix of liquid and solid) from different sites inside the rumen were withdrawn from all dairy cows 1 h before the morning feeding with a pair of rubber fabric tongs inserted through the rumen cannula to form one composite sample. The composite sample was squeezed through four layers of surgical gauze to obtain the rumen fluid. Rumen fluid was kept in a thermos flask pre-warmed at 39 °C and immediately transferred to the laboratory. Rumen fluid from the five dairy cows was mixed in equal proportions kept at 39 °C and filled with carbon dioxide (CO_2) prior to in vitro inoculation.

2.4. In Vitro Batch Culturing and Sample Collection

Samples (500 mg) of each forage ratio, of both the hay and silage mixtures, were transferred into 120 mL glass bottles with Hungate stoppers and screw caps, resulting in a total of 50 samples in glass bottles (5 forage ratios × 2 preservation methods × 5 fermentation bottles). Fifty mL of fresh buffer solution at a pH of 6.85 [18] and 25 mL of filtered rumen fluid were mixed in each bottle, which were all purged with nitrogen gas (N_2) to maintain anaerobic conditions. Thereafter, each bottle was connected to an automated trace gas recording instrument (AGRS-III) [19] to continuously record the cumulative gas production (GP). Five bottles without forage samples were incubated as blanks. Simultaneously, an additional 30 glass bottles (5 forage ratios × 2 preservation methods × 3 fermentation bottles) prepared in the same manner were separately connected to pre-emptied air bags that collected all of the fermentation gases for subsequent determination and calculation. Three fermentations without samples were also included as blanks. All bottles were incubated at 39 °C for 48 h, and these batch cultures were repeated in 3 experimental runs. Two hundred and forty bottles (5 forage ratios × 2 preservation methods × 5 fermentation bottles × 3 experimental replicates + 5 forage ratios × 2 preservation methods × 3 fermentation bottles × 3 experimental replicates) plus 8 bottles in each replicate that were used as blanks (rumen fluid + buffer) were examined in this experiment.

After incubation, the remaining contents in each bottle were separately filtered through preweighed nylon bags (8 × 12 cm, 42 μm pore size). The filtrate was sampled for the determination of ammonia N, microbial protein production (MCP) and volatile fatty acids (VFAs). The nylon bags were thoroughly rinsed and dried at 65 °C for 48 h to a constant weight (the difference between two consecutive weights was no more than 0.20 mg per g) to determine in vitro dry matter disappearance (IVDMD), in vitro organic matter disappearance (IVOMD), in vitro neutral detergent fiber disappearance (NDFD) and in vitro acid detergent fiber disappearance (ADFD) after the chemical analysis of residues remaining in the bags.

2.5. Chemical Analysis

Determination of DM, OM, ether extract (EE), CP, and calcium (Ca), K, and magnesium (Mg) followed the methods of the Association of Official Analytic Chemists (AOAC) [20]. Following the method of Van Soest et al. [21], the contents of neutral detergent fiber (NDF), acid detergent fiber (ADF), and acid detergent lignin (ADL) were determined using a Fiber Analyzer (ANKOM Technology Corporation, Fairport, NY, USA) and expressed including residual ash. The concentrations of ammonia N and organic acids, including lactic acid, acetic acid, propionic acid, and butyric acid, in the silage extract were determined according to the method described by Li and Meng [22].

For each cultured fluid, the concentrations of ammonia N [23] and MCP [24] were determined. The determination of VFAs was conducted according to the method described by Cui et al. [25]. Briefly, the filtrate of each cultured fluid kept in polypropylene tubes (10 mL) was centrifuged at 3500× *g* for 15 min at 4 °C and 1.0 mL of the supernatant sample was mixed with 0.3 mL of 250 g/L orthophosphoric acid solution into a polypropylene microtube for 30 min. Thereafter, the mixture was cooled at 4 °C for 2 h and centrifuged at 15,000× *g* for 10 min at 4 °C. The supernatants were kept at −20 °C for later VFAs analysis. Methane, CO_2, and hydrogen gas (H_2) produced per kg OM digested were measured with a gas chromatograph according to the method described by Cui et al. [25].

2.6. Calculations

Flieg's score was calculated according to the method of Kilic [26], as shown in Equation (1).

$$\text{Flieg's score} = 220 + (2 \times \%\,\text{DM} - 15) - 40 \times \text{pH}, \quad (1)$$

The values at 81–100, 61–80, 41–60, 21–40, and 0–20 represented a very good, good, medium, low and poor silage quality, respectively.

Relative forage quality (RFQ) was calculated according to the method of Moore and Undersander [27], as shown in Equation (2).

$$\text{RFQ} = \text{DMI} \times \text{TDN} / 1.32, \quad (2)$$

where DMI is dry matter intake (% of body weight) and TDN is the total digestible nutrients (% of DM). For the alfalfa and orchardgrass–alfalfa mixtures, TDN and DMI were determined according to the methods described by the NRC [28] and Mertens [29], respectively, with NDFD adjusted according to Oba and Allen [30]. For orchardgrass, TDN and DMI were calculated following the methods of Moore and Undersander [27] and Moore and Kunkle [31], respectively.

Cumulative GP data [$GP_{(t)}$, mL/g DM] at time (t) were exported from the AGRS-III instrument into a Microsoft Excel file and fitted to a nonlinear model [32] by iterative regression analysis, as shown in Equation (3).

$$\text{GP}_t = \text{A} \times [1 - e^{-c \times (t - \text{Lag})}], \quad (3)$$

where A represents the asymptotic GP generated at a constant fractional rate (c) per unit time, t is the time of the gas recording, *e* is the base of the natural logarithm, and Lag represents the lag time before GP commences. The average GP rate (AGPR) at half of A was calculated according to the equation of García-Martínez et al. [33], as shown in Equation (4).

$$\text{AGPR} = \frac{\text{A} \times c}{2 \times (\text{Ln2} + c \times \text{Lag})}, \quad (4)$$

Single-factor associative effects index (SFAEI, %) and multiple-factor associative effects index (MFAEI, %) were calculated according to the methods described by Niderkorn et al. [34] and Wang [35], as shown in Equations (5)–(7).

$$\text{SFAEI} = \frac{\text{observed value} - \text{calculated value}}{\text{calculated value}} \times 100 \quad (5)$$

$$\begin{aligned}\text{Calculated value} &= \text{the measured value of sole orchardgrass} \times \text{the proportion of orchardgrass} \\ &\text{in mixtures} + \text{the measured value of sole alfalfa} \times \text{the proportion of alfalfa in mixtures}\end{aligned} \quad (6)$$

$$\text{MFAEI} = \text{the sum of each single} - \text{factor associative effect value}, \quad (7)$$

In our current study, MFAEI is the sum of the six SFAEI of RFQ, OMD, NDFD, GP_{48}, c, and MCP.

2.7. Statistical Analysis

Data were analyzed using a randomized complete block design with two preservation methods (hay vs. silage) and five orchardgrass to alfalfa ratios (100:0, 75:25, 50:50, 25:75, and 0:100) for each block. One-way ANOVA was used to determine the linear and quadratic responses to the proportion of alfalfa inclusion on chemical composition and ensiling characteristics of the hay and the silage mixtures, respectively. The in vitro experimental data consisted of five fermentation replicates per forage mixture sample in each of the three experimental runs, for a total of 150 observations of the mixed ratios and preservation methods treatments. After averaging fermentation replicates within an experimental run, the data of 30 observations were subjected to ANOVA using a general linear model in which the fixed effects of preservation methods and mixed ratios of orchardgrass to alfalfa were considered. All statistical analyses were performed using R software (version 3.2.3). The least squares means and standard error (SEM) were calculated, and orthogonal contrasts were performed to examine the linear and quadratic effects of the proportion of alfalfa in hay and silage mixtures. Significance was declared at $p < 0.05$ unless otherwise noted.

3. Results

3.1. Chemical Composition of Hay and Silage Mixtures

Increasing the proportion of alfalfa increased the contents of CP, ADL, and Ca and decreased NDF, ADF, ash, and Mg in both hay and silage mixtures (Figure 1). Crude protein, NDF, ADF, and ADL were greater in the hay vs. silage mixtures. The TDN, DMI, and RFQ showed a quadratic increase with increasing alfalfa proportions, and these values were reduced in the hay mixtures compared to the silage mixtures (Figure 1).

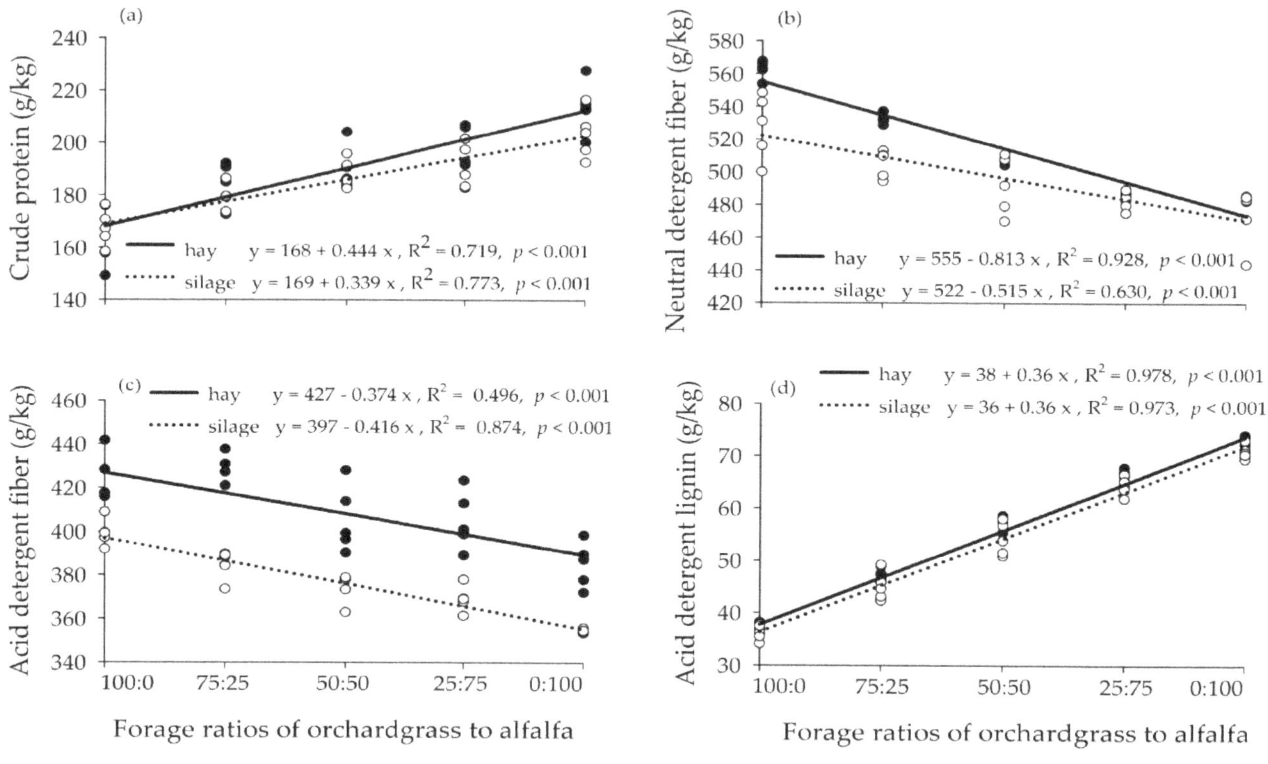

Figure 1. *Cont.*

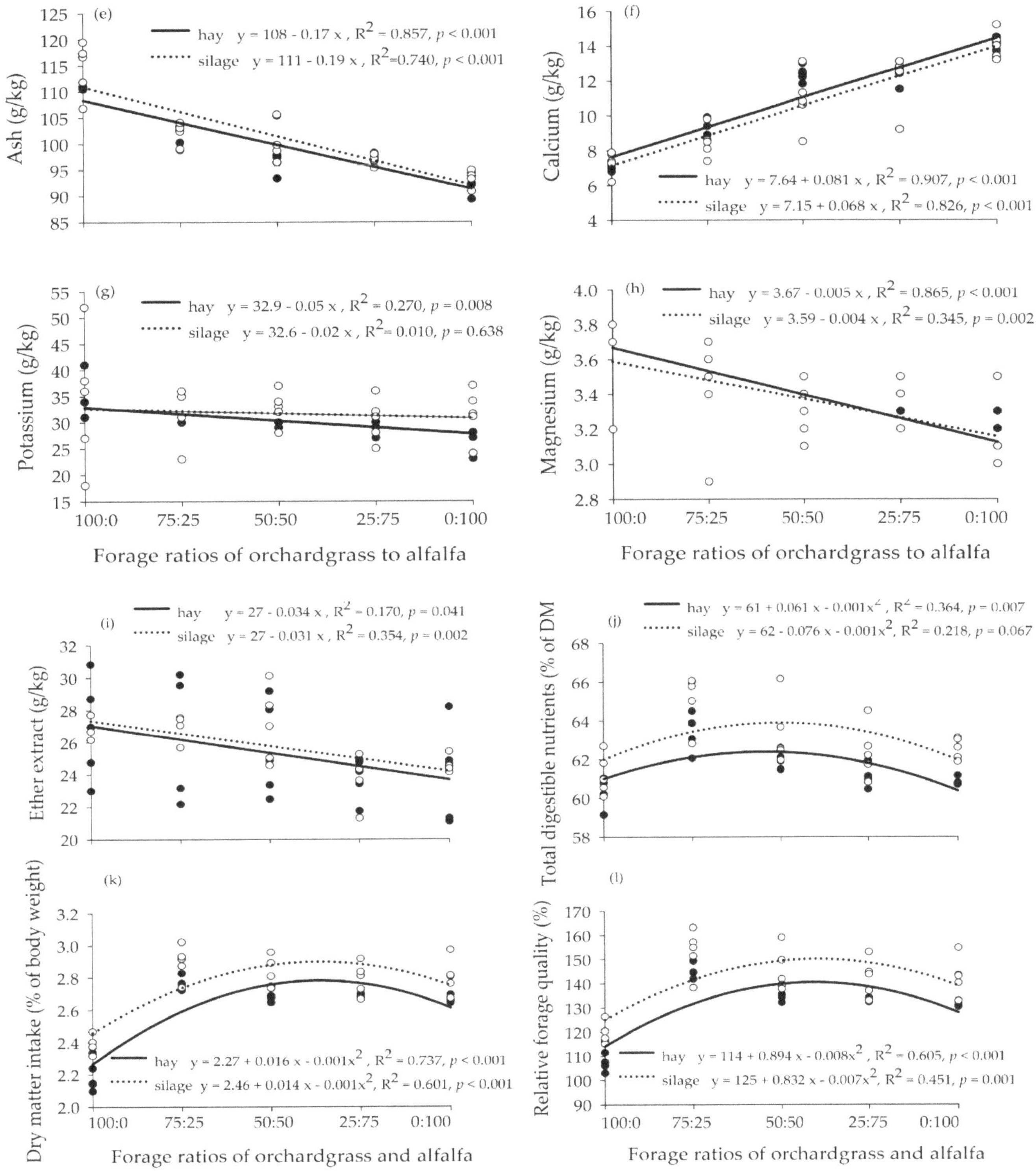

Figure 1. Chemical composition of orchardgrass and alfalfa mixtures preserved as hay and silage: (**a**) crude protein; (**b**) neutral detergent fiber; (**c**) acid detergent fiber; (**d**) acid detergent lignin; (**e**) ash; (**f**) calcium; (**g**) potassium; (**h**) magnesium; (**i**) ether extract; (**j**) total digestible nutrients; (**k**) dry matter intake; (**l**) relative forage quality.

3.2. Ensiling Characteristics of Orchardgrass and Alfalfa Silage Mixtures

The silage pH and the concentrations of ammonia N and acetic acid increased, whereas the concentration of lactic acid decreased as the proportion of alfalfa increased in silage mixtures (Table 2). The concentration of ammonia N in alfalfa was approximately twice that of orchardgrass (11.1 mmol/L and 5.36 mmol/L, respectively). However, the concentration of lactic acid in alfalfa was 355 mmol/L lower than that in orchardgrass (Table 2). According to the Flieg's scores for silage quality, mixtures of orchardgrass and alfalfa produced excellent quality silages when alfalfa proportion was ≤ 50% (Table 2). The silage mixtures of orchardgrass and alfalfa resulted in higher SFAEI of lactic acid and lower SFAEI of pH, compared with the sole orchardgrass and alfalfa (Table 2).

Table 2. Ensiling characteristics of orchardgrass and alfalfa silage mixtures.

Item	Forage Ratios of Orchardgrass to Alfalfa					SEM [2]	p^3	
	100:0	75:25	50:50	25:75	0:100		Linear	Quadratic
pH	4.50	4.83	4.75	5.03	4.96	0.049	<0.001	<0.001
Ammonia N (mmol/L)	5.36	7.96	9.96	11.2	11.1	0.585	<0.001	<0.001
Lactic acid (mmol/L)	963	998	969	754	608	50.7	0.004	0.005
Acetic acid (mmol/L)	1.58	2.55	7.54	7.88	8.63	0.860	<0.001	<0.001
Propionic acid (mmol/L)	0.38	0.30	0.29	0.31	0.14	0.023	0.002	0.005
Butyric acid (mmol/L)	0.12	0.10	0.13	0.14	0.15	0.008	0.081	0.159
Flieg's score [1]	95.6	89.2	87.6	79.3	76.8	1.67	<0.001	<0.001
Single-factor associative effects indices (%) of ensiling characteristics [4]								
pH	-	4.6	0.5	3.8	-	0.76		
Ammonia N (mmol/L)	-	17.0	20.6	15.4	-	4.09		
Lactic acid (mmol/L)	-	14.2	23.4	8.2	-	5.76		
Acetic acid (mmol/L)	-	−23.7	12.8	14.8	-	10.2		
Propionic acid (mmol/L)	-	−6.4	11.0	51.1	-	11.93		

[1] Flieg's score was calculated by means of the pH values and DM contents of the silages: Flieg's score = 220 + (2 × % DM − 15) − 40 × pH. Flieg's score: < 20, very bad; 21–40, bad; 41–60, medium; 61–80, good; 81–100, very good. Since there was no difference on DM contents (ranging from 376 g/kg to 383 g/kg) among silage mixtures, the data were not shown. [2] SEM, standard error of the mean. [3] Linear and quadratic represent the effects of alfalfa inclusion on different silage mixtures. [4] Single-factor associative effects indices (SFAEI, %) = 100 × [(observed value − calculated value)/calculated value]. Calculated value was determined based on the proportional contribution of sole forage (orchardgrass vs. alfalfa) in the silage mixture. Positive or negative values indicated positive or negative associative effects of orchardgrass and alfalfa mixtures.

3.3. Degradation Characteristics and Gas Production Kinetics of Hay and Silage Mixtures

The IVDMD increased linearly (Figure 2a, $p < 0.001$), whereas ADFD decreased linearly (Figure 2d, $p < 0.001$) with increasing proportions of alfalfa in hay and silage mixtures. Positive associative effects were observed on IVOMD (Figure 2b, $p < 0.001$), cumulative GP_{48} (Table 3, $p < 0.05$), and asymptotic GP (A) (Table 3, $p < 0.05$); furthermore, the greatest values of these measurements were observed in the 50:50 orchardgrass-alfalfa mixtures. Negative associative effects on NDFD (Figure 2c, $p < 0.05$) were observed when the proportion of alfalfa increased in hay and silage mixtures. The IVDMD (Figure 2a, $p < 0.001$) and IVOMD (Figure 2b, $p < 0.05$) were greater in the silage mixtures than in the hay mixtures, but no effects of preservation method were observed for NDFD (Figure 2c), ADFD (Figure 2d), GP_{48} (Table 3) or other parameters of GP kinetics (Table 3). Positive associative effects were also observed for the fractional GP rate (c) and AGPR as the proportion of alfalfa increased in hay and silage mixtures ($p < 0.05$). Regarding fermentation end-product gases, no changes were observed in CO_2 or H_2 production; however, increasing the proportion of alfalfa linearly increased CH_4 production ($p < 0.05$). Greater CH_4 production was observed in the silage mixtures than in the hay mixtures ($p < 0.05$; Table 3). No interactions between forage ratios and preservation method were observed on degradation or gas production parameters ($p > 0.05$; Table 3; Figure 2).

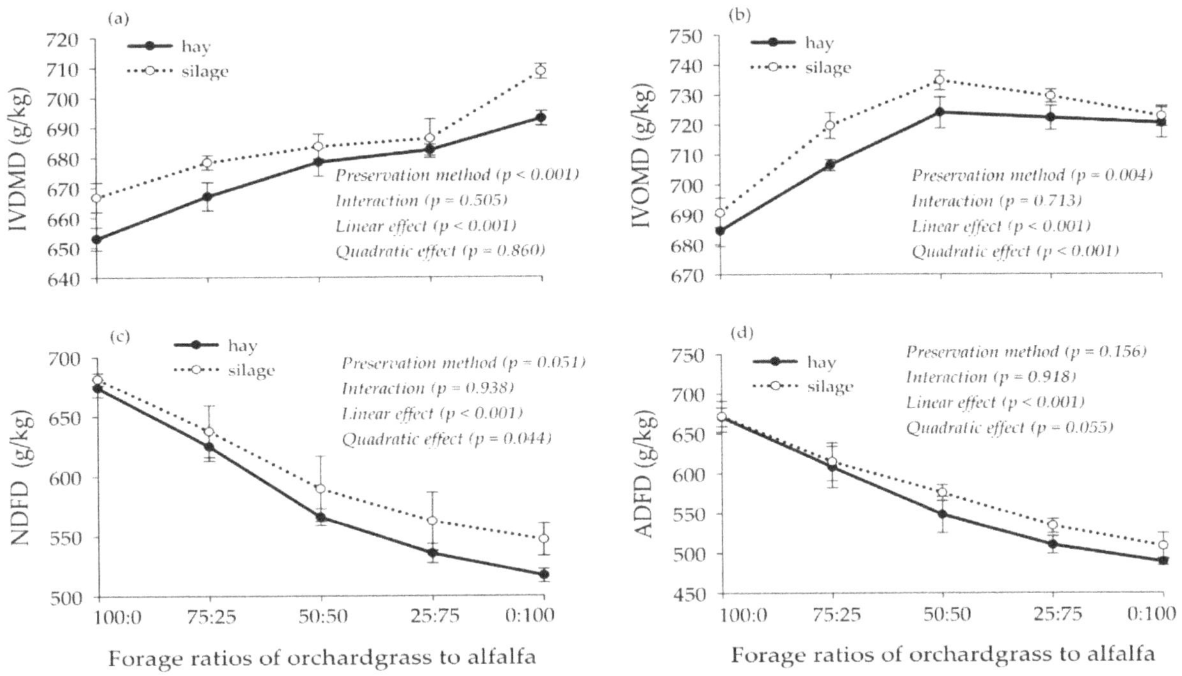

Figure 2. In vitro degradation characteristics of orchardgrass and alfalfa mixtures preserved as hay and silage: (**a**) IVDMD, in vitro dry matter disappearance; (**b**) IVOMD, in vitro organic matter disappearance; (**c**) NDFD, in vitro neutral detergent fiber disappearance; (**d**) ADFD, in vitro acid detergent fiber disappearance. The preservation method indicates whether the forage was preserved as hay or silage, and the interaction indicates the interaction between the preservation method and the ratio of orchardgrass to alfalfa. Orthogonal contrasts were used to examine the linear and quadratic effects of alfalfa inclusion on different hay and silage mixtures. Bars show the mean ± standard error.

Table 3. Gas production kinetics of orchardgrass and alfalfa mixtures preserved as hay and silage.

Item [1]	Preservation Method	Forage Ratios of Orchardgrass to Alfalfa					SEM [2]	p^3		
		100:0	75:25	50:50	25:75	0:100		Preservation Method	Linear	Quadratic
GP_{48} (mL/g DM)	Hay	122	128	133	134	131	1.1	0.847	<0.001	0.026
	Silage	124	124	136	134	134				
GP kinetics										
A (mL/g DM)	Hay	123	128	134	135	131	1.1	0.837	<0.001	0.025
	Silage	124	124	136	134	134				
c	Hay	0.13	0.14	0.16	0.16	0.15	0.002	0.165	0.011	0.004
	Silage	0.13	0.14	0.16	0.14	0.14				
Half-time (h)	Hay	2.71	2.69	2.55	2.60	2.60	0.012	0.339	0.005	0.005
	Silage	2.72	2.66	2.57	2.65	2.65				
AGPR (mL/h)	Hay	23.4	25.1	28.4	29.2	29.8	0.37	0.319	<0.001	0.025
	Silage	24.2	24.1	28.4	28.5	29.1				
Fermentation gas pattern (mL/g OM digested)										
CO_2	Hay	174	176	178	180	176	0.8	0.081	0.150	0.675
	Silage	176	166	178	175	177				
CH_4	Hay	31.6	34.5	36.2	36.4	37.1	0.62	0.045	0.006	0.287
	Silage	34.6	37.5	37.8	39.0	39.2				
H_2	Hay	0.59	0.36	0.82	0.76	1.51	0.121	0.476	0.513	0.164
	Silage	2.00	0.84	0.94	0.66	0.40				

[1] GP_{48}, cumulative gas production at 48 h; A, asymptotic gas production; c, fractional rate of the gas production of 'A'; half-time, the time when 'A' reached its half-maximal value; AGPR, average gas production rate when half of 'A' occurred. [2] SEM, standard error of the mean. [3] The preservation method indicates whether the forage was preserved as hay or silage. Linear and quadratic represent the effects of alfalfa inclusion in the different hay and silage mixtures. Since there were no interactions between the preservation method and the ratios of orchardgrass to alfalfa, we did not present *p*-values on interaction in the Table 3.

3.4. Fermentation Characteristics of Hay and Silage Mixtures

Concentration of ammonia N increased linearly ($p < 0.05$) when the proportion of alfalfa increased in hay and silage mixtures (Table 4). A positive associative effect on MCP was detected as the proportion of alfalfa increased ($p < 0.05$). No effects of alfalfa inclusion on the molar proportions of acetate, propionate, butyrate, and isobutyrate were observed ($p > 0.05$). However, the concentrations of total VFAs, valerate and isovalerate increased linearly with increasing alfalfa proportion ($p < 0.05$). No effects of preservation method were observed on these fermentation parameters ($p > 0.05$; Table 4). There were no interactions between forage ratios and preservation method on fermentation parameters ($p > 0.05$; Table 4).

Table 4. Fermentation characteristics of orchardgrass and alfalfa mixtures preserved as hay and silage.

Item [1]	Preservation Method	Forage Ratios of Orchardgrass to Alfalfa					SEM [2]	p^3		
		100:0	75:25	50:50	25:75	0:100		Preservation Method	Linear	Quadratic
Ammonia N (mg/dL)	Hay	20.0	20.0	20.4	21.5	21.5	0.31	0.635	0.029	0.996
	Silage	20.0	20.2	21.2	21.7	21.8				
MCP (μg/mL)	Hay	270	276	329	345	338	5.2	0.555	<0.001	0.034
	Silage	282	296	315	338	303				
Total VFAs (mmol/L)	Hay	110	116	118	117	120	1.2	0.208	0.049	0.688
	Silage	113	110	115	110	119				
				VFA pattern (mmol/L)						
Acetate	Hay	71.2	76.6	78.4	77.4	77.8	1.10	0.085	0.309	0.809
	Silage	74.5	70.1	72.6	68.8	76.2				
Propionate	Hay	18.8	17.6	17.4	18.9	19.0	0.17	0.297	0.136	0.631
	Silage	17.2	17.9	19.5	17.4	18.4				
Butyrate	Hay	9.09	9.76	9.51	9.33	10.2	0.31	0.620	0.300	0.887
	Silage	9.45	9.56	9.82	10.3	10.4				
Isobutyrate	Hay	2.14	2.63	2.56	2.14	2.20	0.084	0.257	0.900	0.811
	Silage	2.36	2.34	2.64	2.63	2.83				
Valerate	Hay	2.61	3.02	3.10	2.81	3.50	0.093	0.801	0.015	0.710
	Silage	2.80	2.88	2.94	3.25	3.40				
Isovalerate	Hay	6.04	6.20	6.68	6.54	7.16	0.185	0.097	0.024	0.704
	Silage	6.60	6.76	7.18	7.23	7.99				

[1] Ammonia N, ammonia nitrogen; MCP, microbial protein production; VFAs, volatile fatty acids. [2] SEM, standard error of the mean. [3] The preservation method indicates whether the forage was preserved as hay or silage. Linear and quadratic represent the effects of alfalfa inclusion in different hay and silage mixtures. Since there were no interactions between the preservation method and the ratios of orchardgrass to alfalfa, we did not present p-values on interaction in the Table 4.

3.5. SFAEI and MFAEI of Hay and Silage Mixtures

The SFAEI of RFQ, IVOMD, c, MCP, and GP_{48} was positive in both hay and silage mixtures of orchardgrass and alfalfa (Figure 3). The SFAEI of RFQ was greatest at 75:25 orchardgrass to alfalfa hay and silage mixtures. Increasing the proportion of alfalfa increased the SFAEI of IVOMD, GP_{48}, c, and MCP by +5.14%, +5.52%, +16.1%, +13.4%, respectively, and decreased the SFAEI of NDFD (−1.79%) when compared with the calculated value. Positive MFAEI was observed in all mixtures, and the highest MFAEI (up +40%) was observed in 50:50 of orchardgrass to alfalfa hay and silage mixtures.

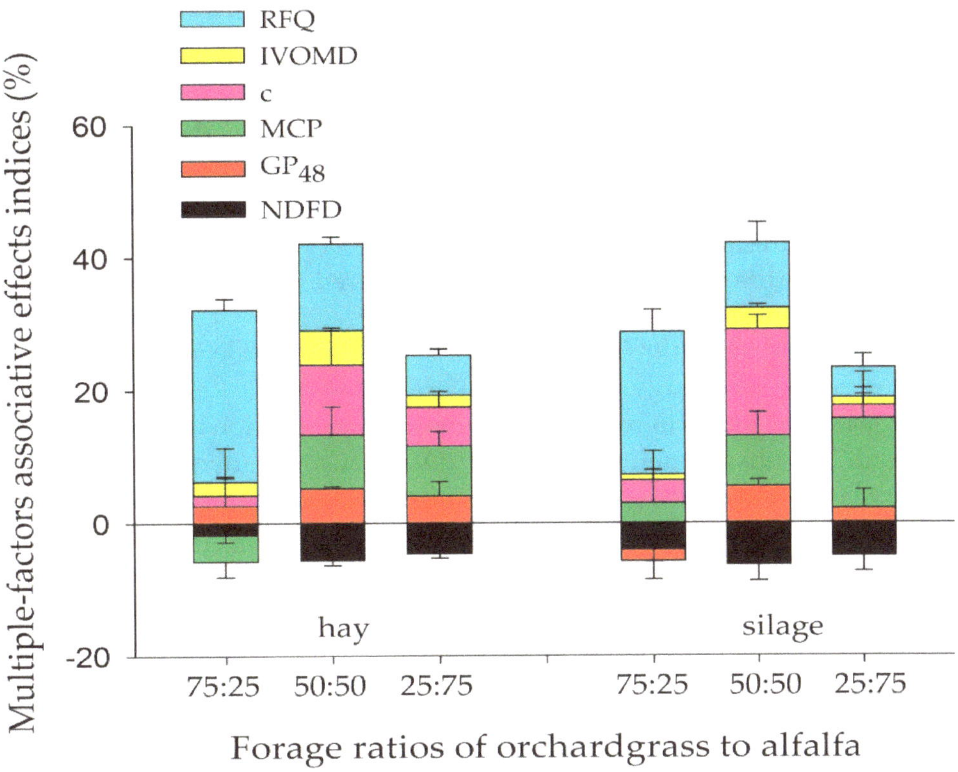

Figure 3. Single-factor associative effects indices (SFAEI, %) and multiple-factors associative effects indices (MFAEI, %) of hay and silage mixtures. RFQ, relative forage quality; IVOMD, in vitro dry matter disappearance; NDFD, in vitro neutral detergent fiber disappearance; c, gas production rate; GP_{48}, gas production at 48 h; MCP, microbial protein production. Bars show the mean ± standard error. Single-factor associative effects indices (SFAEI, %) = 100 × [(observed value − calculated value)/calculated value]. Calculated value was determined based on the proportional contribution of sole forage (orchardgrass vs. alfalfa) in the mixture. Positive or negative values indicated positive or negative associative effects of orchardgrass and alfalfa mixtures.

4. Discussions

As hypothesized, combining orchardgrass with alfalfa optimized the ruminal degradation and fermentation efficiency through complementary effects of nutrients in forage mixtures and the greatest MFAEI (up about +40%) was obtained at 50:50 of orchardgrass to alfalfa in both hay and silage mixtures. Although mixtures of orchardgrass and alfalfa decreased NDFD, this antagonistic effect was quite little for the total MFAEI, which presumably has been compensated for by synergistic effects of other nutrients. Considering the observed associative effects of mixing orchardgrass and alfalfa, it is possible to optimize forage use efficiency in ruminants.

4.1. Chemical Composition of Hay and Silage Mixtures

Greater CP and reduced NDF and ADF contents are usually observed in legumes compared to grasses, due to N fixation from the atmosphere and less cell wall material in legumes [36,37], similar to the current study. A reduction in CP, NDF, and ADF contents were observed in the silage mixtures than in the hay mixtures, probably because a smaller fraction of the protein was broken down into peptides, amino acids, and ammonia and some of the cellulose and hemicellulose were hydrolyzed during the ensiling process [13,38]. Greater values of RFQ were observed in alfalfa-containing mixtures compared with the orchardgrass alone, suggesting that alfalfa inclusion has the potential to improve the overall forage quality of mixtures and the subsequent animal performance [39,40].

4.2. Ensiling Characteristics of Orchardgrass and Alfalfa Silage Mixtures

Legumes are usually more difficult to preserve as silage than are grasses due to a reduction in water-soluble carbohydrates and a greater buffering capacity [7]. The results from the current study indicate that alfalfa silage had increased pH and ammonia N concentrations and decreased lactic acid concentrations compared to orchardgrass silage and silage mixtures of orchardgrass and alfalfa showed suitable ensiling characteristics and high associative effects indices. In accordance with previous studies [41,42], silage mixtures could be a feasible way to preserve nutrition and improve fermentation quality through the role of dominated Lactobacillus, compared with the sole silage crops.

4.3. Associative Effects on the Degradation Characteristics and Gas Production Kinetics of Hay and Silage Mixtures

The IVDMD linearly increased in response to increasing proportions of alfalfa in the present study due to its high CP and low fiber contents [43]. The IVDMD of silage mixtures was greater than that of hay mixtures, which agrees with previous results that the apparent digestibility of DM and OM in alfalfa silage was greater than that in alfalfa hay [44].

Negative associative effects on NDFD were detected as the proportion of alfalfa increased in mixtures, which is in agreement with other studies [10,45]. These results may be attributed to the antagonistic effects that occur in the fermentation extent and degradation rate of fiber fractions. Similarly, these antagonistic effects were more evident when the proportion of alfalfa was greater than 25% in alfalfa–Caucasian bluestem hay mixtures [46]. The dominant fibrolytic bacteria, that are not proteolytic, require NH_3 and branched-chain VFAs that are usually derived from active proteolytic bacteria and amino acid fermentation bacteria that do not ferment carbohydrates [47,48]. The degradable and rapidly digested N from legumes compensated for the slow release of N from grasses and coincided with associations and shifts in the ruminal microbial community [15,49] to better degrade the fiber fractions in grass-legume mixtures [46,50].

However, the NDF of legumes is more resistant to degrade than that of grasses because legumes have higher proportions of indigestible NDFs [51] with more complicated structures (e.g., cuticle sheets and vascular bundles) and physicochemical properties (e.g., alkali solubility and ether links), which limit the attachment of fibrolytic bacteria and the enzymatic hydrolysis of forage polysaccharides [37] and decrease the potential for fiber digestion [52]. In the current study, it appears that the positive effect of alfalfa inclusion on NDFD by supplying a rapidly digested N source for fibrolytic bacteria was offset by the negative effect of increased indigestible fiber fractions, subsequently decelerating the fermentation extent and degradation rate of NDF. High levels of fat, starch, and sugars that are rapidly fermented to VFAs can also reduce rumen pH and hence depress the activity of fibrolytic bacteria, decreasing NDFD [53–55].

The volume of GP was positively correlated with OM degradability, which is indirectly reflected in the potential of microbes to degrade feed and the extent of microbial fermentation [56]. The current results indicate positive associative effects on cumulative GP and IVOMD when the proportion of alfalfa increased in hay and silage mixtures, implying that complementary digestible nutrients in orchardgrass-alfalfa mixtures provided favorable substrates for microbes, resulting in improved fermentation. The presence of positive associative effects on the asymptotic fractional GP rate (c) but negative associative effects on the time to reach half of the asymptotic GP (half-time) suggests that the inclusion of alfalfa in mixtures accelerated microbial fermentation of digestible components and/or resulted in a shift in the site of digestion of cell contents, including N [9]. In summary, no differences were observed in IVDMD and IVOMD for either hay or silage mixtures containing 50, 75, or 100% alfalfa, however, all mixtures at these proportions had greater IVDMD and IVOMD and reduced NDFD and ADFD values than did mixtures with 25% or 0% alfalfa. If the DM and OM degradability of a mixture needs to be increased, at least 50% alfalfa is required based on the results obtained in this study.

There was no effect on CO_2 and H_2 with increasing alfalfa proportions, this could be explained by similar responses in acetate, propionate, and butyrate [57]. The amount of CH_4 produced per unit of

OM digested increased linearly with increasing alfalfa proportion, and the higher nitrogen-free extract (NFE) (44.07% vs. 43.75%) and lower crude fiber (CF) (21.0% vs. 33.93%) in alfalfa compared with orchardgrass could explain the comparatively high CH_4 production with alfalfa [58,59]. Feedstuffs with higher GP and IVDMD tended to have greater CH_4 production [60]; this trend is in accordance with the current study where silage mixtures, which had greater IVDMD and IVOMD, had increased CH_4 production per gram of OM digested than did the hay mixtures.

4.4. Associative Effects on the Fermentation Characteristics of Hay and Silage Mixtures

Ammonia N showed a linear increase in cultured fluids as the proportion of alfalfa increased, which was partially due to the proteolysis in alfalfa that occurred during in vitro incubation [61]. Rumen fermentation and microbial protein synthesis proceeded at a lower than maximal rate when the concentrations of ammonia N in the rumen fluid were less than 23.5 mg/dL and 5 mg/dL, respectively [62,63]. In the current study, the concentration of ammonia N in the cultured fluid was above the requirements to maximize MCP synthesis (>5 mg/dL) and thus the sufficient energy should match the synthetic abilities of microorganisms to effectively utilize these N resources [64]. Positive associative effects on MCP were observed as the proportion of alfalfa increased in our study, implying that mixtures of orchardgrass and alfalfa contributed more to the balance of readily degradable carbohydrates and readily degradable protein than did the monocultures, subsequently improving microbial protein synthesis and increasing bacterial N flow and microbial growth.

The concentration of total VFAs linearly increased with increasing proportions of alfalfa, and similarly, increased VFAs production was also observed in sweet sorghum-alfalfa mixtures [15]. In our study, alfalfa inclusion did not change the molar proportions of acetate, propionate, and butyrate, probably because the ruminal microbial community and metabolic pathways by which specific microbes utilize substrates to produce VFAs end-products were not altered [65]. The concentrations of valerate and isovalerate increased linearly with increasing alfalfa proportions, which was in accordance with a previous study [17] showing that the CP level was significantly correlated with valerate and isovalerate production.

5. Conclusions

The forage ratio of 50:50 for orchardgrass and alfalfa hay and silage mixtures favored the growth of rumen microorganisms and improved nutrient digestion and ruminal fermentability. Increments of rumen degradability and methane production were more pronounced in silage mixtures than hay mixtures. This study provided potential evidence on how to maximize forage use efficiency in ruminant systems, and further research could be performed to assess the synergistic effects of this practice on agronomic characteristics and ruminant production performance.

Author Contributions: Conceptualization, Z.X. and Y.Z.; Methodology, Z.X., Y.Z. and H.Y.; Formal analysis, Z.X. and Y.Z.; Investigation, Y.W. (Yanlu Wang) and Y.W. (Yuqi Wei). Data curation, Z.X. and Y.Z.; Writing—original draft preparation, Z.X.; Writing—review and editing, N.L., H.Y., P.M., E.P. and Y.Z.; Supervision, Y.Z.; Project administration, Y.Z.; Funding acquisition, Y.Z. All authors have read and agreed to the published version of the manuscript.

References

1. Stewart, A.V.; Ellison, N.W. Dactylis. In *Wild Crop Relatives: Genomic and Breeding Resources*; Kole, C., Ed.; Springer: Berlin/Heidelberg, Germany, 2011; pp. 73–87.
2. Turner, L.; Donaghy, D.; Lane, P.; Rawnsley, R. A comparison of the establishment, productivity, and feed quality of four cocksfoot (*Dactylis glomerata* L.) and four brome (*Bromus* spp.) cultivars, under leaf stage based defoliation management. *Aust. J. Agric. Res.* **2007**, *58*, 900–906. [CrossRef]

3. Butkutė, B.; Lemežienė, N.; Kanapeckas, J.; Navickas, K.; Dabkevičius, Z.; Venslauskas, K. Cocksfoot, tall fescue and reed canary grass: Dry matter yield, chemical composition and biomass convertibility to methane. *Biomass Bioenergy* **2014**, *66*, 1–11. [CrossRef]
4. Bourquin, L.D.; Titgemeyer, E.C.; Merchen, N.R.; Fahey, G.C. Forage level and particle size effects on orchardgrass digestion by steers.1. Site and extent of organic matter, nitrogen, and cell wall digestion. *J. Anim. Sci.* **1994**, *72*, 746. [CrossRef] [PubMed]
5. Niderkorn, V.; Martin, C.; Rochette, Y.; Julien, S.; Baumont, R. Associative effects between orchardgrass and red clover silages on voluntary intake and digestion in sheep: Evidence of a synergy on digestible dry matter intake. *J. Anim. Sci.* **2015**, *93*, 4967–4976. [CrossRef] [PubMed]
6. Lee, M.R.; Olmos Colmenero, J.D.J.; Winters, A.L.; Scollan, N.D.; Minchin, F.R. Polyphenol oxidase activity in grass and its effect on plant-mediated lipolysis and proteolysis of Dactylis glomerata (cocksfoot) in a simulated rumen environment. *J. Sci. Food Agric.* **2006**, *86*, 1503–1511. [CrossRef]
7. Phelan, P.; Moloney, A.P.; McGeough, E.J.; Humphreys, J.; Bertilsson, J.; O'Riordan, E.G.; O'Kiely, P. Forage Legumes for Grazing and Conserving in Ruminant Production Systems. *Crit. Rev. Plant Sci.* **2015**, *34*, 281–326. [CrossRef]
8. Peyraud, J.; Astigarraga, L. Review of the effect of nitrogen fertilization on the chemical composition, intake, digestion and nutritive value of fresh herbage: Consequences on animal nutrition and N balance. *Anim. Feed Sci. Technol.* **1998**, *72*, 235–259. [CrossRef]
9. Patra, A. Effects of supplementing low-quality roughages with tree foliages on digestibility, nitrogen utilization and rumen characteristics in sheep: A meta-analysis. *J. Anim. Physiol. Anim. Nutr.* **2010**, *94*, 338–353. [CrossRef]
10. Johansen, M.; Søegaard, K.; Lund, P.; Weisbjerg, M.R. Digestibility and clover proportion determine milk production when silages of different grass and clover species are fed to dairy cows. *J. Dairy Sci.* **2017**, *100*, 8861–8880. [CrossRef]
11. INRA. *INRA Feeding System for Ruminants*; Wageningen Academic Publishers: Wageningen, The Netherlands, 2018; pp. 44–59.
12. Niderkorn, V.; Martin, C.; Bernard, M.; Le Morvan, A.; Rochette, Y.; Baumont, R. Effect of increasing the proportion of chicory in forage-based diets on intake and digestion by sheep. *Animal* **2019**, *13*, 718–726. [CrossRef]
13. Hristov, A.N.; Sandev, S.G. Proteolysis and rumen degradability of protein in alfalfa preserved as silage, wilted silage or hay. *Anim. Feed Sci. Technol.* **1998**, *72*, 175–181. [CrossRef]
14. Plaizier, J.C. Replacing chopped alfalfa hay with alfalfa silage in barley grain and alfalfa-based total mixed rations for lactating dairy cows. *J. Dairy Sci.* **2004**, *87*, 2495–2505. [CrossRef]
15. Chen, L.; Dong, Z.; Li, J.; Shao, T. Ensiling characteristics, in vitro rumen fermentation, microbial communities and aerobic stability of low-dry matter silages produced with sweet sorghum and alfalfa mixtures. *J. Sci. Food Agric.* **2019**, *99*, 2140–2151. [CrossRef]
16. Dal Pizzol, J.; Ribeiro-Filho, H.; Quereuil, A.; Le Morvan, A.; Niderkorn, V. Complementarities between grasses and forage legumes from temperate and subtropical areas on in vitro rumen fermentation characteristics. *Anim. Feed Sci. Technol.* **2017**, *228*, 178–185. [CrossRef]
17. Getachew, G.; Robinson, P.; DePeters, E.; Taylor, S. Relationships between chemical composition, dry matter degradation and in vitro gas production of several ruminant feeds. *Anim. Feed Sci. Technol.* **2004**, *111*, 57–71. [CrossRef]
18. Menke, K.H.; Steingass, H. Estimation of the energetic feed value obtained from chemical analysis and in vitro gas production using rumen fluid. *Anim. Res. Dev.* **1988**, *28*, 7–55.
19. Zhang, D.; Yang, H. In vitro ruminal methanogenesis of a hay-rich substrate in response to different combination supplements of nitrocompounds; pyromellitic diimide and 2-bromoethanesulphonate. *Anim. Feed Sci. Technol.* **2011**, *163*, 20–32. [CrossRef]
20. AOAC. *Official Methods of Analysis of AOAC International*, 17th ed.; Association of Official Analytical Chemists: Gaithersburg, MD, USA, 2005.
21. Van Soest, P.J.; Robertson, J.B.; Lewis, B.A. Methods for dietary fiber, neutral detergent fiber, and nonstarch polysaccharides in relation to animal nutrition. *J. Dairy Sci.* **1991**, *74*, 3583–3597. [CrossRef]
22. Li, Y.; Meng, Q. Effect of different types of fibre supplemented with sunflower oil on ruminal fermentation and production of conjugated linoleic acids in vitro. *Arch. Anim. Nutr.* **2006**, *60*, 402–411. [CrossRef]

23. Verdouw, H.; Vanechteld, C.J.A.; Dekkers, E.M.J. Ammonia determination based on indophenol formation with sodium salicylate. *Water Res.* **1978**, *12*, 399–402. [CrossRef]
24. Makkar, H.; Sharma, O.; Dawra, R.; Negi, S. Simple determination of microbial protein in rumen liquor. *J. Dairy Sci.* **1982**, *65*, 2170–2173. [CrossRef]
25. Cui, J.; Yang, H.; Yu, C.; Bai, S.; Wu, T.; Song, S.; Sun, W.; Shao, X.; Jiang, L. Effect of urea fertilization on biomass yield, chemical composition, in vitro rumen digestibility and fermentation characteristics of straw of highland barley planted in Tibet. *J. Agric. Sci.* **2016**, *154*, 151–164. [CrossRef]
26. Kilic, A. *Silo Feed (Instruction, Education and Application Proposals)*; Bilgehan Press: Izmir, Turkey, 1986.
27. Moore, J.E.; Undersander, D.J. Relative forage quality: An alternative to relative feed value and quality index. In Proceedings of the 13th Annual Florida Ruminant Nutrition Symposium, Gainesville, FL, USA, 10–11 January 2002.
28. National Research Council. *Nutrient Requirements of Dairy Cattle*; National Academies Press: Washington, DC, USA, 2001.
29. Mertens, D. Predicting intake and digestibility using mathematical models of ruminal function. *J. Anim. Sci.* **1987**, *64*, 1548–1558. [CrossRef] [PubMed]
30. Oba, M.; Allen, M. Evaluation of the importance of the digestibility of neutral detergent fiber from forage: Effects on dry matter intake and milk yield of dairy cows. *J. Dairy Sci.* **1999**, *82*, 589–596. [CrossRef]
31. Moore, J.; Kunkle, W. Evaluation of equations for estimating voluntary intake of forages and forage-based diets. *J. Anim. Sci.* **1999**, *1*, 204.
32. France, J.; Dijkstra, J.; Dhanoa, M.; Lopez, S.; Bannink, A. Estimating the extent of degradation of ruminant feeds from a description of their gas production profiles observed in vitro: Derivation of models and other mathematical considerations. *Br. J. Nutr.* **2000**, *83*, 143–150. [CrossRef] [PubMed]
33. García-Martínez, R.; Ranilla, M.; Tejido, M.; Carro, M. Effects of disodium fumarate on in vitro rumen microbial growth, methane production and fermentation of diets differing in their forage: Concentrate ratio. *Br. J. Nutr.* **2005**, *94*, 71–77. [CrossRef]
34. Niderkorn, V.; Baumont, R.; Le Morvan, A.; Macheboeuf, D. Occurrence of associative effects between grasses and legumes in binary mixtures on in vitro rumen fermentation characteristics. *J. Anim. Sci.* **2011**, *89*, 1138–1145. [CrossRef]
35. Wang, J. *Methods in Ruminant Nutrition Research*; Modern Education Press: Beijing, China, 2011. (In Chinese)
36. Krawutschke, M.; Kleen, J.; Weiher, N.; Loges, R.; Taube, F.; Gierus, M. Changes in crude protein fractions of forage legumes during the spring growth and summer regrowth period. *J. Agric. Sci.* **2013**, *151*, 72–90. [CrossRef]
37. Buxton, D.R.; Redfearn, D.D. Plant limitations to fiber digestion and utilization. *J. Nutr.* **1997**, *127*, 814S–818S. [CrossRef]
38. Kondo, M.; Shimizu, K.; Jayanegara, A.; Mishima, T.; Matsui, H.; Karita, S.; Goto, M.; Fujihara, T. Changes in nutrient composition and in vitro ruminal fermentation of total mixed ration silage stored at different temperatures and periods. *J. Sci. Food Agric.* **2016**, *96*, 1175–1180. [CrossRef] [PubMed]
39. Salama, H.S.A.; Zeid, M.M.K. Hay quality evaluation of summer grass and legume forage monocultures and mixtures grown under irrigated conditions. *Aust. J. Crop Sci.* **2016**, *11*, 1543. [CrossRef]
40. Favre, J.R.; Castiblanco, T.M.; Combs, D.K.; Wattiaux, M.A.; Picasso, V.D. Forage nutritive value and predicted fiber digestibility of Kernza intermediate wheatgrass in monoculture and in mixture with red clover during the first production year. *Anim. Feed Sci. Technol.* **2019**, *258*, 114298. [CrossRef]
41. Wang, M.; Wang, L.; Yu, Z. Fermentation dynamics and bacterial diversity of mixed lucerne and sweet corn stalk silage ensiled at six ratios. *Grass Forage Sci.* **2019**, *74*, 264–273. [CrossRef]
42. Ni, K.; Zhao, J.; Zhu, B.; Su, R.; Pan, Y.; Ma, J.; Zhou, G.; Tao, Y.; Liu, X.; Zhong, J. Assessing the fermentation quality and microbial community of the mixed silage of forage soybean with crop corn or sorghum. *Bioresour. Technol.* **2018**, *265*, 563–567. [CrossRef] [PubMed]
43. Jung, H.G.; Allen, M.S. Characteristics of plant-cell walls affecting intake and digestibility of forages by ruminants. *J. Anim. Sci.* **1995**, *73*, 2774–2790. [CrossRef]
44. Hristov, A.N.; Broderick, G.A. Synthesis of microbial protein in ruminally cannulated cows fed alfalfa silage, alfalfa hay, or corn silage. *J. Dairy Sci.* **1996**, *79*, 1627–1637. [CrossRef]
45. Reid, R.L.; Templeton, W.C., Jr.; Ranney, T.S.; Thayne, W.V. Digestibility, intake and mineral utilization of combinations of grasses and legumes by lambs. *J. Anim. Sci.* **1987**, *64*, 1725–1734. [CrossRef]

46. Bowman, J.; Asplund, J. Evaluation of mixed lucerne and caucasian bluestem hay diets fed to sheep. *Anim. Feed Sci. Technol.* **1988**, *20*, 19–31. [CrossRef]
47. Hobson, P.N.; Stewart, C.S. *The Rumen Microbial Ecosystem*, 2nd ed.; Blackie Academic & Professional: London, UK, 1997.
48. Millen, D.D.; Arrigoni, M.D.B.; Pacheco, R.D.L. *Rumenology*, 1st ed.; Springer International Publishing: Cham, Switzerland, 2016.
49. Liu, J.; Zhang, M.; Xue, C.; Zhu, W.; Mao, S. Characterization and comparison of the temporal dynamics of ruminal bacterial microbiota colonizing rice straw and alfalfa hay within ruminants. *J. Dairy Sci.* **2016**, *99*, 9668–9681. [CrossRef]
50. Kammes, K.L.; Allen, M.S. Nutrient demand interacts with forage family to affect digestion responses in dairy cows. *J. Dairy Sci.* **2012**, *95*, 3269–3287. [CrossRef] [PubMed]
51. Bhatti, S.A.; Bowman, J.G.P.; Firkins, J.L.; Grove, A.V.; Hunt, C.W. Effect of intake level and alfalfa substitution for grass hay on ruminal kinetics of fiber digestion and particle passage in beef cattle. *J. Anim. Sci.* **2008**, *86*, 134–145. [CrossRef] [PubMed]
52. Van Soest, P.J. *Nutritional Ecology of the Ruminant*, 2nd ed.; Cornell University press: Ithaca, NY, USA, 1994.
53. Chen, K.J.; Jan, D.F.; Chiou, P.W.S.; Yang, D.W. Effects of dietary heat extruded soybean meal and protected fat supplement on the production, blood and ruminal characteristics of Holstein cows. *Asian-Australas. J. Anim. Sci.* **2002**, *15*, 821–827. [CrossRef]
54. Brask, M.; Lund, P.; Hellwing, A.L.F.; Poulsen, M.; Weisbjerg, M.R. Enteric methane production, digestibility and rumen fermentation in dairy cows fed different forages with and without rapeseed fat supplementation. *Anim. Feed Sci. Technol.* **2013**, *184*, 67–79. [CrossRef]
55. Oba, M.; Mewis, J.L.; Zhining, Z. Effects of ruminal doses of sucrose, lactose, and corn starch on ruminal fermentation and expression of genes in ruminal epithelial cells. *J. Dairy Sci.* **2015**, *98*, 586–594. [CrossRef]
56. Menke, K.; Raab, L.; Salewski, A.; Steingass, H.; Fritz, D.; Schneider, W. The estimation of the digestibility and metabolizable energy content of ruminant feedingstuffs from the gas production when they are incubated with rumen liquor in vitro. *J. Agric. Sci.* **1979**, *93*, 217–222. [CrossRef]
57. Pen, B.; Sar, C.; Mwenya, B.; Kuwaki, K.; Morikawa, R.; Takahashi, J. Effects of Yucca schidigera and Quillaja saponaria extracts on in vitro ruminal fermentation and methane emission. *Anim. Feed Sci. Technol.* **2006**, *129*, 175–186. [CrossRef]
58. Lee, H.; Lee, S.; Kim, J.; Oh, Y.; Kim, B.; Kim, C.; Kim, K. Methane production potential of feed ingredients as measured by in vitro gas test. *Asian Australas. J. Anim. Sci.* **2003**, *16*, 1143–1150. [CrossRef]
59. Cronje, P.B. *Ruminant Physiology: Digestion, Metabolism, Growth and Reproduction*; CABI Publishing: Wallingford, UK, 2000.
60. Jayanegara, A.; Wina, E.; Soliva, C.; Marquardt, S.; Kreuzer, M.; Leiber, F. Dependence of forage quality and methanogenic potential of tropical plants on their phenolic fractions as determined by principal component analysis. *Anim. Feed Sci. Technol.* **2011**, *163*, 231–243. [CrossRef]
61. Zhang, S.J.; Chaudhry, A.S.; Osman, A.; Shi, C.Q.; Edwards, G.R.; Dewhurst, R.J.; Cheng, L. Associative effects of ensiling mixtures of sweet sorghum and alfalfa on nutritive value, fermentation and methane characteristics. *Anim. Feed Sci. Technol.* **2015**, *206*, 29–38. [CrossRef]
62. Mehrez, A.; Ørskov, E.; McDonald, I. Rates of rumen fermentation in relation to ammonia concentration. *Br. J. Nutr.* **1977**, *38*, 437–443. [CrossRef] [PubMed]
63. Satter, L.; Slyter, L. Effect of ammonia concentration on rumen microbial protein production in vitro. *Br. J Nutr.* **1974**, *32*, 199–208. [CrossRef] [PubMed]
64. Wang, D.; Fang, J.; Xing, F.; Yang, L. Alfalfa as a supplement of dried cornstalk diets: Associative effects on intake, digestibility, nitrogen metabolisation, rumen environment and hematological parameters in sheep. *Livest. Sci.* **2008**, *113*, 87–97. [CrossRef]
65. Eun, J.S.; Beauchemin, K.; Schulze, H. Use of exogenous fibrolytic enzymes to enhance in vitro fermentation of alfalfa hay and corn silage. *J. Dairy Sci.* **2007**, *90*, 1440–1451. [CrossRef]

9

Rumen Inoculum Collected from Cows at Slaughter or from a Continuous Fermenter and Preserved in Warm, Refrigerated, Chilled or Freeze-Dried Environments for In Vitro Tests

Mauro Spanghero [1,*], Maria Chiaravalli [2], Stefania Colombini [2], Carla Fabro [1], Federico Froldi [3], Federico Mason [4], Maurizio Moschini [3], Chiara Sarnataro [1], Stefano Schiavon [5] and Franco Tagliapietra [5]

1. Dipartimento di Scienze Agroalimentari, Ambientali e Animali, University of Udine, 33100 Udine, Italy; carla.fabro@uniud.it (C.F.); sarnataro.chiara@spes.uniud.it (C.S.)
2. Dipartimento di Scienze Agrarie e Ambientali, University of Milan, 20122 Milano, Italy; maria.chiaravalli@unimi.it (M.C.); stefania.colombini@unimi.it (S.C.)
3. Dipartimento di Scienze Animali, della Nutrizione e degli Alimenti, Università Cattolica del Sacro Cuore of Piacenza, 29122 Piacenza, Italy; federico.froldi@unicatt.it (F.F.); maurizio.moschini@unicatt.it (M.M.)
4. Department of Biodiversity Protection, Institute of Animal Reproduction and Food Research of Polish Academy of Sciences (IARFR PAS), 10-748 Olsztyn, Poland; f.mason@pan.olsztyn.pl
5. Dipartimento di Agronomia, Animali, Alimenti, Risorse naturali e Ambiente, University of Padova, 35122 Padova, Italy; stefano.schiavon@unipd.it (S.S.); franco.tagliapietra@unipd.it (F.T.)

* Correspondence: mauro.spanghero@uniud.it;

Simple Summary: The utilization of animal donors of rumen fluid for laboratory experiments can raise ethical concerns due to invasive methods of collection (rumen cannulated or intubated animals). Societies are strongly oriented to support cruelty free experiments and alternatives to the collection of rumen fluids from live animals are urgently requested from the scientific community. Thus, in order to attenuate the dependence of laboratories on animal donors, this study compared the rumen inoculum collected at slaughter with the fermentation liquid from a rumen continuous fermenter and both rumen inoculum were used fresh or preserved (by refrigeration, chilling and freeze-drying). The results support the possibility of using continuous fermenters to generate inoculum for in vitro purposes, and short-term refrigeration is confirmed to be a valuable storage system to facilitate transfer inoculum from the collection sites. These findings should attenuate the need for laboratories' frequent collections from animals while continuing research in ruminant nutrition.

Abstract: The utilization of animal donors of rumen fluid for laboratory experiments can raise ethical concerns, and alternatives to the collection of rumen fluids from live animals are urgently requested. The aim of this study was to compare the fresh rumen fluid (collected at slaughter, W) with that obtained from a continuous fermenter (RCF) and three methods of rumen fluid preservation (refrigeration, R, chilling, C, and freeze-drying, FD). The fermentability of different inoculum was evaluated by three in vitro tests (neutral detergent fiber (NDF) and crude protein (CP) degradability and gas production, NDFd, RDP and GP, respectively) using six feeds as substrates. Despite the two types of inoculum differed in terms of metabolites and microbiota concentration, the differences in vitro fermentability between the two liquids were less pronounced than expected (−15 and 20% for NDFd and GP when the liquid of fermenter was used and no differences for RDP). Within each in vitro test, the data obtained from rumen and from fermenter liquids were highly correlated for the six feeds, as well as between W and R (r: 0.837–0.985; $p < 0.01$). The low fermentative capacity was found for C and, particularly, FD for liquids. RCF could be used to generate inoculum for in vitro purposes and short-term refrigeration is a valuable practice to manage inoculum.

Keywords: rumen liquid; in vitro fermentation; rumen degradability; gas production

1. Introduction

Rumen fluid sampled from live animals is used in laboratory experiments (e.g., in vitro rumen fermentation) to evaluate the nutritive value and gaseous emissions of ruminant feeds [1,2] or to inoculate continuous fermenters for studies of rumen fermentation [3].

In general, the utilization of animals as donors of rumen fluid can raise some ethical concerns, because the collection of rumen fluid is an invasive practice, which requires donor animals that are surgically modified (e.g., rumen cannulated), or immobilised and intubated with esophageal probes. An alternative is to collect rumen fluid from animals slaughtered for production purposes in commercial slaughterhouses, but it is difficult to monitor feeding before slaughter. Overall, societies are strongly oriented to support cruelty free experiments and alternatives to collection of rumen fluids from live animals are urgently requested of the scientific community to continue research activity in ruminant nutrition.

Rumen continuous fermenters (RCFs) are laboratory apparatus developed to simulate the rumen conditions for studies of rumen metabolism. They generate a fermenting fluid by starting from an initial rumen inoculum and by a continuous influx of artificial saliva, an output of fermentation products and a constant supply of nutrients (substrates). However, they could also be modified and adapted to be used as artificial generators of rumen fermentation fluid, which could be standardised with respect to several conditions (type and amount of fermentable substrate, pH, dilution, etc.). There are studies which have compared fermentation liquids from different RCFs or between fluids collected from rumens and fermenters [3–7]. The liquid from fermenters is less concentrated in terms of volatile fatty acids (VFA) and protozoa, and the cellulosolitic bacterial strains seem reduced in some types of RCF (e.g., Rusitec, [8]) while in other fermenters, the bacteria microbiota was comparable to that measured directly on the rumen inoculum collected in vivo [6,7]. However, no experiments have simultaneously compared rumen and RCFs fluids as inoculum in terms of results of different in vitro tests.

Overall, independently from the mode of inoculum provision, the possibility of preserving rumen fluid and to create stocks would allow the concentration of the collection in specialized centers to facilitate the transfer of the inoculum and reduce the need of laboratories to frequently collect liquids from live animals. The preservation techniques of rumen fluid have been investigated in several studies, which have mainly considered the usage of low temperatures (e.g., refrigeration and chilling at −20, at −80 °C, or in liquid N), the addition of cryoprotectants, and also freeze-drying [9–13]. In general, there are encouraging findings, but also a scarce homogeneity among trials in terms of measurement of the maintenance of the fermentative capacity of liquid after preservation.

The present research has the general aim to attenuate the needs of direct collection of rumen fluid from animals by: (i) artificially generating rumen fermenting fluids; (ii) evaluating preservation methods for stock rumen fluid. These technologies should reduce the need for laboratories to frequently collect from animals, with the potential advantage of reducing inoculum variability. The specific aim is to compare the fresh rumen fluid (collected at slaughter) with that obtained from a stratified single-flow RCF system [14] and three methods of rumen fermenting fluid preservation (refrigeration, chilling and freeze-drying).

Unlike other research, the novelty of this study is to compare the different inoculum in terms of the results obtained by three widely utilised in vitro methods (degradability of neutral detergent fiber, NDFd, degradability of protein (RDP), and gas production, (GP) respectively). The chosen in vitro tests allow a wide evaluation of the fermentation potential of inoculum, because they quantify the fermentation of main dietary components of ruminant rations, such as fibers (NDFd), protein (RDP) and (mainly) non-fibrous carbohydrates (GP).

2. Materials and Methods

2.1. Trial Organization

The experiment was organised in two subsequent identical fermentation trials (runs) and was carried out by four Italian research groups from the University of Milano, Padova, Piacenza, and Udine (Labs 1, 2, 3, and 4, respectively).

For each trial, 4 dry multiparous Holstein Friesian cows were slaughtered for production purposes in a commercial slaughterhouse, after being fed for 3 weeks a basal diet (32.3% meadow hay, 34.0% corn silage, 27.3 % compound feed and 6.4 % soybean meal). The cows of both runs were in good health, housed in the same barn located 50 km from the slaughterhouse, fed the last feeding in the morning 2 h before being moved to the slaughterhouse, and had free access to fresh water until slaughter. After slaughter, the rumen liquors, collected in equal amounts from the 4 cows by Lab 4, were coarsely filtered, the liquids bulked together during continuous flushing with CO_2, and then divided into 2 main amounts (see Figure 1). The first amount represented the rumen inoculum for the in vitro tests (NDFd, RDP and GP test), while the second amount was used to inoculate the RCF system. Both inoculum were used as: warm at 39 °C (W), refrigerated at 4 °C (R), chilled at −80 °C (C), and freeze-dried (FD). After the collection at slaughter, W (kept inside pre-warmed thermic bottles flushed with CO_2) and R (kept inside bottles flushed with CO_2, immersed in ice water within a portable fridge to quickly lower the temperature to 4 °C) were divided in aliquots of 250–300 mL each and immediately delivered to Lab 1, 2, and 3 to start the in vitro tests within 6 h from slaughter. The amounts to be preserved as C, FD or to be used in the RCF fermenter were immediately brought to Lab 4 (maintained warm at 39 °C). Here, the C or FD, amounts were divided in aliquots of 250–300 mL each. The C aliquots were chilled at −80 °C, while the FD aliquots were processed according to Luchini et al. [15]. Briefly, after centrifugation at 5000 g for 30 min at 4 °C, the supernatant was discarded, and the residue obtained was chilled at −80 °C, and subsequently freeze-dried. The C and FD aliquots were delivered to Lab 1, 2, 3 after 30 d of conservation. Before being used as inoculum, the C liquids were thawed in a bain-marie at 39 °C within 2 h and kept at the same temperature for another 2 h, while the FD liquids were reconstituted to the original volume with the artificial saliva used for the in vitro tests and kept at 39 °C for 2 h.

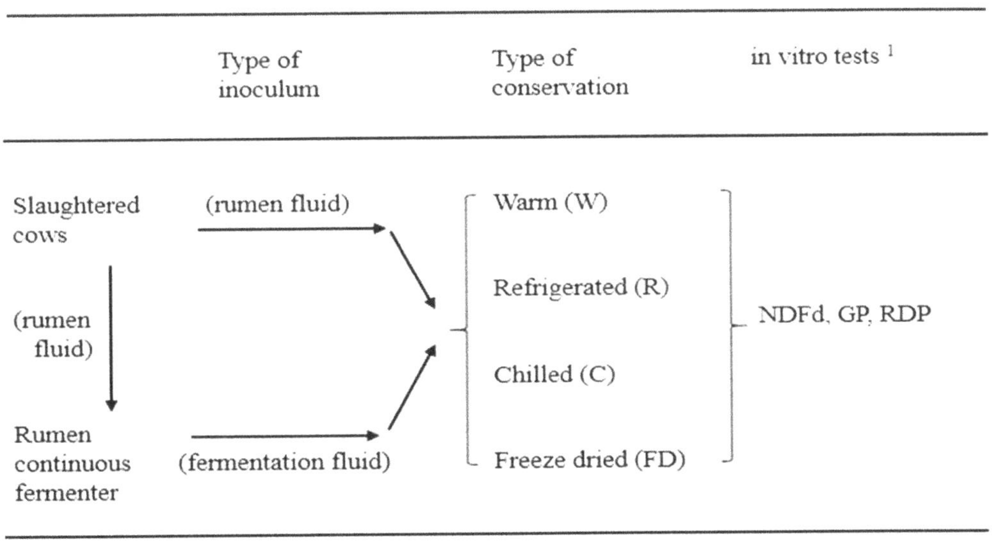

[1] NDFd : NDF degradability; GP : gas production; RDP : rumen degradable protein.

Figure 1. Schematic representation of the trial organization.

The remaining amount of rumen liquid was immediately used to inoculate the RCF fermenter in Lab 4. On the 9th day of incubation in the RCF, the fermentation fluid was collected, and divided in aliquots for each liquid type (W, R, C and FD). The W and R aliquots, prepared as previously described for the rumen inoculum, were immediately transferred to Lab 1, 2 and 3 to perform the in vitro tests within 6 h from the collection of the liquid from the RCF system. The C and FD aliquots were delivered to Labs 1, 2, and 3 after 30 d of conservation.

As incubations of liquids C and FD were delayed by 30 d with respect to W and R and the incubations of liquid from the fermenter were delayed by 9 d from the incubation of rumen liquids. There were in total 4 incubation sessions within each fermentation run.

2.2. Preparation of the Substrates for the In Vitro Tests

All the in vitro tests were performed on the following six ingredients: meadow hay, corn silage, wheat bran, distillers, soybean and barley meal. The samples of corn silage were dried (48 h at 60 °C) and all feed samples were milled through a 1 mm sieve and then analysed for chemical composition.

2.3. The Rumen Continuous Fermenter (RCF) System and the In Vitro Tests

The RCF system, used in Lab 4, was described in detail by Mason et al. [14]. In brief, the system consists of 8 × 2 L glass bottles, immersed in a water bath at 39 °C. A peristaltic pump supplies the buffer solution [16] from a reservoir to the fermenters (dilution rate of 5 %/h) and the outflow, located at the bottom of the bottles, allows stratification of the feeding material. The bottles were inoculated with 600 mL of strained rumen fluid and 800 mL of artificial saliva, and each bottle received a total of 15 g/d of dry matter (DM), in two equal doses, at 09:00–17:00, of the same diet used to feed the donor cows before slaughtering. Each fermentation lasted 9 d and on the last day, the fermentation fluids of the 8 bottles were collected, pooled, and processed to prepare the different inoculum, as previously described.

In all 3 in vitro tests, the inoculum was strained through 4 layers of cheesecloth into pre-warmed (39 °C) flasks with CO_2 and mixed with the buffer solutions ([17] for NDFd and GP and Van Soest buffer [18] for RDP, in a 1:2 and 1:4 ratio, respectively).

The NDFd was tested by Lab 1. Each feed sample was weighed (0.250 ± 0.005 g) in duplicate in Ankom F57 bags (Ankom Technology, Macedon, NY, USA). The bags were incubated in a pre-warmed 100 mL Erlenmeyer flask closed by a rubber stopper with a Bunsen valve for gas release and maintained at 39 °C in a water bath with shaking for 48 h. Each flask was inoculated with 90 mL of the solution under anaerobic conditions, flushing the flask with CO_2. At the end of the incubations, the bags were rinsed with cold water until the water ran clear and then placed for 3 h in a 105 °C forced-air oven to dry. Subsequently, the NDF concentration was determined for each bag using the fiber analyser (Ankom Technology, Macedon, NY, USA).

The RDP was tested by Lab 3 according to the rumen step of the Ross method [19]. Briefly, each sample was weighed (1.000 ± 0.020 g) into a 120 mL glass bottle equipped with a cap and placed into a 39 °C water bath 1 h before the in vitro fermentation. The neutral detergent residue from corn silage was used for microbial contamination correction of the post fermentation feed residues. Each bottle was inoculated with 100 mL solution under CO_2 flushing, closed with the cap, and incubated for 16 h at 39 °C in a shaking water bath. After the incubation, the bottle content was vacuum filtered (110 mm diameter Whatman Filter Papers 54) and the residue was analyzed for Kjeldahl nitrogen.

The GP was measured after 24 h incubation by Lab 2 by using a commercial GP apparatus (Ankom Technology, Macedon, NY, USA; [20]). The system consists of 50 bottles hermetically sealed, equipped with a wireless pressure sensor connected to a computer. Each bottle (317 mL) was filled with 0.500 ± 0.010 g of feed and 75 mL of fermenting solution, obtaining a headspace volume of 242 mL. The bottles were placed in a ventilated incubator at 39 ± 0.4 °C and automatically vented at a fixed pressure (6.8 kPa), to prevent overpressure.

All the tests were performed for each feed substrate and for each inoculum type in duplicate (8 inoculum types × 6 feeds in duplicate) and were replicated in a second fermentation run. The between-run determinations were considered as experimental repetitions. To account for the incubation session effect, a standard rumen fluid was included in each incubation session by each laboratory. Before conducting the whole experiment, Lab 4 prepared a rumen fluid to be used as the control by other labs. The liquid was collected at slaughter from 4 dairy cows (culled in good health) and delivered to Lab 4 within 30 min in airtight glass-bottles refluxed with carbon dioxide and maintained at 39 °C. The whole amount was divided in small aliquots (200 mL), frozen at −20 °C and distributed to Lab 1, 2 and 3. The frozen-thawed inoculum (in a bain-marie at 39 °C within 2 h and kept at the same temperature for another 2 h) were added by each Lab into each incubation run (two incubation bottles added with corn silage as substrate) to detect anomalies between runs.

2.4. *Analysis*

2.4.1. Inoculum Sample Preparation

The samples of the two inoculum for each preservation treatment were collected before performing the in vitro tests for the following analyses: pH, VFA, NH_3, and microbial population. The samples for the VFA analysis were acidified with H_2SO_4 0.1 N and stored at −20 °C while the samples for NH_3 were directly stored at −20 °C. The samples for bacterial DNA analysis were chilled in liquid nitrogen (N) and stored at −80 °C.

2.4.2. Chemical Analysis

The DM content of the feeds was determined by heating at 105 °C for 3 h (method 930.15; [21]), and the ash content was subsequently determined after incineration at 550 °C for 2 h (method 942.05; [21]). The neutral and acid detergent fiber (NDF and ADF, respectively) analysis was performed with a fiber analyzer (Ankom Technology, Macedon, NY, USA) following the procedure of Van Soest et al. [18] without correction for residual ash. The ether extract (EE) and the N contents were respectively determined by the solvent extraction and by the Kjeldahl methods (methods 954.02 and 976.05, [21]).

The pH and the NH_3 content of the inoculum were measured with a glass electrode pH meter (GLP 22, Crison Instruments, S.A. Barcelona, Spain) and an ammonia electrode (Ammonia Gas Sensing Combination Electrode, ©Hach Company, Loveland, CO, USA, 2001).

For the VFA analysis, the aliquots of the inoculum were centrifuged at 20,000 g for 30 min at 20 °C and the supernatant was then filtered using polypore 0.45 μm filters (Alltech Italia, Milan, Italy). The filtrate was injected into a high-performance liquid chromatography instrument (Perkin-Elmer, Norwalk, CN, USA), set to 220 nm according to the method described by Martillotti and Puppo [22].

2.4.3. Microbial Analysis

Genomic DNA was extracted from about 700 μL of inoculum samples using the PowerSoil DNA extraction kit (MoBio Laboratories Inc., Carlsbad, CA, USA) with some modifications as described by Kong et al. [23] (2010). DNA concentration, eluted in 50 μL, was determined by NanoDrop One (Thermo Fisher Scientific Inc., Wilmington, DE, USA).

The quantitative polymerase chain reaction analysis was performed by CFX96 Real Time System (Bio-Rad Technologies Inc, Hercules, CA, USA) using iQ SYBR Green Supermix (Bio-Rad Technologies Inc, Hercules, CA, USA) mixed with 0.3 μL of each forward and reverse primer (0.3 μM), 8.4 μL of sterile water and 1 μL of gDNA to obtain a reaction volume of 20 μL. The amplification program included the denaturation step at 98 °C for 3 min followed by 40 cycles at 98 °C for 15 s, annealing at 60 °C for 30 s and elongation at 72 °C for 30 s.

To determine the specificity of the amplification of each primer, the melting curve was performed. The amplification efficiency was calculated using the formula: $E = 10^{(-1/\text{slope})} - 1$. The relative abundance of the target bacterial groups or species was expressed in a proportion of the total bacteria 16S rRNA gene and was calculated using the following formula: $2^{-\Delta CT}$.

2.4.4. Statistical Analyses

The data from the in vitro tests (NDFd, RDP and GP, within each test and within each feed substrate) and the chemical and microbial composition of fermentation fluids differing for origin and preservation (pH values, VFA and ammonia contents and bacteria abundance) were statistically analysed as a factorial 2 × 4 completely randomised block (2 fermentation runs as blocks) design:

$$Y_{ijk} = \mu + \alpha_i + \beta_j + \gamma_k + (\beta\gamma)_{jk} + \varepsilon_{ijk},$$

where: y_{ijk} is the experimental data; μ is the overall mean; α_i is the random effect (block) of the fermentation run (i = 1, 2); β_j is the fixed effect of the type of inoculum (j = 1, 2); γ_k is the fixed effect of the inoculum preservation method (k = 1,4); and ε_{ijk} is the random error.

Within each in vitro test data, the Pearson correlation coefficient (r) was calculated for the 6 feeds tested in 2 runs with fermentation liquids, differing for origin (liquids from fermenter versus liquids from rumen, 12 data points) or for the type of conservation (W versus R, W versus C and W versus FD liquids, 12 data points for each correlation).

For all statistical analyses, the probability significance levels (p) were 0.05 and 0.01 ($p < 0.05$ and $p \leq 0.01$, respectively).

3. Results

The chemical composition of feeds used as substrates for the in vitro tests and the characteristics of inocula are reported in Tables 1 and 2, respectively.

Table 1. Chemical composition of feeds used as substrates in the in vitro tests.

Feeds	DM	Ash	CP	EE	NDF	ADF
	%	% DM	% DM	% DM	% DM	% DM
Corn silage [1]	91.88	4.16	7.67	3.10	36.79	20.74
Wheat bran	89.71	5.98	16.74	3.47	47.80	13.40
Meadow hay	95.54	11.30	7.34	1.45	58.38	34.82
Distillers	89.15	5.87	34.32	7.79	42.87	11.86
Soya meal, extr.	88.88	6.95	46.09	1.19	21.11	8.88
Barley	89.79	3.08	10.65	1.55	31.89	8.30

[1] pre-dried samples at 60 °C.

Among the parameters reported in Table 2, the concentration of ammonia and the relative abundance of *Fibrobacter succinogenes* in fermentation fluids showed a significant interaction ($p < 0.05$) between the origin of the liquid and the preservation method. However, for both parameters the impact of the interaction in terms of contribution to the mean square of the model was much lower than that of the main effects as well the level of significance ($p < 0.05$ vs. $p \leq 0.01$). Therefore, this study decided to show and to discuss the main effects of the model.

The liquid from the fermenter had a lower total VFA content, acetic percentage on total VFA, acetic:propionic ratio and ammonia concentration compared with the liquid from rumen (34.9 versus 122.1 mMol, 57.8 versus 63.3%, 2.9 versus 5.7 and 15.4 versus 26.1 mg/dL, respectively, $p \leq 0.01$). On the contrary, the valeric and isovaleric acid percentages on total VFA were higher in fluid from the fermenter than the rumen liquid (1.8 versus 1.1% and 2.7 versus 1.4%, respectively, $p \leq 0.01$). The FD preservation differed significantly ($p \leq 0.01$) from the others for the lowest proportion of acetic, propionic and isovaleric acids and for the very high proportion of isobutyric acid. The FD also had a very low content of ammonia in comparison with W and R (6.4 versus 21.8–22.3 mg/dL, $p \leq 0.01$), while the C liquid showed the highest concentration (32.6 mg/dL, $p \leq 0.01$).

Table 2. Volatile fatty acid (VFA) content, pH, ammonia (NH_3) content and bacteria abundance of fermentation liquids from inoculum differing for origin and for conservation.

Items	Type of Inoculum (TI)		Type of Conservation [1] (TC)				TI [2]	TC [2]	TI × TC [2]	RMSE [2]
	From Rumen	From Fermenter	Warm (W)	Refrigerated (R)	Chilled (C)	Freeze—Dried (FD)				
Samples, *n*	8	8	4	4	4	4				
Total VFA (mMol)	122.1	34.9	76.4	74.1	100.3	63.4	**	ns	ns	25.98
VFA composition (% of total VFA)										
Acetic (A)	63.3	57.8	67.3 ^{A}	66.0 ^{A}	65.7 ^{A}	43.4 ^{B}	*	**	ns	3.72
Propionic (P)	12.5	20.2	18.2 ^{A}	18.5 ^{A}	16.9 ^{A}	11.9 ^{B}	**	**	ns	1.84
Isobutyric	10.9	9.6	0.4 ^{B}	0.4 ^{B}	4.2 ^{B}	36.0 ^{A}	ns	**	ns	5.77
Butyric	10.5	8.0	10.1	10.8	9.5	6.5	ns	ns	ns	2.35
Isovaleric	1.4	2.7	2.1 ^{A}	2.4 ^{A}	2.3 ^{A}	1.2 ^{B}	**	**	ns	0.34
Valeric	1.1	1.8	1.6	1.6	1.5	1.2	*	ns	ns	0.40
A:P	5.7	2.9	3.9	3.8	4.0	5.4	**	ns	ns	1.27
pH	6.4	6.7	6.5	6.5	6.5	6.8	ns	ns	ns	0.51
NH_3 (mg/dL)	26.1	15.4	21.8 ^{B}	22.3 ^{B}	32.6 ^{A}	6.4 ^{C}	**	**	*	3.28
Relative abundance (% of total bacteria)										
Genus *Prevotella*	45.1	19.7	25.8	31.7	34.6	37.5	**	ns	ns	8.90
Fibrobacter succinogenes	0.39	2.04	2.36 ^{A}	2.10 ^{A}	0.39 ^{B}	0.01 ^{B}	**	**	*	0.63
Ruminococcus albus group	0.017	0.003	0.013	0.010	0.011	0.006	**	ns	ns	0.0056

[1] Means of the type of conservation with different superscripts (A, B, C) are statistically different. [2],*: $p < 0.05$; **: $p \leq 0.01$; ns: not significant; RMSE: root mean square error.

The relative abundances of the genus *Prevotella* and *Ruminococcus albus* group were lower ($p < 0.01$) in the liquid from the fermenter compared to the liquid from rumen, while the preservation method did not show effects. On the contrary, the *Fibrobacter succinogenes* was higher ($p < 0.01$) in the liquid from the fermenter and its relative abundance in W and R liquids was higher ($p < 0.01$) than the others.

The results of the in vitro tests on the six feeds and the correlation coefficients calculated on the six feeds by using the sets of two inocula differing for origin (liquid from rumen versus liquid from fermenter) or preservation (W versus R, W versus C and W versus FD liquids) are reported in Tables 3 and 4, respectively. The in vitro tests used different inoculum and were performed in eight subsequent incubation sessions by each Lab. To correct any possible effect of the incubations session, a standard rumen fluid was included in each fermentation, which was used on the same substrate (corn silage) as the control. The data from the standard allowed the exclusion of any appreciable effect of the incubation session and the in vitro results from the different incubation sessions did not require corrections.

The NDFd data of the corn silage samples for the FD mode of preservation were removed from the analysis, having an abnormal fermentation due to technical problems.

The utilization of the liquid from the fermenter determined a significant NDFd reduction for three feeds (corn silage, wheat bran, and distillers) and the correlation analysis indicated a close relationship for NDFd between the liquid from rumen and diluted liquids from the fermenter ($r = 0.960$, $p \leq 0.01$). Considering the preservation of the inoculum, the W and R liquids were not different for all the tested feeds and the correlation indicated a close correspondence ($r = 0.985$, $p \leq 0.01$). The C and FD liquids depressed ($p \leq 0.01$) the NDFd for two and three feeds, respectively. However, the correlation W versus C was high ($r = 0.905$, $p \leq 0.01$), while the W versus FD was non-significant.

The RDP results indicated a good correspondence between fluid from the fermenter and liquid from the rumen liquids and the correlation between the liquid from rumen and the liquid from the fermenter data was high ($r = 0.837$, $p \leq 0.01$). The R and C liquids gave similar RDP results to W liquid for all feeds and the regression W versus R showed a good correspondence of values ($r = 0.892$), while the W versus C was not significant. The RDP obtained with the FD liquid were quite variable, with the values for two feeds being higher than those obtained with W, while for the soya, the opposite was true, and the correlation W versus FD was not significant.

Unlike NDFd and RDP, the gas production showed a significant interaction—inoculum × preservation ($p \leq 0.01$)—for some feeds (corn silage, wheat bran, and distillers). However, the impact of the interaction in terms of the contribution to the mean square of the model was much lower than that of the main effects as well the level of significance ($p < 0.05$ versus $p \leq 0.01$). Therefore, it was decided to show the results and to discuss the main effects of the model. For all the feeds, the liquid from rumen gave higher gas production than the liquid from the fermenter ($p \leq 0.01$ and, only for soya, $p < 0.05$).

Moreover, the preservation method had a statistical effect on the gas production ($p \leq 0.01$ and, only for soya and barley, $p < 0.05$). In general, the gas production of C was not statistically different from W and R liquids, except for wheat bran and barley where C gas production was not statistically different from W. The FD liquid generated the lowest yields of gas and all the ingredients differed significantly from those of the W liquid. All correlations of preserved liquids with W were statistically significant ($r = 0.797$–0.921, $p \leq 0.01$).

Table 3. In vitro degradability of NDF (NDFd), CP (RDP), and gas production (at 24 h of fermentation, GP) of six substrates using rumen inoculum differing for origin and for conservation method.

Items	Type of Inoculum (TI)		Type of Conservation [1] (TC)				TI^2	TC^2	$TI \times TC^2$	$RMSE^2$
	From Rumen	From Fermenter	Warm (W)	Refrigerated (R)	Chilled (C)	Freeze-Dried (FD)				
NDFd (%)										
Corn silage	37.5	30.1	39.0^{A}	40.0^{A}	22.5^{B}	-	*	**	ns	6.34
Wheat bran	45.9	38.2	46.7^{A}	46.2^{A}	39.3^{B}	36.0^{B}	**	**	ns	3.40
Meadow hay	44.5	40.7	51.8^{A}	$47.9^{A,B}$	$39.0^{B,C}$	31.8^{C}	ns	**	ns	5.36
Distillers	52.5	41.3	49.4	51.7	47.9	38.5	*	ns	ns	6.85
Soya meal, extr.	81.4	78.5	92.2^{A}	93.7^{A}	85.5^{A}	48.5^{B}	ns	**	ns	6.26
Barley	59.4	54.3	58.0	58.8	55.7	55.1	ns	ns	ns	5.39
RDP (%)										
Corn silage	66.8	66.2	57.0^{b}	49.8^{b}	66.0^{b}	93.3^{a}	ns	*	ns	15.21
Wheat bran	59.9	61.9	58.1	58.9	60.4	66.2	ns	ns	ns	10.20
Meadow hay	39.0	45.6	32.5^{b}	27.8^{b}	44.3^{b}	64.7^{a}	ns	*	ns	11.35
Distillers	33.8	34.5	34.8	37.5	33.8	30.6	ns	ns	ns	5.00
Soya meal, extr.	45.1	53.4	52.8^{A}	56.8^{A}	$47.8^{A,B}$	39.7^{B}	*	*	ns	5.10
Barley	59.0	57.4	51.8	49.7	60.4	71.0	ns	ns	ns	17.38
GP (mL)										
Corn silage	223	193	242^{A}	234^{A}	184^{B}	172^{B}	**	**	**	26.13
Wheat bran	183	146	$176^{A,B}$	188^{A}	$157^{B,C}$	136^{C}	**	**	**	18.40
Meadow hay	162	119	177^{A}	198^{A}	124^{B}	64^{C}	**	**	ns	24.81
Distillers	193	153	197^{A}	207^{A}	156^{B}	131^{B}	**	**	**	50.44
Soya meal, extr.	178	134	176^{B}	213^{A}	136^{C}	100^{D}	*	*	ns	23.31
Barley	266	226	$263^{A,B}$	264^{A}	$231^{A,B}$	226^{B}	**	*	ns	27.85

[1] Means of the type of conservation with different superscripts (A, B, C and a, b) are statistically different. [2] *: $p < 0.05$; **: $p \leq 0.01$; ns: not significant; RMSE: root mean square error.

Table 4. Pearson correlation coefficients (r) of in vitro data obtained for the six feeds used as substrates in two fermentation run (12 observations) by using inocula differing for origin and for conservation.

	Type of Inoculum (TI)		Type of Conservation (TC)					
	Rumen vs. Fermenter		Warm vs. Refrigerated		Warm vs. Chilled		Warm vs. Freeze-Dried	
	r	p [1]	r	p [1]	r	p [1]	r	p [1]
NDFd [2]	0.960	**	0.985	**	0.905	**	0.554	ns
RDP [3]	0.837	**	0.892	**	0.345	ns	0.431	ns
GP [4]	0.939	**	0.921	**	0.797	**	0.850	**

[1] statistical significance of r: **: $p \leq 0.01$; ns: not significant. [2] NDFd: NDF degradability; [3] RDP: CP degradability; [4] GP: gas production (at 24 h of fermentation).

4. Discussion

4.1. *Type of Inoculum*

The present work used a rumen fluid collected from cows immediately after slaughter in controlled conditions (animals fed a known diet, not slaughtered in emergency, in good health status, transported from farms located near the slaughterhouse, rumen fluid sampled within 20 min of slaughter) as a reference in vivo rumen liquor. To our knowledge, there are no specific comparisons between rumen fluid collected via cannula, through the esophageal tube and at slaughtering. The authors reason that rumen fluid collected at slaughter in controlled conditions has a limited difference from that sampled by other methods. This is based on our previous work [24] where a close relationship between NDF degradability measured in vitro (average of rumen collected by different methods) and in situ was found. Our recent data [25] showed comparable values of in vitro NDF degradability using rumen fluids from slaughtered cows and from cows with rumen cannula fed similar diets. Moreover, the collection at slaughter is an accepted method of sampling rumen fluid for microbiota studies by the Rumen Microbial Genomics Network [26] and it is mentioned as an alternative to sampling via cannula [2].

A first aim of this study was to compare the fresh rumen inoculum with that obtained from a continuous culture system (RCF, see Figure 1). The data from the literature [4,5] showed that fermentation liquids from continuous cultures are less concentrated than rumen fluids collected directly from the rumen in terms of fermentation metabolites. This is confirmed also by data of this experiment, where a particularly low VFA concentration was found in the continuous fermenter liquids, being approximately 30% of the VFA concentration of rumen fluid. The concentration of VFAs measured in the rumen liquids were comparable to those found in the literature, while those of the RCF liquids were slightly lower than those obtained previously with the same in vitro system [14,27]. In addition, the ammonia reduction in the fermenter liquids was less marked than VFA, being approximately 60% of the ammonia in the rumen fluid. Moreover, a higher proportion of valeric and isovaleric acids was found in RCF liquids. These latter acids mainly originated from the metabolism of amino acids and this agrees with the relatively high ammonia concentration mentioned above.

The microbial population was represented in this study by only some bacterial strains and was affected by the type of inoculum. After the incubation period in the RCF, *Prevotella* genus and *Ruminococcus albus* group decreased the relative abundance, while *Fibrobacter succinogenes* increased, compared to the rumen liquids. *F. succinogenes,* both in the liquid from rumen and in the RCF liquids, was present at higher percentages than the *R. albus* group, as reported also from by Soto et al. [6,7] and by Koike et al. [28]. Our limited observations on the bacterial rumen microbiota confirm that the RCF in vitro environment shifts bacteria composition, because it depresses some strains and is favourable to others.

Despite the previously discussed differences between the two types of inoculum in terms of metabolites and microbiota concentration, the effects on the in vitro tests were less pronounced than expected, at least for NDF and crude protein degradability. On average, compared to the rumen fluid, the utilization of fermenter liquid inoculum determined a small reduction of NDFd (−12%), while the degradability of protein was slightly increased (+5%). Furthermore, the correspondence of NDFd data obtained on the six ingredients by the two inoculum was very close (r: 0.960, $p \leq 0.01$). On the contrary, gas production was depressed by 20% with the usage of RCF, but also in this case, there was a close correlation between the values obtained with the two types of inoculum (r: 0.939, $p \leq 0.01$). Therefore, despite the differences described above in metabolite and microbiota composition, the overall in vitro fermentability and the ranking of feeds in term of in vitro rumen fermentation do not seem to be greatly affected using the RCF fluid in comparison to rumen inoculum. This surprising result could be in part explained because the in vitro tests work at a high ratio of fermentation liquid to substrate (20, 50 and 180 mL/g of DM substrate for RDP, GP and NDFd, respectively). In such conditions, there is an abundance of degrading capacity and it could be speculated that the RCF fermenting fluid is not so diluted to greatly reduce the fermentability measured by the in vitro tests, at least for the fiber and protein fractions.

4.2. Method of Inoculum Preservation

A second aim of this study was to compare the fresh rumen inocula with those preserved by refrigeration, chilling and freeze-drying (Figure 1). The rumen fluid maintained at 39 °C for 5–6 h prior to incubation represented the reference rumen fluid because Robinson et al. [29] demonstrated that the storage in such conditions had no detrimental effects on in vitro NDF fermentation. The results of the present study show that rumen fluid refrigerated for 5–6 h had fermentative parameters very close to that of the fluid preserved at the animal body temperature. On the contrary, freezing at −80 °C determined a high ammonia increment and the freeze-drying process caused pronounced variations in comparison with fresh rumen fluid because both acetic and propionic acids proportions were reduced, and the iso-butyric increased abnormally and the ammonia was severely reduced.

The NDFd values obtained from the fluid preserved by refrigeration were very similar to those attained from the inoculum maintained at 39 °C, as previously found by Robinson et al. [29]. A similar situation was found also for the RDP values, apart from the two forage samples. The same close relationship between refrigerated and fluid from rumen was also the case for gas production and this result confirms those found by Hervás et al. [30], who suggested that the preservation of rumen fluid for up to 6 h in crushed ice is a practical alternative to the use of freshly collected rumen fluid as inoculum for in vitro ruminal fermentation studies. Therefore, refrigerated inoculum can be a more practical system of storing rumen inoculum during transport from the site of collection to the laboratory of utilization within a 4 to 6 h period.

The possibility of storing the inoculum for long periods was evaluated in terms of chilling at −80 °C and by freeze-drying. The chilled inoculum determined a limited reduction of NDF degradation (−14%) in comparison with the warm inoculum and there was a satisfactory relationship between feeds (r: 0.905).

Furthermore, the degradation of protein was, on average, not depressed and this agrees with the findings of Luchini et al. [31] who suggested that rumen inoculum preserved frozen might be used for in vitro rumen protein degradation experiments. However, the regression between feeds showed an insufficient degree of relation.

Finally, less gas was produced by the chilled inocula (−20%) with a satisfactory relationship among feeds (r: 0.797). The protection of the microbiota from freezing damage can be obtained by adding organic chemicals with cryoprotectant properties but these compounds can interfere with the subsequent in vitro tests because they are fermentable substrates (e.g., glycerol, [12]) or can have toxic effects (e.g., dimethyl sulfoxide). A possible improvement of our chilling procedure [9,12] could be to avoid freezing large amounts of inoculum and to divide it into small amounts (e.g., 30–50 mL) because the high surface/volume accelerates the freezing.

The freeze-drying process produced the lowest fermentation activity and a very poor relation with fresh liquid among feeds, confirming similar findings obtained by Belanche et al. [9]. Several methodological factors can explain these poor results, such as the absence of cryoprotectants, the procedure of pellet preparation in terms of centrifugation intensity, and an improper process of rehydration. Systemic experimental work would be necessary to evaluate the role of specific conditions on the microbiota survival throughout whole process. Finally, it is worth noting that both preservation methods based on low temperatures, determined a drastic drop in the relative abundance of the *Fibrobacter succinogenes*, which plays a main role in fiber degradation. This variation could be associated with a depression of degradability and, consequently, with the low NDFD and GP values measured.

5. Conclusions

The results from the present trial indicate that fermentation liquid from rumen continuous fermenters can be used to generate inoculum for in vitro purposes. This result has relevant implications, not only in terms of artificially generating rumen fermenting fluids by RCF systems, but also from the perspective of the standardization of the fermentation liquids by adopting controlled fermentation conditions (e.g., substrate, dilution rate of saliva, fermentation, and pH). Short-term refrigeration is confirmed to be a valuable storage system which has practical advantages in managing rumen fluid, particularly in the case of laboratories distant from collection sites. Finally, low fermentative capacity was found for chilled and, particularly, for freeze-dried liquids, and the procedures adopted to obtain such preserved inoculum probably require substantial improvements.

Author Contributions: Conceptualization, M.S., S.C., F.M., M.M. and F.T.; Data curation, M.S.; Formal analysis, M.C., C.F., F.F., F.M., C.S., S.S. and F.T.; Funding acquisition, M.S.; Project administration, M.S.; Writing—original draft, M.S.; Writing—review & editing, S.C., F.M., M.M., S.S. and F.T.. All Authors participated in critical revision of the manuscript and all have approved the final draft submitted.

References

1. López, S. In vitro and in situ techniques for estimating digestibility. In *Quantitative Aspects of Ruminant Digestion and Metabolism*, 2nd ed.; CABI Publishing: Wallingford, UK, 2005; pp. 87–121.
2. Yáñez-Ruiz, D.R.; Bannink, A.; Dijkstra, J.; Kebreab, E.; Morgavi, D.P.; O'Kiely, P.; Reynolds, C.K.; Schwarm, A.; Shingfield, K.J.; Yu, Z.; et al. Design, implementation and interpretation of in vitro batch culture experiments to assess enteric methane mitigation in ruminants—A review. *Anim. Feed Sci. Technol.* **2016**, *216*, 1–18. [CrossRef]
3. Hristov, A.N.; Lee, C.; Hristova, R.; Huhtanen, P.; Firkins, J.L. A meta-analysis of variability in continuous-culture ruminal fermentation and digestibility data. *J. Dairy Sci.* **2012**, *95*, 5299–5307. [CrossRef] [PubMed]
4. Carro, M.D.; Ranilla, M.J.; Martin-García, A.I.; Molina-Alcaide, E. Comparison of microbial fermentation of high- and low-forage diets in Rusitec, single-flow continuous-culture fermenters and sheep rumen. *Animal* **2009**, *3*, 527–534. [CrossRef] [PubMed]
5. Muetzel, S.; Lawrence, P.; Hoffmann, E.M.; Becker, K. Evaluation of a stratified continuous rumen incubation system. *Anim. Feed Sci. Technol.* **2009**, *151*, 32–43. [CrossRef]
6. Soto, E.C.; Yánez-Ruiz, D.R.; Cantalapiedra-Hijar, G.; Vivas, A.; Molina-Alcaide, E. Changes in ruminal microbiota due to rumen content processing and incubation in single-flow continuous culture fermenters. *Anim. Prod. Sci.* **2012**, *52*, 813–822. [CrossRef]
7. Soto, E.C.; Molina-Alcaide, E.; Khelil, H.; Yáñez-Ruiz, D.R. Ruminal microbiota developing in different in vitro simulation systems inoculated with goats' rumen liquor. *Anim. Feed Sci. Technol.* **2013**, *185*, 9–18. [CrossRef]
8. Martínez, M.E.; Ranilla, M.J.; Tejido, M.L.; Saro, C.; Carro, M.D. Comparison of fermentation of diets of variable composition and microbial populations in the rumen of sheep and Rusitec fermenters. II. Protozoa population and diversity of bacterial communities. *J. Dairy Sci.* **2010**, *93*, 3699–3712. [CrossRef]
9. Belanche, A.; Palma-Hidalgo, J.M.; Nejjam, I.; Serrano, R.; Jiménez, E.; Martín-García, I.; Yáñez-Ruiz, D.R. In vitro assessment of the factors that determine the activity of the rumen microbiota for further applications as inoculum. *J. Sci. Food Agric.* **2018**, *99*, 163–172. [CrossRef]

10. Chaudhry, A.S.; Mohamed, R.A.I. Fresh or frozen rumen contents from slaughtered cattle to estimate in vitro degradation of two contrasting feeds. *Czech J. Anim. Sci.* **2012**, *57*, 265–273. [CrossRef]
11. Denek, N.; Can, A.; Avci, M. Frozen rumen fluid as microbial inoculum in the two-stage in vitro digestibility assay of ruminant feeds. *S. Afr. J. Anim. Sci.* **2010**, *40*, 251–256. [CrossRef]
12. Prates, A.; de Oliveira, J.A.; Abecia, L.; Fondevila, M. Effects of preservation procedures of rumen inoculum on in vitro microbial diversity and fermentation. *Anim. Feed Sci. Technol.* **2010**, *155*, 186–193. [CrossRef]
13. Zeigler, L.D.; Schlegel, M.L.; Edwards, M.S. Development of a rumen fluid preservation technique and application to an in vitro dry matter digestibility assay. In *Proceedings of the Fifth Conference on Zoo and Wildlife Nutrition, Chester, PA, USA, 5–8 October 2003*; Ward, A., Brooks, M., Maslanka, M., Eds.; AZA Nutrition Advisory Group: Minneapolis, MN, USA, 2003.
14. Mason, F.; Zanfi, C.; Spanghero, M. Testing a stratified continuous rumen fermenter system. *Anim. Feed Sci. Technol.* **2015**, *201*, 104–109. [CrossRef]
15. Luchini, N.D.; Broderick, G.A.; Combs, D.K. Preservation of ruminal microorganisms for in vitro determination of ruminal protein degradation. *J. Anim. Sci.* **1996**, *74*, 1134–1143. [CrossRef]
16. Slyter, L.L.; Bryant, M.P.; Wolin, M.J. Effect of pH on population and fermentation in a continuously cultured rumen ecosystem. *J. Appl. Microbiol.* **1966**, *14*, 573–578.
17. Menke, K.H.; Steingass, H. Estimation of the energetic feed value obtained from chemical analysis and gas production using rumen fluid. *Anim. Res. Dev.* **1988**, *28*, 7–55.
18. Van Soest, P.J.; Robertson, J.B.; Lewis, B.A. Methods for dietary fiber, neutral detergent fiber, and nonstarch polysaccharides in relation to animal nutrition. *J. Dairy Sci.* **1991**, *74*, 3583–3597. [CrossRef]
19. Ross, D.A.; Gutierrez-Botero, M.; van Amburgh, M.E. Development of an in-vitro intestinal digestibility assay for ruminant feeds. In Proceedings of the Cornell Nutrition Conference, Syracuse, NY, USA, 22–24 October 2013; pp. 190–202.
20. Tagliapietra, F.; Cattani, M.; Bailoni, L.; Schiavon, S. In vitro rumen fermentation: Effect of headspace pressure on the gas production kinetics of cornmeal and meadow hay. *Anim. Feed. Sci. Technol.* **2010**, *158*, 197–201. [CrossRef]
21. AOAC International. *Official Methods of Analysis of AOAC International*, 17th ed.; AOAC International: Gaithersburg, MD, USA, 2000.
22. Martillotti, F.; Puppo, S. Liquid chromatographic determination of organic acids in silages and rumen fluids. *Ann. dell'Istituto Sper. Zootec.* **1985**, *18*, 1–10.
23. Kong, Y.; Teather, R.; Forster, R. Composition, spatial distribution, and diversity of the bacterial communities in the rumen of cows fed different forages. *FEMS Microbiol. Ecol.* **2010**, *74*, 612–622. [CrossRef]
24. Spanghero, M.; Berzaghi, P.; Fortina, R.; Masoero, F.; Rapetti, L.; Zanfi, C.; Tassone, S.; Gallo, A.; Colombini, S.; Ferlito, J.C. Technical note: Precision and accuracy of in vitro digestion of neutral detergent fiber and predicted net energy of lactation content of fibrous feeds. *J. Dairy Sci.* **2010**, *93*, 4855–4859. [CrossRef]
25. Spanghero, M. (Dipartimento di Scienze Agroalimentari, Ambientali e Animali, University of Udine, Udine, Italy). Unpublished work. 2019.
26. Rumen Microbial Genomics Network. A Report in Support of the Rumen Microbial Genomics (RMG) Network Describing Standard Guidelines and Protocols for Data Acquisition, Analysis and Storage. Available online: http://www.rmgnetwork.org/user/file/37/RMG%20Network%20Report%20standard%20guidelines.pdf (accessed on 14 October 2019).
27. Spanghero, M.; Mason, F.; Zanfi, C.; Nikulina, A. Effect of diets differing in protein concentration (low *vs* medium) and nitrogen source (urea vs soybean meal) on in vitro rumen fermentation and on performance of finishing Italian Simmental bulls. *Livest. Sci.* **2017**, *196*, 14–21. [CrossRef]
28. Koike, S.; Pan, J.; Kobayashi, Y.; Tanaka, K. Kinetics of in sacco fiber-attachment of representative ruminal cellulolytic bacteria monitored by competitive PCR. *J. Dairy Sci.* **2003**, *86*, 1429–1435. [CrossRef]
29. Robinson, P.H.; Campbell Mathews, M.; Fadel, J.G. Influence of storage time and temperature on in vitro digestion of neutral detergent fibre at 48 h, and comparison to 48 h in sacco neutral detergent fibre digestion. *Anim. Feed Sci. Technol.* **1999**, *80*, 257–266. [CrossRef]

30. Hervás, G.; Frutos, P.; Javier Giráldez, F.; Mora, M.J.; Fernández, B.; Mantecón, A.R. Effect of preservation on fermentative activity of rumen fluid inoculum for in vitro gas production techniques. *Anim. Feed Sci. Technol.* **2005**, *123*, 107–118. [CrossRef]
31. Luchini, N.D.; Broderick, G.A.; Combs, D.K. In vitro determination of ruminal protein degradation using freeze-stored ruminal microorganisms. *J. Anim. Sci.* **1996**, *74*, 2488–2499. [CrossRef] [PubMed]

10

Models Based on the Mitscherlich Equation for Describing Typical and Atypical Gas Production Profiles Obtained from In Vitro Digestibility Studies using Equine Faecal *Inoculum*

Christopher D. Powell [1,*], Mewa S. Dhanoa [2], Anna Garber [3], Jo-Anne M. D. Murray [3], Secundino López [4,5,*], Jennifer L. Ellis [1] and James France [1]

1 Department of Animal Biosciences, University of Guelph, Guelph, ON N1G 2W1, Canada; jellis@uoguelph.ca (J.L.E.); jfrance@uoguelph.ca (J.F.)

2 Institute of Grassland and Environmental Research, Plas Gogerddan, Aberystwyth SY23 3EB, UK; dandhanoa@hotmail.com

3 College of Medical, Veterinary and Life Sciences, School of Veterinary Medicine, University of Glasgow, Glasgow G61 1QH, UK; Anna.Garber@glasgow.ac.uk (A.G.); Jo-Anne.Murray@glasgow.ac.uk (J.-A.M.D.M.)

4 Departamento de Producción Animal, Universidad de León, E-24007 León, Spain

5 Instituto de Ganadería de Montaña, CSIC-Universidad de León, Finca Marzanas s/n, 24346 Grulleros, Spain

* Correspondence: cpowell@uoguelph.ca (C.D.P.); s.lopez@unileon.es (S.L.);

Simple Summary: Feedstuff evaluation through animal trials is time consuming and expensive. An alternative, the gas production method, measures the amount of fermentation gas produced from incubating feedstuffs with microbes from ruminal fluid or faecal samples. Models can be applied to gas production profiles to determine extent of feedstuff degradation either in the rumen or in the hindgut. Typical gas production profiles show a monotonically increasing monophasic pattern. However, atypical gas production profiles exist whereby at least two consecutive phases of gas production are present; these profiles are much less well described. Two models are proposed to fit these biphasic profiles, a sum of two Mitscherlich equations, and sum of Mitscherlich + linear equations. Additionally, two models that describe typical monophasic gas production curves, the simple Mitscherlich and the generalised Mitscherlich (root-*t*) model, were assessed for comparison. Models were fitted to 25 gas production profiles resulting from incubating feedstuffs with faecal *inocula* from equines. Of these 25 profiles, 17 displayed atypical biphasic patterns, and 8 displayed typical monophasic patterns. The two biphasic models were found to describe both the atypical and typical gas production profiles accurately. These models allow for the evaluation of feedstuffs using cost- and time-efficient methods.

Abstract: Two models are proposed to describe atypical biphasic gas production profiles obtained from in vitro digestibility studies. The models are extensions of the standard Mitscherlich equation, comprising either two Mitscherlich terms or one Mitscherlich and one linear term. Two models that describe typical monophasic gas production curves, the standard Mitscherlich and the France model [a generalised Mitscherlich (root-*t*) equation], were assessed for comparison. Models were fitted to 25 gas production profiles resulting from incubating feedstuffs with faecal *inocula* from equines. Seventeen profiles displayed atypical biphasic patterns while the other eight displayed typical monophasic patterns. Models were evaluated using statistical measures of goodness-of-fit and by analysis of residuals. Good agreement was found between observed atypical profiles values and fitted values obtained with the two biphasic models, and both can revert to a simple Mitscherlich allowing them to describe typical monophasic profiles. The models contain kinetic fermentation parameters that can be used in conjunction with substrate degradability information and digesta passage rate to calculate extent of substrate degradation in the rumen or hindgut. Thus, models link the in vitro gas production technique to nutrient supply in the animal by providing information relating to digestion and nutritive value of feedstuffs.

Keywords: gas production technique; in vitro digestibility; Mitscherlich equation; feedstuff evaluation; fermentation kinetics; substrate degradation

---

1. Introduction

The in vitro gas production technique [1,2] is widely applied in animal nutrition for ranking and evaluating feedstuffs. This technique is based upon the assumption that the gas produced from incubating a feedstuff with a microbial *inoculum* is the consequence of the anaerobic fermentation of that feedstuff [3]. In ruminant nutrition, gas production profiles generated have been used in conjunction with the retention time of digesta (derived from the rate of passage) to determine extent of degradation in the rumen [3–9]. In equine nutrition, the technique has been proposed as an in vitro surrogate for determining the digestibility and nutritive value of feedstuffs using in vivo methods [10–14].

Typical gas production profiles are diminishing returns or sigmoidal in shape (see [5] for illustration), and France et al. [4] derived a purpose-built function in the form of a generalised Mitscherlich equation with an additional root-t term to represent a variable fractional rate of degradation for fitting to a wide range of curve shapes. This model is commonly referred to as the "France" model, and this term will be used herein. However, atypical patterns have also been recorded. Groot et al. [15] reported biphasic profiles and selected a function comprising two generalised rectangular hyperbolae to fit them, while other atypical patterns have been observed by research workers though not formally reported in the scientific literature. Interpretation of these atypical patterns include the autonomous fermentation of feed components in incubated feedstuffs, with these feed components representing chemical or nutritional fractions, with total gas produced being a summation of gas produced from each fermented feed component [16].

The Mitscherlich equation has a long history of application in the agricultural sciences and in applied biology generally, both as a response function and as a growth function [17,18]. The Mitscherlich, which is an expression of the principle of the Law of Diminishing Increments as originally applied to the effect of fertilization on crop yields, is a function that reaches an asymptotic maximum and represents diminishing returns behaviour in rising to the asymptote. It is a special case of the function proposed by France et al. [4]. In this paper, we consider four types of gas production profile (diminishing returns, sigmoidal, biphasic and asymptotic, biphasic but non-asymptotic). The profiles considered were obtained from incubating feedstuffs with faecal *inocula* from equines using the gas production method of Theodorou et al. [2]. These data were taken from two experiments with either grazing horses or ponies fed primarily grass hay. The main objective of this paper was to assess the ability of the simple Mitscherlich, and three extensions of this classical function, to describe both typical and atypical gas production profiles. The functions were derived to describe gas production profiles on the basis of substrate degradation, rather than on the basis of gas produced, permitting the estimation of fermentation kinetic parameters. Using relatively simple equations, proposed herein, extent of feedstuff degradation in the hindgut of equines can be calculated using model parameter estimates in conjunction with information regarding substrate degradability and digesta passage rate. Therefore, a secondary objective was to compare how model fits, and by extension model derived parameters, affect extent of feedstuff degradation values when these models are applied to mono- and bi-phasic gas production profiles.

2. Materials and Methods

2.1. Datasets

2.1.1. Experiment 1: Inoculum from Horses

In a study to assess the fermentative capacity of faecal *inocula*, Murray et al. [19] sourced *inoculum* from 14 grass-kept horses (maintained on grass 24 h a day) from the International League for the Protection of Horses in Norfolk, UK. *Inocula* were prepared from these 14 horses—7 of them predisposed to laminitis and the other 7 clinically normal—so that the effect of laminitis on hindgut fermentative activity could be evaluated. Grass hay was the substrate incubated in vitro. Due to the large distance to the laboratory, *inoculum* was stored at −20 °C for transportation on ice. *Inocula* were subsequently thawed and incubated at 38 °C. Gas production was recorded using the method of Theodorou et al. [2] and three replicates per *inoculum* were used. Standard in vitro gas production results were described by Murray et al. [19]. The grass hay data yielded 14 gas production profiles, one for each horse, as the average over the three replicates. Visual inspection of these profiles revealed a predominance of atypical patterns.

2.1.2. Experiment 2: Inoculum from Ponies

The study comprised a total of eleven different *inocula* (see Table 1 for details). Garber et al. [20] sourced eight *inocula* from Welsh Section A geldings arranged in a 4 × 4 Latin square experimental design aiming to investigate the in vitro fermentation of high fibre/high concentrate diets supplemented with yeast (control diets with no yeast). Another 3 faecal *inocula* were obtained in an experiment in which ponies were fed a grass hay only diet (control), or the same grass hay supplemented with increasing concentrations of a fibrolytic enzyme (either 0.75 or 3.75 mL of enzyme solution per kg DM hay). Gas production was recorded using the ANKOM RF gas production system [20] and three replicates per *inoculum* were used. Preliminary results were reported by Garber et al. [21]. The data yielded 11 gas production profiles, one for each treatment, after averaging the three replicates for each *inoculum*. Visual inspection revealed both typical and atypical patterns.

The entirety of the observed gas production values (Dataset 1–25) used in this study can be found in the Supplementary Information section of this paper, Table S1.

Table 1. Details of the eleven treatments used in Experiment 2.

Control and Enzyme Treatments		
Dataset-Substrate-*Inoculum*	Substrate (Chopped)	*Inoculum* from
15-C-I1	**C** Grass hay untreated (same as ponies were fed).	**I1** Ponies fed control diet consisting of grass hay fed at 100%
16-E1-I2	**E1** Grass hay treated with enzyme 0.75 mL/kg DM hay	**I2** Ponies fed grass hay treated with enzyme 0.75 mL/kg DM hay
17-E2-I3	**E2** Grass hay treated with enzyme 3.75 mL/kg DM hay	**I3** Ponies fed grass hay treated with enzyme 3.75 mL/kg DM hay
Other treatments		
Dataset-Substrate-*Inoculum*	Substrate (Ground to pass 1 mm screen)	*Inoculum* from
18-A-I4	**A** Grass hay (50%) + alfalfa (50%)	**I4** Ponies fed 25% alfalfa and 75% grass hay (dry matter (DM) basis).
19-A1-I4	**A1** Grass hay (75%) + alfalfa (25%)	Not supplemented with yeast.
20-B-I5	**B** Grass hay (50%) + alfalfa (50%) + Yeast (0.011 g)	**I5** Ponies fed 25% alfalfa and 75% grass hay (DM basis).
21-B1-I5	**B1** Grass hay (75%) + alfalfa (25%) + Yeast (0.011 g)	Supplemented daily with 30 g yeast per pony was mixed with alfalfa.
22-C-I6	**C** Grass hay (50%) + concentrate (50%)	**I6** Ponies fed 25% concentrate and 75% grass hay (DM basis).
23-C1-I6	**C1** Grass hay (75%) + concentrate (25%)	Not supplemented with yeast.
24-D-I7	**D** Grass hay (50%) + concentrate (50%) + Yeast (0.011 g)	**I7** Ponies fed 25% concentrate and 75% grass hay (DM basis).
25-D1-I7	**D1** Grass hay (75%) + concentrate (25%) + Yeast (0.011 g)	Supplemented daily with 30 g yeast per pony mixed with concentrate.

2.2. Models Fitted

The classical Mitscherlich equation used in crop science has the general form:

$$y = A - (A - B)e^{-ct}$$

where the ordinate y is crop yield, the abscissa t is fertilizer rate, and A, B and c are constants. The parameter A represents the asymptotic value of y (i.e., maximum yield) and B the minimum yield (i.e., no fertilizer application).

For application to the gas production technique, the ordinate becomes cumulative gas production (mL) and the abscissa becomes time since inoculation (h). Cumulative gas production at zero time can be considered negligible and a lag $T \geq 0$ (h) may occur before onset of fermentation, so the Mitscherlich equation becomes:

$$y = A\left(1 - e^{-c(t-T)}\right); t \geq T \tag{1}$$

In this equation, A would represent the asymptotic gas production (mL) and c (h^{-1}) the fractional rate of fermentation. In this paper, we explore Equation (1) and three different extensions (Equations (2)–(4) below) of this classical function for use in describing typical and atypical gas production profiles.

Gas production profiles are typically monophasic, asymptotic and often sigmoidal (e.g., [5]). France et al. [4] derived the following equation from rate:state principles to describe such profiles:

$$y = A\left\{1 - \exp\left[-c(t - T) - d\left(\sqrt{t} - \sqrt{T}\right)\right]\right\}; t \geq T \tag{2}$$

Here, A (mL) is the asymptotic value of y, and c (h^{-1}) and d ($h^{-0.5}$) are fractional rate constants. Equation (1) is a special case of Equation (2) (i.e., $d = 0$). This equation is commonly referred to as the France model.

Biphasic, asymptotic gas production profiles have also been observed (e.g., [15]), and these would appear to lend themselves to description by the sum of two Mitscherlich terms:

$$y = A_1(1 - e^{-c_1(t-T_1)}) + A_2(1 - e^{-c_2(t-T_2)}); t \geq T_1,\ t \geq T_2 \tag{3}$$

The first term in Equation (3) is zero until time T_1 and likewise the second term until time T_2. Equation (1) is also a special case encompassed by Equation (3) (i.e., $A_2 = 0$). This latter equation will be referred to as the double Mitscherlich model.

As mentioned above, instances of profiles that do not exhibit typical asymptotic behaviour have also been observed but not formally reported. Such profile forms suggest a function resulting from the sum of a Mitscherlich term and a linear term might provide an appropriate description:

$$y = A(1 - e^{-c(t-T_1)}) + \beta(t - T_2); t \geq T_1,\ t \geq T_2 \tag{4}$$

where the parameter β (mL h^{-1}) is the slope of an underlying linear trend. As for Equation (3), the first term in Equation (4) is zero until time T_1 and likewise the second term until time T_2. Putting $\beta = 0$ in Equation (4) yields Equation (1). This latter equation will be referred to as the Mitscherlich + linear model.

2.3. Extent of Degradation

The extent of degradation (E) of substrate in a specific compartment or region of the gastro-intestinal tract may be calculated from the gas production curve, provided the *inoculum* used to generate the profile is representative of that compartment. If the profile is diminishing returns in shape, first-order kinetics with a constant fractional rate of degradation describes substrate degradation and Equation (1) can be fitted to the profile. Extent of degradation is then given by:

$$E = S_0 e^{-kT} c / [(c + k)(S_0 + U_0)] \tag{5}$$

where S_0 (g) is the amount of the incubated substrate that is potentially degradable, U_0 (g) the amount that is undegradable, T (h) the lag before commencement of degradation, c (h^{-1}) the fractional rate of fermentation, and k (h^{-1}) is the fractional rate of passage out of the compartment [4].

If the profile is sigmoidal, first-order kinetics with a variable fractional degradation rate would account for substrate degradation and the France model (Equation (2)) can be fitted. Extent of degradation is then given by:

$$E = S_0 e^{-kT} (1 - kI_1) / (S_0 + U_0) \tag{6}$$

$$= kS_0 I_2 / (S_o + U_0) \tag{7}$$

where

$$I_1 = \int_T^{\infty} \exp\{-[(c + k)(t - T) + d(\sqrt{t} - \sqrt{T})]\} dt$$

$$I_2 = \int_T^{\infty} e^{-kt} \{1 - \exp[-c(t - T) - d(\sqrt{t} - \sqrt{T})]\} dt$$

and c (h^{-1}) is the constant portion of the fractional degradation rate and d ($h^{-0.5}$) the coefficient of the variable portion. The integrals I_1 and I_2 are non-analytical and therefore have to be evaluated numerically [4].

If the profile is linear with an abrupt cut-off (i.e., a broken stick), zero-order kinetics with constant rate of degradation independent of substrate remaining can be assumed and a piecewise linear model fitted. The extent of degradation is then given by:

$$E = \beta\left[\ln\left(k\beta^{-1} S_0 + e^{kT}\right) - kT\right] / [k(S_0 + U_0)] \tag{8}$$

where β (mL h^{-1}) is the slope of the line fitting the ascending portion of the profile [22].

If the gas production profile is multiphasic, then the extent of degradation for each phase can be calculated by applying the appropriate equation, and the weighted extents summed to estimate overall extent of degradation. For example, if the profile resolves into two diminishing returns components (1 and 2) as in Equation (3), then Equation (5) can be independently applied to each of the two phases and the overall extent calculated as:

$$E = (w_1 E_1 + w_2 E_2) / (w_1 + w_2) \tag{9}$$

where w_1 and w_2 are the relative weights assigned to the respective phases. If the profile resolves into a diminishing returns and a linear (with abrupt cut-off) component as in Equation (4), then Equations (5) and (8) respectively can be applied to the two phases and the overall extent calculated again using Equation (9). As an arbitrary rule of thumb, the asymptotic gas production values for the two phases (abrupt cut-off value if a phase is linear), viz. A_1 and A_2, can be adopted as the weights w_1 and w_2 respectively.

Thus, for the equine data considered herein, the extent of degradation of substrate in the hindgut can be calculated using Equations (5)–(9) if we assume faecal *inoculum* is representative of that region of the gastro-intestinal tract. Herein, when calculating extent of degradation, the amount of the incubated substrate that is potentially degradable S_0 (g), the amount that is undegradable U_0 (g), and the fractional rate of passage out of the compartment k (h^{-1}), were assumed to be 0.538, 0.465 and 0.019, respectively, for all datasets [23].

2.4. Fitting and Evaluation of Models

Each of the four models (Equations (1)–(4)) was fitted by non-linear regression to the 25 gas production profiles using the NLIN procedure in the statistical software SAS [24]. Initial estimates of parameter values were obtained through visual inspection of the data.

Using various statistical tests, the models were evaluated for goodness-of-fit along with analysis of residuals. Mean square prediction error (MSPE) was calculated as the sum of the squared difference between predicted and observed values divided by the number of observations [25]. The accuracy factor (AF) index is a measure of the average deviation of a model's predictions and is used as a simple index of the level of confidence in these predictions [26]. Agreement between model predictions and observations was further determined using the concordance correlation coefficient (CCC), a single statistic ranging between −1 (perfect disagreement) and +1 (perfect agreement) which contains both accuracy and precision indicators [27,28]. The Akaike information criterion (AIC) is a test for model selection which accounts for goodness-of-fit while penalizing for over-fitting, with the model resulting in the smallest AIC being the most appropriate [29].

The ability of each model to predict gas production without systematically over- or under-estimating was examined using the number of runs test and the Durbin–Watson (DW) test. The runs test examines a sequence of residuals for unusual groupings of positive or negative residuals and tests the null hypothesis that the arrangement of signs (+/−) is random, with too few runs indicating the presence of autocorrelation [30]. The DW test examines dependencies in the error terms by testing for correlations between a residual and the residuals immediately before and after it in the sequence. Compared to the runs test, the DW provides greater information regarding analysis of residuals by not only considering the sign of the residual but also its magnitude. The DW statistic (D), and upper (D_u) and lower (D_l) critical values, were calculated according to [30]. When D is less than the lower critical value D_l, evidence of positive autocorrelation occurs, and when D is greater than the upper critical value D_u, evidence of negative autocorrelation occurs.

3. Results

The ability of the Mitscherlich, and the other derived functions, to describe typical and atypical cumulative gas production profiles was assessed by fitting the four equations (Equations (1)–(4)) to 25 datasets. The profiles examined resulted from incubating forage using faecal *inoculum* from equines following the methodology of Theodorou et al. [2]. Using parameter estimates resulting from fitting these models to the gas production profiles, extents of substrate degradation were calculated and compared.

3.1. Fitting Behaviour

Of the 25 gas production profiles considered, 17 displayed atypical patterns, characterized by more than one phase, while the remaining 8 displayed typical monophasic patterns. No convergence issues were encountered when fitting the simple Mitscherlich (Equation (1)), double Mitscherlich (Equation (3)) and the Mitscherlich + linear (Equation (4)) to any of the datasets. The use of an "if than" statement concerning $t \geq T_2$ and its effect on A_2 and β in SAS allowed both Equations (3) and (4) to revert to the simple Mitscherlich if that resulted in a better fit compared to the extended biphasic equations (i.e., when $A_2 = 0$ in Equation (3) and $\beta = 0$ in Equation (4)). The France equation (Equation (2)) also encompasses the ability to revert to a simple Mitscherlich (Equation (1)) when $d = 0$. When fitted to the 25 gas production profiles, the France equation (Equation (2)) reverted to the simple Mitscherlich (Equation (1)) in four cases as the best fit for these gas production profiles was achieved when $d = 0$. Likewise, the double Mitscherlich (Equation (3)) reverted to the simple Mitscherlich (Equation (1)), i.e., $A_2 = 0$, in five cases as a single Mitscherlich term described these profiles better than two Mitscherlich terms.

When fitting the France model (Equation (2)) to the atypical gas production curves, the convergence criteria had to be relaxed in order to reach successful convergence. When enforcing relaxed convergence criteria, Equation (2) was unable to converge for one of the 25 datasets. In order to achieve biologically meaningful parameters, lag time (T) and fractional rate constant (c) were constrained to be non-negative when fitting each model. Furthermore, in fitting Equation (2) a constraint was placed on parameter d, viz. $d \geq -2c\sqrt{T}$ to ensure the fractional rate of degradation remained non-negative [4].

3.2. Parameter Estimates and Fitted Gas Production Curves

Initial parameter estimates of lag time (T), asymptotic value (A) and slope (β) were determined by visual inspection of the gas production curves, while ranges for the fractional rate constants (c and d) were provided. The final parameter estimates resulting from fitting Equations (1)–(4) to Dataset 1 and 8 of Experiment 1 and Dataset 18 and 22 of Experiment 2 are presented in Tables 2 and 3, respectively. The final estimates resulting from fitting Equations (1)–(4) to the remaining 21 datasets are given in the Supplementary Information section of this paper (Tables S2–S5). Using the parameter estimates in Tables 2 and 3, the gas production profiles resulting from applying Equations (1)–(4) to Dataset 1 and 8 are shown in Figure 1. Both Dataset 1 and 8 show clear atypical biphasic gas production curves which are more faithfully represented by the double Mitscherlich (Equation (3)) and Mitscherlich + linear (Equation (4)) equations than the monophasic simple Mitscherlich (Equation (1)) or the France model (Equation (2)). Examining the four lower panels of Figure 1, the extent to which each phase contributes to the overall gas production curve of the double Mitscherlich (Equation (3)) and Mitscherlich + linear (Equation (4)) are clearly distinguishable.

Table 2. Parameter estimates obtained by fitting Equations (1)–(4) to Dataset 1 (laminitis) and 8 (clinically normal) from Experiment 1. An asterisk (*) denotes the equation that resulted in the best fit, based on AIC, to a particular dataset.

	Simple Mitscherlich (Equation (1))		**France (Equation (2))**		**Double Mitscherlich (Equation (3)) ‡**		**Mitscherlich + linear (Equation (4)) §**	
	Dataset 1	**Dataset 8**	**Dataset 1**	**Dataset 8**	**Dataset 1**	**Dataset 8**	**Dataset 1**	**Dataset 8**
A	163.5	123.7	246.8	159.5	60.3, 84.3 *	60.8, 66.8 *	63.7	64.6
c	0.014	0.025	0.005	0.008	0.093, 0.026 *	0.140, 0.027 *	0.078	0.111
T	0	0	0	0	3.1, 43.4 *	2.8, 34.6 *	2.8, 35.9	2.5, 28.7
d			0.019	0.049				
β							0.936	0.723

‡ The two scale parameters of this equation are entered under A in the order A_1, A_2 in this table. Likewise, the two rate parameters under c in the order c_1, c_2, and the two lags under T in the order T_1, T_2. § The two lag parameters of this equation are entered under T in the order T_1, T_2.

Table 3. Parameter estimates obtained by fitting Equations (1)–(4) to Dataset 18 (50% grass hay + 50% alfalfa) and 22 (50% grass hay + 50% concentrate), from Experiment 2. An asterisk (*) denotes the equation that resulted in the best fit, based on AIC, to a particular dataset.

	Simple Mitscherlich (Equation (1))		**France (Equation (2))**		**Double Mitscherlich (Equation (3)) ‡**		**Mitscherlich + linear (Equation (4)) §**	
	Dataset 18	**Dataset 22**	**Dataset 18**	**Dataset 22**	**Dataset 18**	**Dataset 22**	**Dataset 18**	**Dataset 22**
A	120.6	148.7 *	122.9	148.7 †	109.7, 38.7	148.7 Ψ	109.6 *	147.7
c	0.071	0.078 *	0.057	0.078 †	0.090, 0.016	0.078 Ψ	0.091 *	0.079
T	0	0.5 *	0	0.5 †	0.3, 37.2	0.5 Ψ	0.3, 34.8 *	0.5, 33.1
d			0.037					
β							0.473 *	0.040

‡ The two scale parameters of this equation are entered under A in the order A_1, A_2 in this table. Likewise, the two rate parameters under c in the order c_1, c_2, and the two lags under T in the order T_1, T_2. § The two lag parameters of this equation are entered under T in the order T_1, T_2. † Best fit by France, Equation (2), achieved with $d = 0$, therefore reverting to a simple Mitscherlich, viz. Equation (1). Ψ Best fit by double Mitscherlich, Equation (3), achieved with $A_2 = 0$, therefore reverting to a simple Mitscherlich, viz. Equation (1).

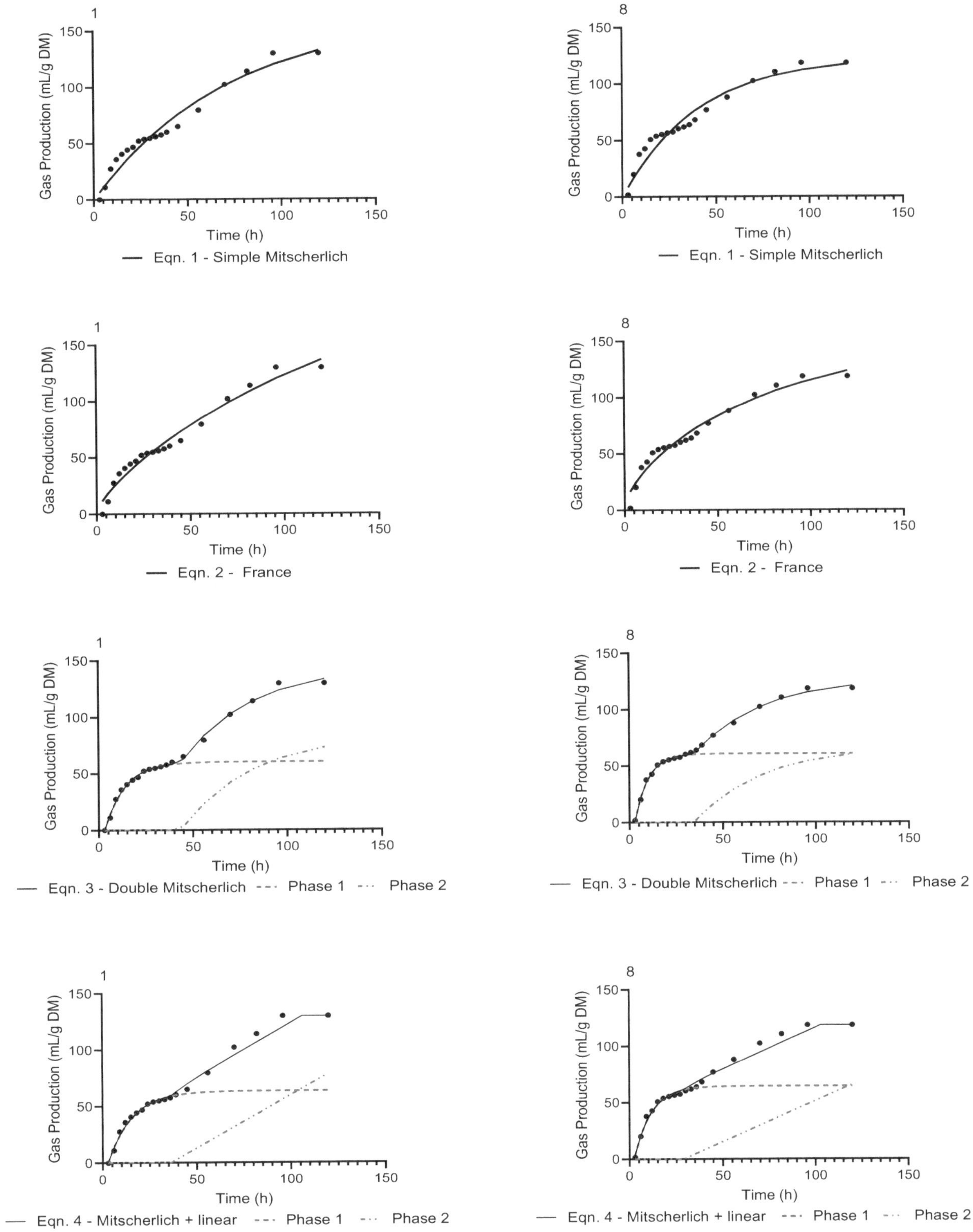

Figure 1. Observed (•) atypical gas production profiles and predicted curves resulting from fitting Equations (1)–(4) to Dataset 1 and 8 of Experiment 1.

In contrast to Dataset 1 and 8, Dataset 18 and 22 of Experiment 2 display more typical monophasic gas production curves as shown in Figure 2. Again, examining the bottom four panels of Figure 2, the second phase of the biphasic models (viz. Equations (3) and (4)) is much less evident, with the second phase being entirely absent when fitting the double Mitscherlich (Equation (3)) to Dataset 22 as

Equation (3) reverts to the simple Mitscherlich with $A_2 = 0$. Additionally, when fitting the equation of France (Equation (2)) to Dataset 22, the best fit was achieved when $d = 0$, and thus Equation (2) reverted to a simple Mitscherlich (Equation (1)) when applied to this dataset. The gas production profiles resulting from fitting Equations (1)–(4) to the remaining datasets are shown in the Supplementary Information (Figures S1–S4).

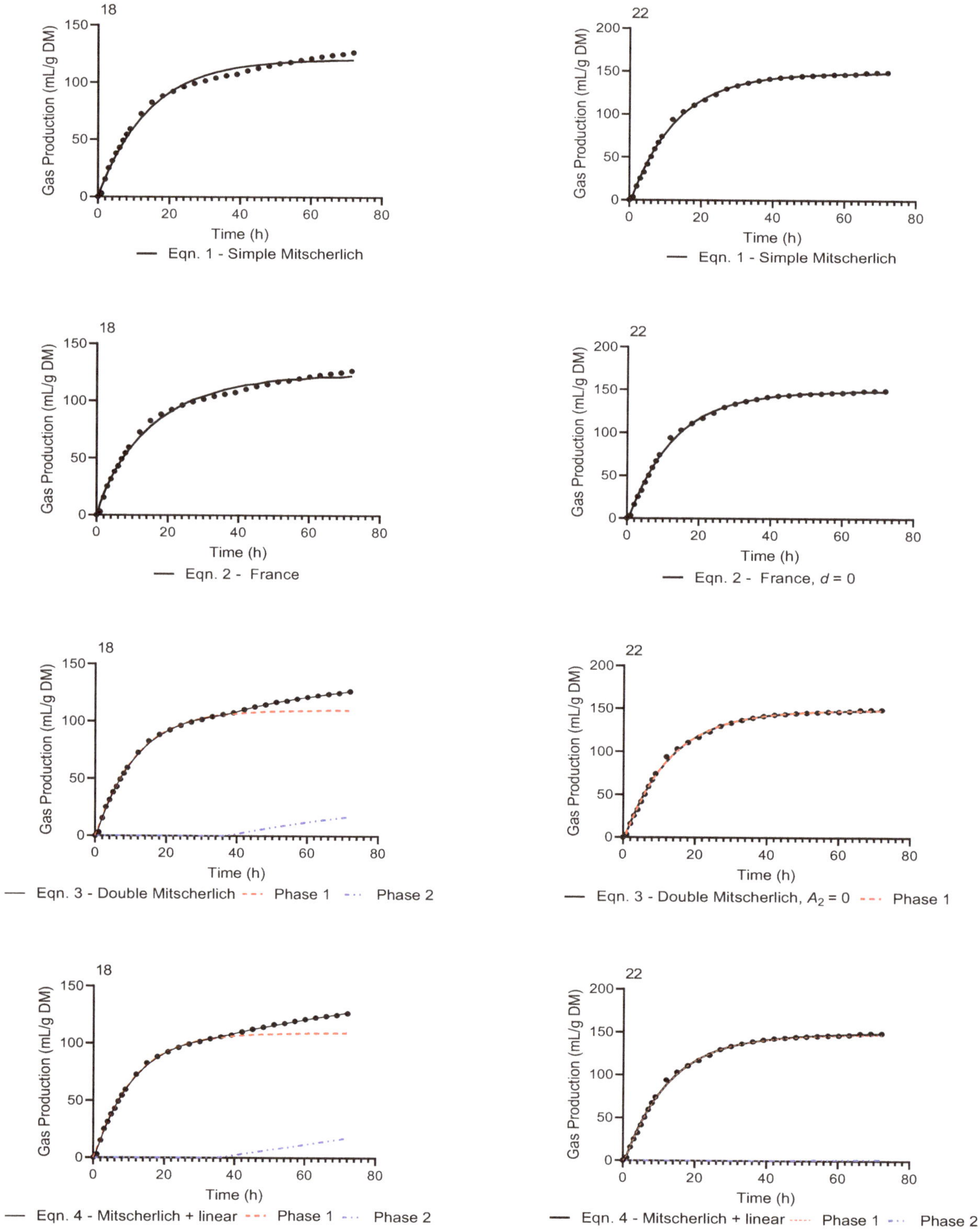

Figure 2. Observed (•) typical gas production profiles and predicted curves resulting from fitting Equations (1)–(4) to Dataset 18 and 22 of Experiment 2.

3.3. Model Evaluation

Goodness-of-fit was assessed using four criteria, namely AIC, MSPE, CCC and AF. The goodness-of-fit values resulting from fitting each of the four models to the 25 datasets were averaged and the models ranked from 1 to 4 based upon their comparative performance with the other models for a given criterion. Individual models averaged goodness-of-fit values, along with their mean rank and the number of times the model ranked first or second under a given criterion, are presented in Table 4.

Table 4. Goodness-of-fit and analysis of residuals from fitting the four equations to the 25 datasets of Experiments 1 and 2.

Criteria	Model			
	Simple Mitscherlich (Equation (1))	France (Equation (2))	Double Mitscherlich (Equation (3))	Mitscherlich + Linear (Equation (4))
Akaike information criterion (AIC)				
Average (±SE)	73.1 (2.5)	66.6 (2.6)	46.7 (2.7)	47.6 (5.2)
Mean rank	3.4	2.9	1.4	1.7
Number of times a model ranked 1st or 2nd	2	5	22	25
Mean square prediction error (MSPE)				
Average (±SE)	31.6 (5.4)	20.9 (3.2)	5.1 (0.5)	10.7 (1.7)
Mean rank	3.4	2.6	1.3	1.6
Number of times a model ranked 1st or 2nd	2	5	22	25
Concordance correlation coefficient (CCC)				
Average (±SE)	0.979 (0.004)	0.985 (0.003)	0.996 (0.001)	0.994 (0.001)
Mean rank	3.5	2.6	1.4	1.7
Number of times a model ranked 1st or 2nd	5	9	22	25
Accuracy factor (AF)				
Average	1.32 (0.04)	1.32 (0.04)	1.18 (0.03)	1.17 (0.03)
Mean rank	2.8	3.2	1.4	1.5
Number of times a model ranked 1st or 2nd	7	3	25	25
Number of runs test				
Too few runs	25	24	7	19
Runs are random	0	0	18	6
Durbin–Watson (DW) test				
Positive correlation	25	23	7	10
No evidence	0	1	2	4
Negative correlation	0	0	16	11

When fitted to the 25 datasets, the double Mitscherlich resulted in the smallest averaged AIC value (46.7 ± 2.7), followed by the Mitscherlich + linear (47.6 ± 5.2) and the France equation (66.6 ± 2.6), with the simple Mitscherlich (73.1 ± 2.5) resulting in the highest average AIC. Based upon AIC, the mean rank of the double Mitscherlich, Mitscherlich + linear, France and the simple Mitscherlich was 1.4, 1.7, 2.9 and 3.4, respectively, with the Mitscherlich + linear being ranked 1st or 2nd 25 times followed by the double Mitscherlich (23), France (5), and the simple Mitscherlich (2). The double Mitscherlich resulted in the lowest average MSPE (5.1 ± 0.5) followed by the Mitscherlich + linear (10.7 ± 1.7), France (20.9 ± 3.2), and the simple Mitscherlich (31.6 ± 5.4). In agreement with AIC, the double Mitscherlich resulted in the highest mean rank (1.3), followed by the Mitscherlich + linear (1.6), France (2.6) and simple Mitscherlich (3.4) with the Mitscherlich + linear ranking 1st or 2nd 25 times compared to the double Mitscherlich (22), France (5), and simple Mitscherlich (2).

Following the same trend as AIC and MSPE, the double Mitscherlich resulted in the highest CCC (0.996 ± 0.001), followed by the Mitscherlich + linear (0.994 ± 0.001), France (0.985 ± 0.003) and simple Mitscherlich (0.979 ± 0.004). Again, the double Mitscherlich yielded the best average rank of 1.4, followed by the Mitscherlich + linear (1.7), France (2.6) and simple Mitscherlich (3.5). The simple Mitscherlich was ranked 1st or 2nd in 5 of the 25 datasets and France in 9 of the 25, whilst the Mitscherlich + linear and double Mitscherlich were ranked 1st or 2nd in 25 and 22 datasets, respectively. On the basis of AF, the Mitscherlich + linear (1.17 ± 0.03) and double Mitscherlich (1.18 ± 0.03) outperformed both the France and simple Mitscherlich (with an averaged AF of 1.32 ± 0.04). The double Mitscherlich was ranked higher than the Mitscherlich + linear, 1.4 vs. 1.5, with both models being ranked 1st or 2nd 25 times. The simple Mitscherlich had a higher rank compared to the France, 2.8 vs. 3.2, with the simple Mitscherlich ranking 1st or 2nd 7 times compared to 3 times for the France equation.

In addition to goodness-of-fit, the runs test and DW test were used for the analysis of residuals. The results of these analyses are presented in Table 4. The runs test determined that too few runs occurred for all datasets when fitting the simple Mitscherlich and France equations. In comparison, runs of residuals were determined to be random in 18 and 6 of the 25 datasets when fitting the double Mitscherlich and Mitscherlich + linear, respectively. Using the DW test there was evidence of positive correlation of the residuals in all of the datasets that the simple Mitscherlich fitted successfully, and in all but one for the France model. In contrast, positive correlation was only found in 7 and 10 of the 25 datasets when fitting the double Mitscherlich and Mitscherlich + linear, with negative correlation being found in 16 and 11 of the 25 datasets, respectively.

3.4. Extent of Degradation

Extent of substrate degradation, calculated from the parameter estimates resulting from fitting the four models to the 25 gas production profiles, is presented in Table 5. When calculating extent of degradation using the biphasic equations, viz. Equations (3) and (4), the relative weights in which each phase contributed to the overall extent of substrate degradation need to be incorporated. Using the double Mitscherlich (Equation (3)) the weights of each phase were simply assumed to be their respective asymptotic gas production values for each phase, A_1 and A_2. For the Mitscherlich + linear, the weight of the first phase was its respective asymptotic value (A_1), while the weight of the second (linear) phase was calculated as the amount of gas produced over the course of this linear segment (i.e., multiplying its slope by the duration of the linear segment). The duration of the linear segment was the difference between time at which the abrupt cut-off value occurred and time at which the linear portion commenced (i.e., T_2, the lag time). The abrupt cut-off was determined in two ways. In Experiment 1, Dataset 1–14, abrupt cut-off values were assumed to occur at the intersection between the linear segment of the equation and the apparent plateau in gas production, which visually occurred between the last two data points. In Experiment 2, Dataset 15–17, a plateau in the observed data was not evident, therefore the abrupt cut-off value was set to the end of incubation. Finally, in Experiment 2, Dataset 18–25, a plateau was eventually reached with the linear segments being horizontal or near horizontal, and abrupt cut-off values were again set to the end of incubation.

Table 5. Calculated extent of degradation (%) from the 14 datasets of Experiment 1 displaying atypical gas production profiles and the 11 datasets of Experiment 2 displaying both typical and atypical profiles. An asterisk (*) denotes the equation that resulted in the best fit, based on AIC, to a particular dataset.

Dataset	Visual Curve Pattern	Simple Mitscherlich (Equation (1))	France (Equation (2))	Double Mitscherlich (Equation (3))	Mitscherlich + Linear (Equation (4))
1	Atypical	22.7	15.5	25.5 *	33.9
2	Atypical	34.3	28.5	35.1 *	40.0
3	Atypical	25.4	12.7	25.1 *	37.5
4	Atypical	26.7	-	25.0 *	35.6
5	Atypical	31.1	21.8	29.1 *	42.0
6	Atypical	30.3	21.7	30.0 *	38.0
7	Atypical	34.9	28.6	32.7 *	39.7
8	Atypical	30.4	24.4	30.0 *	38.0
9	Atypical	23.4	16.0	33.2 *	36.8
10	Atypical	22.4	19.0	24.2 *	34.9
11	Atypical	36.3	32.9	29.1 *	39.6

Table 5. *Cont.*

Dataset	Visual Curve Pattern	Simple Mitscherlich (Equation (1))	France (Equation (2))	Double Mitscherlich (Equation (3))	Mitscherlich + Linear (Equation (4))
12	Atypical	33.4	32.6	30.2 *	38.2
13	Atypical	25.9	22.0	28.1 *	34.5
14	Atypical	33.3	31.3	25.1 *	37.3
15	Atypical	31.6	31.6	31.6 $^{\Psi}$	38.4 *
16	Atypical	34.2	23.5	34.2 $^{\Psi}$	38.9 *
17	Atypical	35.5	19.5	35.5 $^{\Psi}$	42.0 *
18	Typical	42.4	41.6	35.9	41.9 *
19	Typical	42.1	40.5	36.8 *	42.8
20	Typical	43.3 *	43.3 †	43.3 $^{\Psi}$	43.0
21	Typical	43.2	43.2 †	41.2	42.9 *
22	Typical	42.9 *	42.9 †	42.9 $^{\Psi}$	42.8
23	Typical	42.5	42.3	41.9	43.0 *
24	Typical	44.2	44.2 †	43.1	43.8 *
25	Typical	43.0	43.1	41.2	42.8 *

† Best fit by France, Equation (2), achieved when $d = 0$, therefore reverting to a simple Mitscherlich, viz. Equation (1). $^{\Psi}$ Best fit by double Mitscherlich, Equation (3), achieved with $A_2 = 0$, therefore reverting to a simple Mitscherlich, viz. Equation (1).

Although substrate was primarily grass hay, or a grass hay mix, the range of calculated extent of degradation varied widely, from a minimum of 12.7% to a maximum of 44.2%. The wide range in extent of degradation values can be attributed to the use of parameter estimates from a model that fits a given gas production profile poorly. Therefore, Table 5 includes an indicator of which model fitted the particular dataset values best, based upon AIC. Fitting these four models to 25 datasets that encompass both typical and atypical gas production curves resulted in the double Mitscherlich being the best fitting in 15 of these datasets, the Mitscherlich + linear 8, simple Mitscherlich 2 and France 0. Given that extent of degradation is determined using parameter estimates obtained by fitting these models to a dataset, the importance of model fit in calculating extent of degradation is apparent.

4. Discussion

Gas production profiles generated from incubating a substrate with either ruminal or faecal *inocula* have been widely used to provide information regarding the degradability of forages and supplementary feeds in both ruminants and non-ruminants [3,5,6,10–14]. Typical shapes of these profiles range from diminishing returns to strongly sigmoidal [3]. Various models, e.g., Mitscherlich, Michaelis-Menten, Gompertz and logistic, have been proposed to describe these curves, including generalised models such as Richards and that of France which are able to accommodate both diminishing returns and sigmoidal behaviour [4,31–33]. Deriving these models on the basis of substrate degradation rather than amount of gas produced permits the generation of fermentation kinetic parameters [3]. By fitting these models to gas production profiles, such parameters (e.g., fractional rate of degradation and lag time) can be estimated. These model-derived parameters have been used in conjunction with information regarding substrate degradability and digesta passage rate to calculate extent of substrate degradation in the rumen [3–6,34]. This method has been successfully applied to typical monophasic sigmoidal and diminishing returns gas profiles to evaluate substrates based upon the extent of their degradability [3,5,6,8–14,34].

In addition to the typical sigmoidal and diminishing return patterns displayed by gas production profiles, atypical multiphasic curves have been reported [15,16,19,20,35]. Although multiphasic gas production curves have been described by both Groot et al. [15] and Wang et al. [36], proposed models are based upon the amount of gas produced, rather than the amount of substrate degraded, resulting in the model being unable to link the gas production technique to animal performance [3].

4.1. Profile Shapes and Associated Parameters

The diminishing returns behaviour described by the simple Mitscherlich is a result of the interaction between the constant fractional degradation rate (c) and the amount of degradable substrate (S) available for fermentation. The amount of degradable substrate available is time dependent with the maximal value occurring at time zero (S_0). The instantaneous rate of degradation is calculated by multiplying the constant fractional degradation rate (c) by the amount of degradable substrate (S) at time t [3]. As the fractional degradation rate is constant, and S is maximal at the commencement of the incubation, instantaneous rate of degradation is maximal at the start of incubation, following a lag period if present. As fermentation progresses, the amount of degradable substrate decreases while the fractional rate of degradation remains constant. Therefore, the instantaneous rate of degradation declines continuously, from its maximum at the start of the incubation until it finally reaches zero due to available fermentable material being exhausted. When fermentation ceases, due to a lack of degradable substrate, the instantaneous rate of degradation becomes zero, with no additional gas production occurring, having reached an upper asymptote. Therefore, the characteristic diminishing returns pattern of the simple Mitscherlich describes a scenario whereby rate of fermentation, and thus gas production, is initially at a maximum and continuously decreases, as a function of time, until an asymptote is reached.

Unlike the simple Mitscherlich, the France model is capable of describing both diminishing returns and sigmoidal behaviour. This is achieved by assuming that fractional rate of degradation can vary with time. Depending on the values of the fractional rate constants, viz. c and d in this manuscript, the fractional rate of degradation can remain constant, decrease or increase with time [4]. In the France model, when the fractional rate of degradation is constant diminishing return type behaviour is described (as in the simple Mitscherlich). None of the gas production profiles examined in this study showed clear sigmodal behaviour and therefore the flexibility of the France model in its ability to describe both diminishing returns and sigmoidal shapes was not demonstrated. When describing sigmoidal behaviour, initially the rate of degradation increases resulting in exponential-type behaviour. As it continues to increase, a point of inflexion occurs whereby the rate reaches its maximal value. Following inflexion, the rate of degradation decreases resulting in diminishing returns behaviour with an asymptote being approached.

Of the 25 datasets examined in this study, in all but three the substrate was ground and passed through a 1 mm screen prior to incubation, with the remaining three being chopped (length not reported). A clear increase in gas production, and associated increase in extent of substrate degradation, is observed when comparing the ground (Dataset 18–25) with the chopped substrates (Dataset 15–17) of Experiment 2 (see Supplementary Information Figure S3 vs. Figure S4 and Table 5 for associated gas production profiles and extent of degradation values, respectively). However, 17 of these datasets, including both ground and chopped substrates, are atypical in nature and exhibit a second phase of gas production. In these 17 datasets the first phase is well described by the simple Mitscherlich whereby following a lag, rate of degradation and thus gas production are initially at their maximum and continuously decrease until an asymptote is approached. Following this first phase, a second phase occurs. This second phase can show either diminishing returns or a linear pattern. These phases might be attributed to differences in chemical or nutritional fractions of the feedstuffs [16]. Phase 1 may represent the gas produced from the fermentation of sugars or a soluble readily fermentable fraction, while the second phase consists of gas produced from the fermentation of structural carbohydrates or an insoluble potentially fermentable fraction [35,37]. Alternatively, the occurrence of the second gas production phase can potentially be attributed to chemical or structural barriers implicit in the substrate that must be overcome in order to continue degradation [15]. Furthermore, the possibility of microbial turnover in batch cultures, and the small amount of gas produced from 'self-fermentation' may add to the second phase of gas production [35,38]. Many other factors may influence the profile shape including: inter-animal variability, ration of the donor of the microbial *inoculum*, length of time the donor animal was adapted to the ration, time of day the *inoculum* was collected, ruminal vs. faecal sources of *inoculum* and frozen vs. fresh *inoculum* [5,19,35]. This leads to the conclusion that these factors may have the potential to influence microbial diversity and abundance in the *inoculum*, which in turn influences fermentative ability and by extension influences gas production.

4.2. Extent of Degradation

The ability to describe typical diminishing returns and sigmoidal gas production profiles using a variety of models (e.g., Mitscherlich, Michaelis-Menten, logistic, Gompertz and France) is well established [3,34,36]. However, the description of atypical multiphasic gas production curves is much less established, particularly how to link the in vitro gas production technique to the extent of degradation in the animal. Groot et al. [15] proposed a model that fitted multiphasic profiles using two or more generalised rectangular hyperbolae. Applying this model, some authors have estimated the amount of gas produced and rate of gas production of various feedstuffs [35,39]. Likewise, Wang et al. [36] described single- and multi-phase gas production curves using logistic-exponential equations. However, in these studies differences in feedstuffs were identified on the basis of the amount of gas produced rather than on criteria linked to animal performance.

Two biphasic models are presented in this paper that make use of a simple Mitscherlich term when describing the first phase of gas production and a second phase comprising either an additional Mitscherlich or a simple linear term with abrupt cut-off. Depending on the nature of the profile, both the double Mitscherlich (Equation (3)) and Mitscherlich + linear (Equation (4)) fitted the atypical datasets well, resulting in parameter values that can be used to calculate and compare the extent of degradation of respective substrates and substrate treatments. For example, three of the datasets examined in this study (Dataset 15–17) encompass chopped hay treated with increasing levels of an enzyme (0, 0.75 or 3.7 mL enzyme per kg DM hay, respectively). The Mitscherlich + linear fitted these datasets the best and using the associated parameter values, extent of substrate degradation was demonstrated to increase with increasing levels of enzymatic treatment, viz. 38.4%, 38.9% and 42.0% for Dataset 15–17, respectively. When fitting the Mitscherlich + linear to these datasets, it was assumed that following the linear trend an abrupt cut-off is reached (i.e., an asymptote is reached) and gas production ceases. This can be observed by inspecting Dataset 1 and 8 in Figure 1 whereby gas production ceases to increase between the last two data points. In comparison, examining Dataset 15–17 (see Supplementary Information, Figure S3), visually there appears to be potential for further gas production as an asymptote, in the form of an abrupt cut-off following the linear segment, has yet to be reached. If the linear trend continues after the 76 h incubation period, there is potential for continued substrate degradation and associated gas production. Therefore, extent of substrate degradation calculated using the Mitscherlich + linear would be an underestimate if fermentation continued beyond the 76 h incubation period used to generate these gas production profiles.

As previously mentioned, the extent of degradation calculated using the fermentation kinetic parameters generated from applying the four proposed models to 25 datasets ranged widely from a minimum of 12.7% to a maximum of 44.2%. This wide range in values can partially be attributed to the use of parameter values from a model that fits a particular gas production profile poorly. Even in a given dataset, large variation in calculated extent of degradation values existed. For example, in Dataset 3, extent of degradation using the France model was 12.7% compared to 37.5% with the Mitscherlich + linear. These findings are in contrast with those of Dhanoa et al. [34] whereby the model being applied (generalised Mitscherlich, simple Mitscherlich, generalised Michaelis-Menten, simple Michaelis-Menten, Gompertz and logistic) had very little effect on extent of degradation values. However, the gas production profiles of Dhanoa et al. [34] using mixed rumen microorganisms as the *inoculum* were all monophasic in nature and therefore reasonably well described by the aforementioned models. Indeed, when examining the eight typical gas production profiles of this manuscript, Dataset 18–25, there was very little difference in extent of degradation in a given dataset regardless of model applied. When comparing the standard deviation of extent of degradation determined by the four models when applied to the same typical gas production dataset, Dataset 22 had the lowest value of

0.1% while Dataset 18 had the highest deviation at 3.1%. Examining Dataset 22, the simple Mitscherlich fitted the dataset best with an associated extent of degradation of 42.9%. Both the double Mitscherlich and France models reverted to the simple Mitscherlich, as the simple Mitscherlich fitted this dataset better than their generalized forms, and therefore were in agreement with an extent of degradation of 42.9%. The Mitscherlich + linear was also in agreement with this value, 42.8%. In contrast, examining the 17 atypical biphasic gas production profiles, the model applied had considerable ramifications on calculated extent of degradation. In these datasets, the lowest standard deviation of extent of degradation between the four models when applied to a single dataset occurred in Dataset 12 at 3.4% while the largest deviation occurred in Dataset 3 at 10.1%. In the atypical profile of Dataset 12, the double Mitscherlich fitted the gas production profile the best with a calculated extent of degradation of 30.2%, the simple Mitscherlich, France and Mitscherlich + linear overestimated the extent of degradation, viz. 33.4%, 32.6% and 38.2%, respectively. Overall, this discrepancy in extent of degradation values for a given dataset can be attributed to fitting a monophasic equation (viz. Equations (1) and (2)) to a distinctly biphasic profile, or fitting a linear term to a non-linear segment, resulting in poor kinetic parameter estimates and by extension extent of degradation values.

It is important to note that when calculating extent of degradation, the value of S_0, the amount of incubated substrate that is potentially degradable, was taken from the literature. This value was set to 0.538 regardless of dataset and the associated substrate represented by that dataset. The value of 0.538 is the apparent in vivo dry matter digestibility, using ponies, of ground and pelleted hay consisting of a 50:50 mix of Lucerne hay and Cocksfoot hay [23]. When performing the gas production technique of Theodorou et al. [2], the potentially undegradable fraction of the substrate (U_0) can be obtained by weighing the residual matter after gas production has ceased. Likewise, the potentially degradable value (S_0) can be calculated by subtracting U_0 from the quantity of substrate initially incubated. However, these values were not available at the time of this current study and a constant value was assumed. Therefore, greater differences in calculated values of extent of degradation should be expected as S_0 and U_0 will vary between substrates, substrate composition and the treatment received.

5. Conclusions

Two models, a double Mitscherlich and Mitscherlich + linear with abrupt cut-off, were proposed and derived to describe atypical gas production patterns characterized by two distinct phases of gas production. The models fitted these atypical curves well and due to their hybrid nature are also able to describe typical monophasic gas production profiles through their ability to revert to a simple Mitscherlich. These models contain kinetic parameters that can be used to calculate extent of substrate degradation using relatively simple equations. Given that extent of degradation is linked to nutrient supply, these models provide useful information regarding the evaluation of feedstuffs using in vitro methods [34,40].

Supplementary Materials:
Table S1: Observed Gas Production Values of Datasets 1–25 from Experiment 1 and 2; **Table S2**: Final parameter estimates from fitting the simple Mitscherlich (Eqn. 1) to Dataset 1–25; **Table S3**: Final parameter estimates from fitting the France model (Eqn. 2) to Dataset 1–25; **Table S4**: Final parameter estimates from fitting the double Mitscherlich model (Eqn. 3) to Dataset 1–25; **Table S5**: Final parameter estimates from fitting the Mitscherlich + linear (Eqn. 4) to Dataset 1–25. **Figure S1**: Observed (•) and predicted gas production profiles resulting from fitting Equations (1)–(4) to Dataset 1–7, horses displaying clinical signs of laminitis, from Experiment 1; **Figure S2**: Observed (•) and predicted gas production profiles resulting from fitting Equations (1)–(4) to Datasets 8–14, clinically normal horses, from Experiment 1; **Figure S3**: Observed (•) and predicted gas production profiles resulting from fitting Equations (1)–(4) to Dataset 15–17, datasets exhibiting atypical dual-phase gas production curves, from Experiment 2. **Figure S4:** Observed (•) and predicted gas production profiles resulting from fitting Equations (1)–(4) to Dataset 18–25, datasets exhibiting typical single-phase gas production curves, from Experiment 2.

Author Contributions: Conceptualization, J.F. and M.S.D.; methodology, J.F., M.S.D. and S.L.; software, C.D.P.; validation, C.D.P.; formal analysis, C.D.P.; investigation, C.D.P.; resources, A.G. and J.-A.M.D.M.; data curation, C.D.P.; writing—original draft preparation, C.D.P.; writing—review and editing, C.D.P., M.S.D., A.G., J.-A.M.D.M.,

S.L., J.L.E. and J.F.; visualization, C.D.P.; supervision, J.F. and S.L.; project administration, J.F.; funding acquisition, J.F. All authors have read and agreed to the published version of the manuscript.

References

1. Menke, K.H.; Steingass, H. Estimation of the energetic feed value obtained from chemical analysis and in vitro gas production using rumen fluid. *Anim. Res. Dev.* **1988**, *28*, 7–55.
2. Theodorou, M.K.; Williams, B.A.; Dhanoa, M.S.; McAllan, A.B.; France, J. A simple gas production method using a pressure transducer to determine the fermentation kinetics of ruminant feeds. *Anim. Feed Sci. Technol.* **1994**, *48*, 185–197. [CrossRef]
3. France, J.; Dijkstra, J.; Dhanoa, M.S.; López, S.; Bannink, A. Estimating the extent of degradation of ruminant feeds from a description of their gas production profiles observed *in vitro*: Derivation of models and other mathematical considerations. *Br. J. Nutr.* **2000**, *83*, 143–150. [CrossRef] [PubMed]
4. France, J.; Dhanoa, M.S.; Theodorou, M.K.; Lister, S.J.; Davies, D.R.; Isac, D. A model to interpret gas accumulation profiles associated with in vitro degradation of ruminant feeds. *J. Theor. Biol.* **1993**, *163*, 99–111. [CrossRef]
5. Mauricio, R.M.; Owen, E.; Mould, F.L.; Givens, I.; Theodorou, M.K.; France, J.; Davies, D.R.; Dhanoa, M.S. Comparison of bovine rumen liquor and bovine faeces as inoculum for an in vitro gas production technique for evaluating forages. *Anim. Feed Sci. Technol.* **2001**, *89*, 33–48. [CrossRef]
6. Dhanoa, M.S.; France, J.; Crompton, L.A.; Mauricio, R.M.; Kebreab, E.; Mills, J.A.N.; Sanderson, R.; Dijkstra, J.; López, S. Technical note: A proposed method to determine the extent of degradation of a feed in the rumen from the degradation profile obtained with the in vitro gas production technique using feces as the inoculum. *J. Anim. Sci.* **2004**, *82*, 733–746. [CrossRef]
7. López, S. In vitro and in situ techniques for estimating digestibility. In *Quantitative Aspects of Ruminant Digestion and Metabolism*; Dijkstra, J., Forbes, J.M., France, J., Eds.; CAB International: Wallingford, UK, 2005; pp. 87–121.
8. Calabrò, S.; Cutrignelli, M.I.; Piccolo, G.; Bovera, F.; Zicarelli, F.; Gazaneo, M.P.; Infascelli, F. In vitro fermentation kinetics of fresh and dried silage. *Anim. Feed Sci. Technol.* **2005**, *123–124*, 129–137. [CrossRef]
9. Yao, K.Y.; Gu, F.F.; Liu, J.X. In vitro rumen fermentation characteristics of substrate mixtures with soybean meal partially replaced by microbially fermented yellow wine lees. *Ital. J. Anim. Sci.* **2020**, *19*, 18–24. [CrossRef]
10. Lowman, R.S.; Theodorou, M.K.; Hyslop, J.J.; Dhanoa, M.S.; Cuddeford, D. Evaluation of an in vitro batch culture technique for estimating the in vivo digestibility and digestible energy content of equine feeds using equine faeces as the source of microbial inoculum. *Anim. Feed Sci. Technol.* **1999**, *80*, 11–27. [CrossRef]
11. Murray, J.A.M.D.; Longland, A.; Moore-Colyer, M. In vitro fermentation of different ratios of high-temperature dried lucerne and sugar beet pulp incubated with an equine faecal inoculum. *Anim. Feed Sci. Technol.* **2006**, *129*, 89–98. [CrossRef]
12. Elghandour, M.M.Y.; Vázquez Chagoyán, J.C.; Salem, A.Z.M.; Kholif, A.E.; Martínez Castañeda, J.S.; Camacho, L.M.; Buendía, G. In vitro fermentative capacity of equine fecal inocula of 9 fibrous forages in the presence of different doses of Saccharomyces cerevisiae. *J. Equine Vet. Sci.* **2014**, *34*, 619–625. [CrossRef]
13. Murray, J.M.D.; McMullin, P.; Handel, I.; Hastie, P.M. Comparison of intestinal contents from different regions of the equine gastrointestinal tract as inocula for use in an in vitro gas production technique. *Anim. Feed Sci. Technol.* **2014**, *187*, 98–103. [CrossRef]
14. Garber, A.; Hastie, P.M.; Handel, I.; Murray, J.M.D. In vitro fermentation of different ratios of alfalfa and starch or inulin incubated with an equine faecal inoculum. *Livest. Sci.* **2018**, *215*, 7–15. [CrossRef]
15. Groot, J.C.J.; Cone, J.W.; Williams, B.A.; Debersaques, F.M.A.; Lantinga, E.A. Multiphasic analysis of gas production kinetics for in vitro fermentation of ruminant feeds. *Anim. Feed Sci. Technol.* **1996**, *64*, 77–89. [CrossRef]
16. López, S.; Dijkstra, J.; Dhanoa, M.S.; Bannink, A.; Kebreab, E.; France, J. A generic multi-stage compartmental model for interpreting gas production profiles. In *Modelling Nutrient Digestion and Utilization in Farm Animals*; Sauvant, D., Van Milgen, J., Faverdin, P., Friggens, N., Eds.; Wageningen Academic Publishers: Wageningen, The Netherlands, 2010; pp. 139–147.
17. Mitscherlich, E. Das gesetz des minimums und das gesetz des abnehmenden bodenertrages. *Landwirstsch Jahrb* **1909**, *38*, 537–552.

18. Thornley, J.; France, J. Growth functions. In *Mathematical Models in Agriculture: Quantitative Methods for the Plant, Animal and Ecological Sciences*, 2nd ed.; CAB International: Wallingford, UK, 2007; pp. 136–169.
19. Murray, J.A.M.D.; Scott, B.; Hastie, P.M. Fermentative capacity of equine faecal inocula obtained from clinically normal horses and those predisposed to laminitis. *Anim. Feed Sci. Technol.* **2009**, *151*, 306–311. [CrossRef]
20. Garber, A.; Hastie, P.M.; Farci, V.; Murray, J.M.D. Yeast increases in vitro gas production of high fibre substrate when incubated with equine faecal inoculum. In Proceedings of the 8th European Equine Health & Nutrition Congress, Antwerp, Belgium, 23–24 March 2017; p. 119.
21. ANKOM^{RF}. *Gas Production System–Operator's Manual (Rev F 1/26/15)*; Ankom Technology: Macedon, NY, USA, 2015.
22. Dhanoa, M.S.; López, S.; Sanderson, R.; France, J. Simplified estimation of forage degradability in the rumen assuming zero-order degradation kinetics. *J. Agric. Sci.* **2009**, *147*, 225–240. [CrossRef]
23. Drogoul, C.; Poncet, C.; Tisserand, J.L. Feeding ground and pelleted hay rather than chopped hay to ponies 1. Consequences for in vivo digestibility and rate of passage of digesta. *Anim. Feed Sci. Technol.* **2000**, *87*, 117–130. [CrossRef]
24. SAS Institute Inc. *SAS/STAT® 9.2 User's Guide, Second Edition*; SAS Institute Inc.: Cary, NC, USA, 2009.
25. Bibby, J.; Toutenburg, H. *Prediction and Improved Estimation in Linear Models*; John Wiley and Sons: London, UK, 1977.
26. Ross, T. Indices for performance evaluation of predictive models in food microbiology. *J. Appl. Bacteriol.* **1996**, *81*, 501–508. [CrossRef]
27. Lin, L.I.-K. A concordance correlation coefficient to evaluate reproducibility. *Biometrics* **1989**, *45*, 255–268. [CrossRef]
28. Lin, L.; Torbeck, L.D. Coefficient of accuracy and concordance correlation coefficient: New statistics for methods comparison. *PDA J. Pharm. Sci. Technol.* **1998**, *52*, 55–59. [PubMed]
29. Akaike, H. A new look at the statistical model identification. *IEEE Trans. Automat. Contr.* **1974**, *19*, 716–723. [CrossRef]
30. Draper, N.R.; Smith, H. Fitting a straight line by least squares. In *Applied Regression Analysis*; John Wiley & Sons, Inc.: Hoboken, NJ, USA, 1998; pp. 15–96.
31. Krishnamoorthy, U.; Soller, H.; Steingass, H.; Menke, K.H. A comparative study on rumen fermentation of energy supplements in vitro. *J. Anim. Physiol. Anim. Nutr.* **1991**, *65*, 28–35. [CrossRef]
32. Blümmel, M.; Ørskov, E.R. Comparison of in vitro gas production and nylon bag degradability of roughages in predicting feed intake in cattle. *Anim. Feed Sci. Technol.* **1993**, *40*, 109–119. [CrossRef]
33. Schofield, P.; Pitt, R.E.; Pell, A.N. Kinetics of fiber digestion from in vitro gas production. *J. Anim. Sci.* **1994**, *72*, 2980–2991. [CrossRef]
34. Dhanoa, M.S.; López, S.; Dijkstra, J.; Davies, D.R.; Sanderson, R.; Williams, B.A.; Sileshi, Z.; France, J. Estimating the extent of degradation of ruminant feeds from a description of their gas production profiles observed *in vitro*: Comparison of models. *Br. J. Nutr.* **2000**, *83*, 131–142. [CrossRef]
35. Cone, J.W.; Van Gelder, A.H.; Visscher, G.J.W.; Oudshoorn, L. Influence of rumen fluid and substrate concentration on fermentation kinetics measured with a fully automated time related gas production apparatus. *Anim. Feed Sci. Technol.* **1996**, *61*, 113–128. [CrossRef]
36. Wang, M.; Tang, S.X.; Tan, Z.L. Modeling in vitro gas production kinetics: Derivation of Logistic-Exponential (LE) equations and comparison of models. *Anim. Feed Sci. Technol.* **2011**, *165*, 137–150. [CrossRef]
37. Stefanon, B.; Pell, A.N.; Schofield, P. Effect of maturity on digestion kinetics of water-soluble and water-insoluble fractions of alfalfa and brome hay. *J. Anim. Sci.* **1996**, *74*, 1104–1115. [CrossRef]
38. Theodorou, M.K.; Davies, D.R.; Nielsen, B.B.; Lawrence, M.I.G.; Trinci, A.P.J. Determination of growth of anaerobic fungi on soluble and cellulosic substrates using a pressure transducer. *Microbiology* **1995**, *141*, 671–678. [CrossRef]
39. Van Gelder, A.H.; Hetta, M.; Rodrigues, M.A.M.; De Boever, J.L.; Den Hartigh, H.; Rymer, C.; Van Oostrum, M.; Van Kaathoven, R.; Cone, J.W. Ranking of in vitro fermentability of 20 feedstuffs with an automated gas production technique: Results of a ring test. *Anim. Feed Sci. Technol.* **2005**, *123–124*, 243–253. [CrossRef]
40. Tamminga, S.; Van Straalen, W.M.; Subnel, A.P.J.; Meijer, R.G.M.; Steg, A.; Wever, C.J.G.; Blok, M.C. The Dutch protein evaluation system: The DVE/OEB-system. *Livest. Prod. Sci.* **1994**, *40*, 139–155. [CrossRef]

11

Relations of Ruminal Fermentation Parameters and Microbial Matters to Odd- and Branched-Chain Fatty Acids in Rumen Fluid of Dairy Cows at Different Milk Stages

Keyuan Liu [1,2,†], Yang Li [2,†], Guobin Luo [2,3], Hangshu Xin [2], Yonggen Zhang [2,*] and Guangyu Li [1,*]

1 Institute of Special Economic Animal and Plant Science, Chinese Academy of Agricultural Sciences, Changchun 130112, China; liukeyuan0212@163.com
2 Department of Animal Science and Technology, Northeast Agricultural University, Harbin 150030, China; liyang1405053@sina.com (Y.L.); guobinluo@126.com (G.L.); laura_liuky@foxmail.com (H.X.)
3 Zhejiang NHU Company Ltd., Shaoxing 312500, China
* Correspondence: zhangyonggen@sina.com (Y.Z.); liguangyu@caas.cn (G.L.)
† These authors contributed equally.

Simple Summary: The objective of this study was to determine the relationships between milk odd- and branched-chain fatty acids (OBCFAs) and ruminal fermentation parameters, microbial populations, and base contents. Significant relationships existed between the concentrations of C11:0, *iso*-C15:0, *anteiso*-C15:0, C15:0, and *anteiso*-C17:0 in rumen and milk. The total OBCFA content in milk was positively related to the acetate molar proportion but negatively correlated with isoacid levels. The adenine/N ratio was negatively related to milk OBCFA content but positively associated with the *iso*-C15:0/*iso*-C17:0 ratio.

Abstract: The purpose of this research was to evaluate whether relationships exist between odd- and branched-chain fatty acids (OBCFAs) originating from milk fat and the corresponding data of ruminal fermentation parameters, microbial populations, and base contents that were used to mark microbial protein in rumen. Nine lactating Holstein dairy cows with similar body weights and parity were selected in this study, and the samples of rumen and milk were collected at the early, middle, and late stages, respectively. The rumen and milk samples were collected over three consecutive days from each cow, and the ruminal and milk OBCFA profiles, ruminal fermentation parameters, bacterial populations, and base contents were measured. The results showed that the concentrations of OBCFAs, with the exception of C11:0 and C15:0, were significantly different between milk and rumen ($p < 0.05$). The concentrations of *anteiso*-fatty acids in milk were higher than those in rumen, and the contents of linear odd-chain fatty acids were higher than those of branched-chain fatty acids in both milk and rumen. Significant relationships that existed between the concentrations of C11:0, *iso*-C15:0, *anteiso*-C15:0, C15:0, and *anteiso*-C17:0 in rumen and milk ($p < 0.05$). The total OBCFA content in milk was positively related to the acetate molar proportion but negatively correlated with isoacid contents ($p < 0.05$). The populations of *Ruminococcus albus*, *R. flavefacients*, and *Eubacterium ruminantium* were significantly related to milk C13:0 contents ($p < 0.05$). The adenine/N ratio was negatively related to milk OBCFA content ($p < 0.05$) but positively associated with the *iso*-C15:0/*iso*-C17:0 ratio ($p < 0.05$). Milk OBCFAs were significantly correlated with ruminal fermentation parameters, ruminal bacterial populations, and base contents. Milk OBCFAs had the potential to predict microbial nitrogen flow, and the prediction equations for ruminal microbial nitrogen flow were established for OBCFAs in dairy milk.

Keywords: fermentation parameters; microbial populations; microbial bases; odd- and branched-chain fatty acids; lactation stages

1. Introduction

There is growing awareness that milk fat content can respond to physiological and metabolic health situations [1,2]. The milk fatty acid profile is a dynamic pattern influenced by many factors, such as lactational stage, season, and dietary composition [3–5]. The microbial processes of the rumen confer the ability to convert feeds into available nutrients for the ruminant animal [6]. Odd- and branched-chain fatty acids (OBCFAs) of ruminant milk generally originate from rumen bacteria [7] and are mainly present in bacterial membrane lipids [8]. Some studies have discussed the potential of OBCFAs as markers of rumen fermentation and ruminal bacteria [9–12].

Lactating dairy cow digestion is strongly determined by the microbial population in the rumen. In the rumen, microbial fermentation of feedstuffs produces volatile fatty acids (VFAs), which are the main energy supply substances in ruminants. There is a significant relationship between ruminal pH and the profile of VFAs available for absorption [13]. Hence, the composition and amount of milk fatty acids are determined by the proportions and the total amounts of fermentation end-products in ruminants [14]. Many studies have examined the effects of microbial protein synthesis and microbial nucleic acid composition in the rumen on protein nutrition [15]. The composition of the ruminal microbial ecosystem in the forestomach of ruminants is known to be affected by the type and quantity of the ration, feeding intervals, specific additives (e.g., antibiotics), and the host animal itself [16]. To identify the relationships between bacterial populations and milk OBCFA concentrations, seven kinds of bacteria species (cellulolytic or amylolytic bacteria) were selected for the current study. *Ruminococcus albus*, *R. flavefaciens*, *Fibrobacter succinogenes* [17], and *Eubacterium ruminantium* [18] are the predominant ruminal cellulolytic bacteria. The genus *Butyrivibrio fibrisolvens* is a heterogeneous bacterial taxon [19]. *Selenomonas ruminantium* [20] and *Streptococcus bovis* [21] are important for the degradation of starch and lactate, which are abundant in high-grain diets.

However, there are few studies on the relationships between milk OBCFAs during different lactation stages and ruminal bacterial populations. The objective of this research was to estimate the potential use of OBCFAs in milk to predict ruminal fermentation parameters and rumen microbial matter. First, we investigated whether there were relationships between the contents of OBCFA in milk and rumen, and the fermentation parameters, bacterial populations, and bases of milk during different stages of lactation. Second, we developed equations for fermentation parameters, bacterial populations, and bases using the independent datasets of OBCFAs and identified the best OBCFA combination.

2. Materials and Methods

2.1. Animals and Basal Diets

Nine lactating Holstein dairy cows of similar body weights (650 ± 33 kg body weight) were examined at the same fetal time, and samples of rumen and milk were collected at the early, middle, and late lactation stages. The milk yields were 35.44 ± 2.63, 37.62 ± 2.85, and 26.98 ± 2.79 kg/d in the early, middle, and late stages, respectively. The cows were fed total mixed rations (TMRs), whose composition and nutrition levels are shown in Table 1, according to the dairy nutrient requirements of the NRC (national research council) (2001) [22].

Table 1. Feed ingredients and chemical composition (g/kg DM) of the rations.

Ingredients	Content
Alfalfa hay	73
Chinese wildrye	43
Corn silage	334
Corn	220
Soybean meal	41
DDGS	214
Cottonseed meal	58
Molasses	5.0
NaCl	3.0
$CaHPO_4$	1.6
Limestone	5.4
Premix *	3.0
Nutrient levels	
NEL(MJ/kg) †	8.71
CP	162
NDF	312
ADF	183
Ca	8.7
TP	4.6

* The premix provided the following per kg of diets: Vitamin A 330,000 IU, Vitamin D 60,000 IU, Vitamin E 1000 IU, Zn 2100 mg, Mn 1500 mg, Cu 535 mg, Se 12 mg, I 45 mg. † NEL was an estimated value [22]. DM, dry matter; DDGS, Corn distillers dried grains; CP, crude protein; NDF, neutral detergent fiber; ADF, acid detergent fiber; TP, total phosphorus.

2.2. Samples Collection and Analysis Method

All the TMR samples were analyzed for DM, nitrogen (N) (AOAC 968.06), calcium (Ca) (AOAC 927.02), and total phosphorus (TP) (AOAC 965.17) according to the procedures of the AOAC (Association of Official Analytical Chemists) (1990) after air-drying at 60 ± 5 °C [23]. The content of crude protein (CP) was calculated as N × 6.25. The acid detergent fiber (ADF) and neutral detergent fiber (NDF) concentrations were analyzed based on the procedures described by Van Soest et al. [24] using the Ankom system (Ankom 220 fiber analyzer; Ankom, New York, USA) with a heat-stable α-amylase and expressed exclusive of residual ash. Net energy for lactation (NE_L) at a production level was calculated using an NRC summative approach from the dairy nutrient requirement [22].

The rumen contents were evacuated via the gastric canal over 3 consecutive days. One part of the rumen contents was filtered through four layer of cheesecloth and the filtrates were preserved at −20 °C for the analysis of the concentrations of VFAs and ammonia nitrogen (NH_3-N), after determining the pH that was obtained from samples by a pH meter (Sartorius Basic pH Meter, Gottingen, Germany). The filtered samples were treated according to the description of Li and Meng [25]. Then, the contents of VFAs were analyzed by a gas chromatography (GC 2010, Shimadzu, Tokyo, Japan) with an FFAP capillary column (HP-INNOWAX, 30 m × 0.25 mm × 0.2 μm, Agilent, California, USA), and using an ammonia-sensing electrode (Expandable Ion Analyzer EA 940, Orion, Massachusetts, USA) to determine the concentration of NH_3-N.

The other part of the rumen contents were squeezed through two layers of cheesecloth, and about 5 mL was preserved at −80 °C for the extraction of DNA [26] and nearly 100 mL was preserved at −20 °C for the analysis of OBCFA, microbial bases, and the total nitrogen (N). The samples of OBCFA were treated according to the description of Zhang et al. [27] and analyzed by a gas chromatography (GC 2010, Shimadzu, Tokyo, Japan) with an SP-2560™ column for fatty acid methyl esters (100 m × 0.25 mm × 0.2 μm, Supelco, Pennsylvania, USA). The carrier gas was highly pure, and the injector pressure was held constant at 266.9 Kpa. The initial oven temperature was held at 170 °C for 30 min, and increased at 1.5 °C/min to 200 °C and held for 20 min, and then increased by 5 °C/min to 230 °C and held for 5 min. The rumen bases were extracted from freeze-dried samples using perchloric acid, as described by Vlaeminck et al. [9]. The standards of individual bases (≥99.5%, Aladdin, Shanghai, China) were formulated to a concentration of 50 mg/L. The mixed solution mixed by the standard base solutions with the same volume was serially diluted into 5 gradients and subsequently analyzed by HPLC (high performance liquid chromatography) using a C18 column (5 μ, 250 × 4.6 mm, Diamonsil, Guangzhou, China). The buffer solution (20 mM $NH_4H_2PO_4$) was run isocratically at 1 mL/min, and the effluent was monitored at 254 nm.

DNA was extracted from the rumen contents by the bead-beating procedure described previously by Reilly and Attwood [28]. In detail, 1.5 mL of rumen liquid was placed into a 2-mL centrifuge tube and later centrifuged for 5 min at 12,000 g/min, and the supernatant was subsequently removed. Phosphate buffer saline (PBS; 1.5 mL; pH = 8.0) was then added, and the sample was mixed and centrifuged for 5 min at 12,000 g/min. The supernatant was then removed. CTAB (hexadecyl trimethyl ammonium bromide) buffer (800 μL; sterilized solution containing 4 g of CTAB, 16.364 g of NaCl, 20 mL of 1 mol/L Tris-HCl, pH = 8.0, and 8 mL of 0.5 mol/L EDTA (Ethylene Diamine Tetraacetic Acid) to a final volume of 200 mL) was then added, and the sample was mixed, cultured for 20 min in 70 °C, and centrifuged for 10 min at 10,000 g/min. The supernatant was then transferred to a new centrifuge tube, and 5 μL of RNA enzymes (10 mg/mL) were added. The sample was mixed and then incubated for 30 min at 37 °C. An equal volume of phenol/chloroform/isoamyl alcohol 25:24:1 solution was then added, and the sample was mixed for 15 to 30 s until a white emulsion appeared and then centrifuged for 10 min at 13,000 rpm. The supernatant was added to a new centrifuge tube, and the last step was then repeated until the interface became clear. After a clear interface was obtained, an 0.8-fold volume of the isopropyl alcohol mixture was added into the tube, and the sample was gently mixed, placed at room temperature for 5 min, and then placed at −20 °C overnight. The tube was then centrifuged for 15 to 20 min at 13,000 g/min, and the supernatant was carefully decanted to leave a white DNA precipitation in the bottom of the tube. Ice-cold 70% ethanol (750 μL) was then added to the tube to gently resuspend the DNA precipitation, and the sample was then centrifuged for 10 to 15 min at 12,000 g/min. The supernatant was decanted, and the DNA was allowed to air dry. The DNA was then dissolved with 100 μL (depending on the precipitation volume) of TE (Tris-EDTA) buffer and incubated at 70 °C for 5 min. The DNA solution was then stored at −20 °C until use. The air-dried DNA pellet was redissolved in TE buffer (10 mmol/L Tris-HCl, 1 mmol/L EDTA, pH 8.0) and diluted to concentrations of 10 ng/μL and stored at 4 °C until real-time PCR amplification. The bacterial 16SrRNA genes were amplified using absolute quantification PCR (qPCR). The standard DNA in real-time PCR used plasmid DNA containing the respective target gene sequence, which was obtained by PCR cloning using the species-specific primer set. The specific method was described by Singha et al. [29]. The primer sequences for the 16SrRNA genes and specific amplifications of the correct size are shown in Table 2. The qPCR protocol was performed with ABI 7500 system software (ABI 7500, Massachusetts, USA) using TOYOBO (Osaka, Japan) DNA Master SYBR Green II.

Table 2. Primers for real time-PCR.

Bacteria species	Primer Sequence (5′-3′)	Product Size (bp)	Reference
Fibrobacter succinogenes	F-GGCGGGATTGAATGTACCTTGAGA R-TCCGCCTGCCCCTGAACTATC	204	Yang (2007) [30]
Ruminococcus albus	F-GTTTTAGGATTGTAAACCTCTGTCTT R-CCTAATATCTACGCATTTCACCGC	273	Yang (2007) [30]
Ruminococcus flavefaciens	F-GATGCCGCGTGGAGGAAGAAG R-CATTTCACCGCTACACCAGGAA	278	Yang (2007) [30]
Butyrivibrio fibrisolvens	F-TAACATGAGTTTGATCCTGGCTC R-CGTTACTCACCCGTCCGC	113	Yang (2007) [30]
Eubacterium ruminantium	F-CTCCCGAGACTGAGGAAGCTTG R-GTCCATCTCACACCACCGGA	184	Stevenson, et al. (2007) [31]
Streptococcus bovis	F-TTCCTAGAGATAGGAAGTTTCTTCGG R-ATGATGGCAACTAACAATAGGGGT	127	Stevenson, et al. (2007) [31]
Selenomonas ruminantium	F-CAATAAGCATTCCGCCTGGG R-TTCACTCAATGTCAAGCCCTGG	138	Stevenson, et al. (2007) [31]

Nearly 100-mL milk samples were collected and preserved at −20 °C, and treated the samples that referred to a description of Vlaeminck et al. [28]. The OBCFA compositions were analyzed by a gas chromatography (GC 2010, Tokyo, Shimadzu, Japan) and the analysis program was the same as the rumen samples.

2.3. Statistical Analyses

All data statistical analyses were performed using SAS 9.2 (SAS Institute Inc.,Cary, NC, USA).

The data of variable analysis was done by the MIXED procedure. The MIXED model procedure with the inclusion of the random effect of the study was as described by St-Pierre [32]. The effect of different dietary F:C ratios on OBCFAs were estimated following:

$$Y_{ijk} = \mu + T_i + P_j + C_k + \epsilon_{ijk}, \tag{1}$$

where Y_{ijk} is the individual observation, μ is the overall mean, T_i is the effect of the dietary treatment ($i = 3$; F:C = 30:70, 50:50, and 70:30), P_j is the effect of the experimental period, C_k is the effect of the cow, and ϵ_{ijk} is the residual error. The effect of the cow was treated as a random effect. For all statistical analyses, significance was declared at $p < 0.05$.

The correlation between the milk OBCFA profile, and rumen OBCFA concentrations, and data, which were obtained from the VFAs, NH_3-N, and pH, were analyzed by CORR PROC using the Pearson correlation method. The correlations were determined to be significant at $p \leq 0.05$.

The data of milk OBCFA were considered as an independent data set and the data of VFAs, NH_3-N, and pH were a dependent dataset. Multiple regression was applied using the STEPWISE method of the REG procedure. The SLENTRY and START values were all 0.05. The equations were determined by least squares estimation ($p \leq 0.05$). The regression equations were evaluated based on the root mean square error (RMSE) and coefficient of multiple determinations (R^2) of the regression model:

$$\text{RMSE} = \sqrt{\frac{1}{n} \times \sum\nolimits_{i=1}^{n} (y_i - \hat{y}_i)^2}, \tag{2}$$

where n is the number of observations, and y_i and $\hat{y}_i$ are the observed and predicted values, respectively.

3. Results

3.1. Changes during Different Milk Stages in OBCFA Production in Rumen Fluid and Milk

The changes in OBCFAs in rumen fluid and milk fat are shown in Table 3. The C11:0 and C13:0 contents were significantly higher in the late milk stage than in the early and middle stages, while the concentrations of C15:0 and *anteiso*-C17:0 were lower in the late stage than in the other stages in the rumen fluid ($p < 0.05$). The contents of *anteiso*-C15:0 and *iso*-C16:0 were most abundant in the late milk stage, and *iso*-C17:0 concentrations were lowest in milk fat ($p < 0.05$).

The total OBCFA and odd *anteiso*-chain fatty acids contents contained in milk were higher than those in rumen, and the linear odd-chain fatty acids were more abundant both in milk and rumen (Figure 1). Except for C15:0, the concentrations of other OBCFAs were significantly different in milk and rumen. Fatty acids with 15 carbon atoms were the major kinds in rumen, but the *anteiso*-chain fatty acids were much more abundant than other kinds in milk, especially concentrations of *anteiso*-C17:0.

Table 3. Changes of different milk stages on OBCFA production in rumen fluid and milk (g/100 g fatty acids).

OBCFA Profile	Rumen Fluid					Milk				
	Early Stage	Middle Stage	Late Stage	SEM	*p*	Early Stage	Middle Stage	Late Stage	SEM	*p*
C11:0	0.03 [b]	0.04 [b]	0.05 [a]	0.003	0.001	0.03	0.03	0.04	0.01	0.17
C13:0	0.06 [b]	0.06 [b]	0.08 [a]	0.004	0.004	0.01 [b]	0.02 [a]	0.01 [b]	0.002	0.04
Iso-C15:0	0.39	0.33	0.36	0.04	0.61	0.14	0.15	0.15	0.01	0.12
Anteiso-C15:0	0.58	0.55	0.56	0.04	0.79	0.38	0.39	0.38	0.03	0.93
C15:0	1.37 [a]	1.01 [b]	1.04 [b]	0.08	0.004	0.84	0.84	0.78	0.04	0.44
Iso-C16:0	0.30	0.30	0.33	0.02	0.62	0.25 [a]	0.19 [b]	0.16 [b]	0.02	0.02
Iso-C17:0	0.27	0.26	0.29	0.02	0.33	0.44 [b]	0.52 [a]	0.55 [a]	0.03	0.04
Anteiso-C17:0	0.07 [b]	0.09 [a]	0.07 [b]	0.004	0.002	0.83 [b]	1.13 [a]	1.01 [b]	0.07	0.02
C17:0	0.25	0.24	0.23	0.02	0.51	0.57	0.43	0.47	0.06	0.23
TOBCFA	3.32	2.87	3.00	0.14	0.08	3.54	3.57	3.69	0.09	0.42

OBCFA, odd- and branched-chain fatty acids; TOBCFA, the total odd- and branched-chain fatty acids. a, b, c, means with different letters within the same line and the same item differ significantly ($p < 0.05$). SEM = standard error of mean.

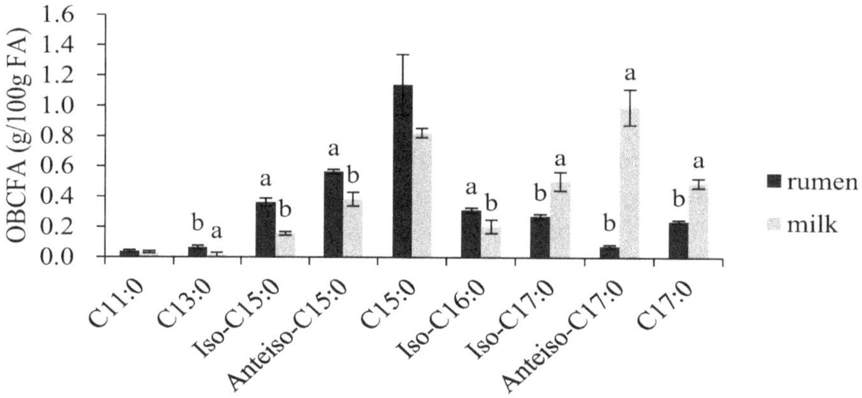

Figure 1. Comparison of rumen fluid and milk OBCFA of different milk stages. The data are means and deviations of three milk stages and error bars the show standard error of the mean. Bars without a common letter (a and b) differ ($p < 0.05$). OBCFA, odd- and branched-chain fatty acids.

3.2. Correlation between Rumen and Milk OBCFA during Different Milk Stages

The contents of OBCFA in rumen fluid were significantly related to that in milk fat (Table 4). Significant relationships of concentrations of C11:0, *iso*-C15:0, *anteiso*-C15:0, C15:0, and *anteiso*-C17:0 were found between rumen fluid and milk fat (r = 0.39~0.66, p = 0.0002–0.04). The concentrations of C11:0 in milk were significantly related to the concentrations of *anteiso*-C15:0 in rumen (r = 0.39, p = 0.04). As for the milk, C13:0 contents were significantly correlated with the ruminal *anteiso*-C15:0, C15:0, and total OBCFA concentrations (r = 0.48~0.52, p = 0.005–0.01). The concentrations of *iso*-C15:0 in milk were positively associated with the contents of C11:0, C13:0, and *anteiso*-C15:0 in rumen fluid (r = 0.39~0.41, p = 0.03–0.04). The contents of *anteiso*-C15:0 in milk were significantly related to the contents of C11:0, C13:0, and total OBCFA in rumen fluid (r = 0.46–0.65, p = 0.0002–0.02). The milk C15:0 contents were positively correlated with the concentrations of *anteiso*-C15:0 and total OBCFA in rumen (r = 0.58–0.62, p = 0.001–0.002). The contents of *iso*-C16:0 in milk were positively correlated with the ruminal C11:0 contents (r = 0.44, p = 0.02); however, they were negatively related to the ruminal *anteiso*-C17:0 contents (r = −0.39, p = 0.04). There was significantly relationship between the concentrations of *iso*-C17:0 in milk and the concentrations of C11:0 in rumen (r = −0.59, p = 0.001). Besides, the contents of *anteiso*-C17:0 in milk were negatively linked with the contents of *iso*-C15:0, C15:0, and total OBCFA in rumen (r = −0.39 to −0.49, p = 0.01–0.046). A negative relationship that existed between the milk C17:0 contents and ruminal C11:0 contents (r = −0.44, p = 0.02).

Table 4. Correlation between rumen and milk OBCFA of different milk stages.

Rumen Fluid	Milk Fat																			
	C11:0		C13:0		*Iso*-C15:0		*Anteiso*-C15:0		C15:0		*Iso*-C16:0		*Iso*-C17:0		*Anteiso*-C17:0		C17:0		TOBCFA	
	r	*p*	*r*	*p*	*r*	*p*	*r*	*p*	*r*	*p*	*r*	*p*	*r*	*p*	*r*	*p*	*r*	*p*	*r*	*p*
C11:0	0.40	0.04	0.13	0.53	0.39	0.04	0.50	0.01	0.16	0.43	0.44	0.02	−0.59	0.001	−0.32	0.11	−0.44	0.02	−0.03	0.87
C13:0	0.26	0.19	0.14	0.47	0.41	0.04	0.65	0.0002	−0.01	0.96	0.37	0.06	−0.36	0.07	−0.25	0.21	−0.23	0.25	0.03	0.89
Iso-C15:0	−0.02	0.93	0.34	0.08	0.42	0.03	0.25	0.20	0.36	0.07	−0.06	0.76	−0.15	0.46	−0.39	0.046	0.16	0.43	−0.03	0.88
Anteiso-C15:0	0.39	0.04	0.48	0.01	0.43	0.03	0.60	0.001	0.58	0.002	−0.24	0.23	−0.004	0.98	−0.30	0.13	0.09	0.65	0.31	0.11
C15:0	0.20	0.33	0.52	0.005	0.20	0.31	0.12	0.54	0.66	0.0002	−0.14	0.50	0.13	0.51	−0.44	0.02	0.22	0.27	0.06	0.76
Iso-C16:0	0.08	0.68	−0.07	0.72	0.17	0.39	0.18	0.38	−0.03	0.90	0.38	0.052	−0.19	0.34	−0.14	0.48	−0.05	0.81	−0.02	0.90
Iso-C17:0	0.29	0.14	0.18	0.36	0.27	0.18	0.37	0.06	0.45	0.02	−0.07	0.74	0.003	0.99	−0.22	0.26	−0.14	0.47	0.13	0.52
Anteiso-C17:0	−0.09	0.65	−0.12	0.57	−0.13	0.52	0.06	0.78	−0.02	0.93	−0.39	0.04	0.31	0.12	0.39	0.04	0.14	0.49	0.34	0.08
C17:0	−0.04	0.83	0.31	0.12	0.02	0.90	−0.03	0.89	−0.01	0.96	−0.29	0.15	0.07	0.72	−0.16	0.44	0.17	0.40	−0.16	0.43
TOBCFA	0.29	0.14	0.49	0.01	0.36	0.07	0.46	0.02	0.62	0.001	−0.06	0.77	−0.07	0.73	−0.49	0.01	0.11	0.57	0.09	0.64

OBCFA, odd- and branched-chain fatty acids; TOBCFA, the total odd- and branched-chain fatty acids. *r* = correlation coefficient.

3.3. Simple Statistics of Experimental Data Used for Model Development

Simple statistical data were gathered from of the present experimental conditions shown in Table 5. According to the CV%, the *Butyrivibrio flarisolvens* population changed less than the other bacterial populations. There was a wide range of rumen fermentation parameters in this research. With regard to rumen VFAs, larger differences were observed in isoacids than in linear-chain VFAs in the rumen. Compared to the variation of NH_3-N, the pH was more stable. With regard to the rumen bacterial bases, the variations were similar to each other.

Table 5. Simple statistics of experimental data used for model development.

Variables	N	Mean	SD	Minimum	Maximum	CV%
Milk odd and branched-chain fatty acids (g/100 g fatty acids)						
C11:0	27	0.03	0.02	0.01	0.08	50.53
C13:0	27	0.02	0.01	0.01	0.04	46.25
Iso-C15:0	27	0.16	0.03	0.09	0.21	18.39
Anteiso-C15:0	27	0.38	0.07	0.23	0.58	19.33
C15:0	27	0.82	0.12	0.64	1.02	14.26
Iso-C16:0	27	0.20	0.07	0.10	0.37	34.76
Iso-C17:0	27	0.50	0.10	0.23	0.62	19.46
Anteiso-C17:0	27	0.99	0.23	0.51	1.44	23.35
C17:0	27	0.49	0.18	0.34	1.08	36.52
TOBCFA	27	3.61	0.25	3.01	4.03	6.94
Rumen fermentation parameters						
Acetate (mmol/mol)	27	523.81	25.31	474.55	577.33	4.83
Propionate (mmol/mol)	27	294.02	18.24	255.62	327.79	6.21
Isobutyrate (mmol/mol)	27	14.84	2.20	10.06	19.35	14.81
Butyrate (mmol/mol)	27	120.65	4.81	112.00	127.81	3.99
Isovalerate (mmol/mol)	27	20.84	2.37	15.89	25.26	11.36
Valerate (mmol/mol)	27	25.84	2.32	21.25	30.63	8.98
TVFA (mmol/l)	27	63.12	13.50	48.76	110.34	21.39
NH3-N (mmol/l)	27	11.69	3.67	4.84	19.95	31.39
pH	27	6.35	0.24	5.87	6.70	3.77
Ruminal bacterial populations (log_{10} copies/mL)						
Fibrobacter succinogenes	27	7.99	0.32	7.33	8.56	4.00
Ruminococcus albus	27	7.03	0.30	6.55	7.55	4.29
Ruminococcus flavafaciens	27	7.73	0.14	7.42	8.02	1.85
Butyrivibro flarisolvens	27	8.91	0.07	8.79	9.08	0.80
Eubacterium ruminantium	27	6.62	0.28	5.88	7.23	4.19
Streptococcus bovis	27	5.27	0.30	4.45	5.74	5.61
Selenomonas ruminantium	27	7.35	0.19	6.97	7.62	2.61
Ruminal bacterial bases (g/kg DM)						
Cytosine	27	1.21	0.11	0.98	1.38	8.94
Uracil	27	0.98	0.14	0.63	1.16	14.10
Guanine	27	2.37	0.29	1.60	2.80	12.20
Adenine	27	1.86	0.16	1.43	2.16	8.57
Ruminal bacterial bases/N (g/100 g N)						
Cytosine/N	27	3.11	0.24	2.75	3.66	0.08
Uracil/N	27	2.52	0.33	1.80	3.01	0.13
Guanine/N	27	6.09	0.62	4.52	7.16	0.10
Adenine/N	27	4.80	0.44	4.13	6.11	0.09

TVFA, the total contents of acetate, propionate, isobutyrate, butyrate, isovalerate, and valerate in rumen. OBCFA, odd- and branched-chain fatty acids; TOBCFA, the total odd- and branched-chain fatty acids. SD = standard deviation; CV = coefficient of variation.

3.4. Relationship of Milk OBCFA Pattern during Different Milk Stages to Ruminal Fermentation Parameters, Bacteria Populations, and Microbial Bases

Relationships existed between the fermentation parameters in rumen fluid and the OBCFA patterns (Table 6). Although there no significant correlations that were found between the molar proportions of acetate and individual OBCFA concentrations, the molar proportions of acetate were positively related to total OBCFA concentrations ($r = 0.40, p = 0.04$). A negative correlation existed between the propionate molar proportions and *iso*-C17:0 concentrations ($r = -0.39, p = 0.045$). The total OBCFA concentrations were negatively correlated with the molar proportions' isobutytate ($r = -0.49, p = 0.01$) while the concentrations of *anteiso*-C15:0 were positively associated with the molar proportions' butyrate ($r = 0.47, p = 0.01$). There was a negative relationship between the milk C15:0 concentration and isovalelate proportion in rumen ($r = -0.41, p = 0.03$). The total OBCFA concentrations were negatively related to isovalelate and valelate proportions ($r = -0.48$ to $-0.49, p = 0.01$) but were positively correlated with total VFA contents ($r = 0.51, p = 0.007$) and the *iso*-C15:0/*iso*-C17:0 ratio and *anteiso*-C15:0/anteiso-C17:0 ratio ($r = -0.56$ to $-0.60, p = 0.001$–0.002). The concentrations of NH_3-N were negatively related to *iso*-C15:0 and *anteiso*-C15:0 concentrations ($r = -0.50$ to $-0.56, p = 0.003$–0.008), but positively correlated with *anteiso*-C17:0 concentrations ($r = 0.47, p = 0.01$). However, there were no apparent relationships between pH and OBCFA concentrations.

The relationships between bacterial populations and OBCFA concentrations are shown in Table 7. The copy numbers of *Fibrobacter succinogenes* were not related to single OBCFAs concentrations, but positively linked with the *iso*-C15:0/*iso*-C17:0 ratio ($r = 0.47, p = 0.01$). The concentrations of C11:0 and C13:0 were positively correlated with the populations of *Ruminococcus albus* ($r = 0.53$–$0.57, p = 0.002$–0.005). The copies of *Ruminococcus flavefaciens* were significantly related to the C13:0 and total OBCFA concentrations ($r = 0.38$–$0.50, p = 0.009$–0.049). A positive correlation was found between C13:0 concentrations and *Eubacterium ruminantium* copy numbers ($r = 0.43, p = 0.03$). The ratio of *anteiso*-C15:0 to *anteiso*-C17:0 was positively correlated with *Streptococcus bovis* copies ($r = 0.48, p = 0.01$); however, no significant relationships that were found between *Selenomonas ruminantium* copies and individual OBCFA concentrations. The populations of *Selenomonas ruminantium* were positively associated with the total OBCFA and branched-chain fatty acid contents ($r = 0.39$–$0.57, p = 0.002$–0.04).

Several relationships were found between the microbial bases and OBCFA patterns (Table 8). Cytosine concentrations were negatively related to *iso*-C17:0 and total OBCFA concentrations ($r = -0.39$ to $-0.50, p = 0.007$–0.04). The concentrations of uracil were significantly correlated with the *iso*-C17:0 content, the sum of *iso*-C15:0 and *iso*-C17:0 contents ($r = -0.51$ to $-0.54, p = 0.003$–0.004). The concentrations of guanine were negatively related to the concentrations of *iso*-C17:0, the total OBCFA contents, and the sum of *iso*-C15:0 and *iso*-C17:0 contents ($r = -0.41$ to $-0.54, p = 0.004$–0.03). A negative relation was presented by the concentrations of adenine and C17:0 ($r = -0.39, p = 0.04$), while the contents of *anteiso*-C15:0 and the C15:0/C17:0 ratio were positively linked with the adenine concentrations ($r = 0.44$–$0.49, p = 0.01$–0.02). The correlations of microbial protein marked by different microbial bases/N ratio with the OBCFAs are also shown in Table 8. The cytosine/N ratio was negatively related to total OBCFA concentrations in milk ($r = -0.40, p = 0.04$). In addition, some negative relationships were found between the guanine/N ratio and some milk OBCFAs, which were *iso*-C17:0 and the sum of *iso*-C15:0 and *iso*-C17:0 ($r = -0.49$ to $-0.55, p = 0.003$–0.01). The adenine/N ratio was negatively connected with *iso*-C17:0, *anteiso*-C17:0, the sum of *iso*-C15:0 and *iso*-C17:0, the sum of *anteiso*-C15:0 and *anteiso*-C17:0, the total *iso*-branched chain fatty acids, the total *anteiso*-branched chain fatty acids, and the total branched chain fatty acids in milk ($r = -0.42$ to $-0.59, p = 0.001$–0.03). However, a positive relationship was found between the adenine/N ratio and the ratio of *iso*-C15:0 and *iso*-C17:0 in milk ($r = 0.50, p = 0.01$).

Table 6. Correlation between ruminal fermentation parameters and milk OBCFA pattern.

Item	Acetate		Propionate		Isobutyrate		Butyrate		Isovalerate		Valerate		TVFA		NH_3-N		pH	
	r	*p*	*r*	*p*	*r*	*p*	*r*	*p*	*r*	*p*	*r*	*p*	*r*	*p*	*r*	*p*	*r*	*p*
C11:0	−0.22	0.28	0.12	0.56	0.36	0.06	0.17	0.41	0.19	0.33	0.26	0.18	−0.31	0.11	0.06	0.75	0.22	0.26
C13:0	0.14	0.48	−0.22	0.28	0.08	0.69	−0.21	0.30	0.16	0.44	−0.05	0.82	−0.20	0.33	0.19	0.34	0.29	0.14
Iso-C15:0	−0.07	0.73	0.19	0.35	−0.08	0.69	0.05	0.79	−0.10	0.63	−0.07	0.74	0.18	0.37	−0.50	0.008	−0.05	0.81
Anteiso-C15:0	−0.34	0.08	0.31	0.11	0.32	0.11	0.47	0.01	0.11	0.60	0.28	0.16	−0.22	0.27	−0.56	0.003	0.21	0.29
C15:0	0.18	0.37	−0.11	0.60	−0.31	0.11	−0.05	0.81	−0.41	0.03	−0.33	0.097	0.37	0.06	−0.28	0.16	0.20	0.32
Iso-C16:0	−0.12	0.54	0.20	0.32	0.19	0.35	0.03	0.90	0.02	0.93	0.22	0.28	−0.06	0.76	−0.21	0.29	−0.31	0.12
Iso-C17:0	0.36	0.06	−0.39	0.045	−0.37	0.06	−0.24	0.22	−0.25	0.20	−0.35	0.08	0.11	0.59	0.22	0.27	0.12	0.56
Anteiso-C17:0	0.26	0.19	−0.35	0.07	−0.19	0.35	−0.12	0.55	−0.17	0.39	−0.21	0.28	0.17	0.40	0.47	0.01	0.18	0.37
C17:0	0.21	0.28	−0.06	0.77	−0.37	0.06	−0.32	0.0996	−0.28	0.16	−0.38	0.05	0.23	0.26	0.12	0.55	−0.05	0.81
TOBCFA	0.40	0.04	−0.33	0.09	−0.49	0.01	−0.26	0.18	−0.48	0.01	−0.49	0.01	0.51	0.007	0.06	0.76	0.08	0.71
C15:0 + C17:0	0.12	0.57	0.03	0.89	−0.29	0.14	−0.11	0.58	−0.37	0.06	−0.32	0.10	0.37	0.06	−0.25	0.22	0.07	0.74
Iso-C15:0 + *iso*-C17:0	0.30	0.13	−0.36	0.06	−0.32	0.10	−0.12	0.55	−0.23	0.26	−0.27	0.17	0.23	0.26	0.02	0.90	0.17	0.39
Anteiso-C15:0 + *anteiso*-C17:0	0.14	0.48	−0.25	0.21	−0.07	0.72	0.04	0.83	−0.11	0.58	−0.09	0.65	0.09	0.65	0.30	0.12	0.22	0.27
Total *iso*-fatty acids	0.32	0.0997	−0.34	0.08	−0.28	0.16	−0.24	0.23	−0.18	0.38	−0.19	0.35	0.22	0.27	−0.02	0.94	−0.23	0.24
Total *anteiso*-fatty acids	0.14	0.48	−0.25	0.21	−0.07	0.71	0.05	0.82	−0.11	0.57	−0.09	0.64	0.09	0.64	0.31	0.12	0.22	0.27
Total odd-chain fatty acids	0.09	0.67	0.05	0.81	−0.26	0.19	−0.09	0.65	−0.33	0.096	−0.30	0.13	0.34	0.08	−0.26	0.20	0.10	0.63
Total branched-chian fatty acids	0.16	0.44	−0.25	0.22	−0.08	0.68	−0.03	0.90	−0.19	0.35	−0.09	0.66	0.12	0.55	0.22	0.28	0.15	0.47
C15:0/C17:0	−0.13	0.51	0.04	0.85	0.17	0.40	0.30	0.13	−0.03	0.90	0.18	0.37	−0.04	0.83	−0.31	0.11	0.27	0.17
Iso-C15:0/*iso*-C17:0	−0.25	0.21	0.35	0.08	0.14	0.49	0.22	0.28	0.05	0.80	0.16	0.42	0.09	0.65	−0.56	0.002	−0.13	0.52
Anteiso-C15:0/*anteiso*-C17:0	−0.32	0.10	0.36	0.06	0.31	0.12	0.30	0.12	0.14	0.50	0.30	0.13	0.07	0.74	−0.60	0.001	0.04	0.84

TVFA, the total contents of acetate, propionate, isobutyrate, butyrate, isovalerate, and valerate in rumen. OBCFA, odd- and branched-chain fatty acids; TOBCFA, the total odd- and branched-chain fatty acids. *r* = correlation coefficient.

Table 7. Correlation between bacteria populations and milk OBCFA pattern.

Item	*Fibrobacter succinogenes*		*Ruminococcus albus*		*Ruminococcus flavafaciens*		*Butyrivibro flarisolvens*		*Eubacterium ruminantium*		*Streptococcus bovis*		*Selenomonas ruminantium*	
	r	*p*	*r*	*p*	*r*	*p*	*r*	*p*	*r*	*p*	*r*	*p*	*r*	*p*
C11:0	0.17	0.38	0.57	0.002	0.18	0.37	0.12	0.54	0.36	0.06	0.19	0.34	0.01	0.95
C13:0	0.11	0.60	0.53	0.005	0.50	0.009	0.15	0.45	0.43	0.03	0.22	0.27	−0.01	0.97
Iso-C15:0	0.37	0.055	−0.03	0.87	0.35	0.07	−0.09	0.65	0.05	0.79	0.02	0.94	−0.05	0.79
Anteiso-C15:0	0.18	0.37	0.35	0.07	−0.04	0.83	0.14	0.47	0.22	0.26	0.33	0.09	−0.25	0.20
C15:0	0.21	0.29	0.25	0.21	0.17	0.39	−0.28	0.16	0.05	0.81	0.22	0.28	0.27	0.18
Iso-C16:0	0.14	0.48	−0.17	0.39	−0.18	0.36	0.14	0.48	−0.07	0.71	−0.21	0.29	−0.09	0.65
Iso-C17:0	−0.25	0.21	−0.03	0.88	0.13	0.53	0.05	0.81	−0.01	0.94	−0.02	0.92	0.33	0.09
Anteiso-C17:0	−0.14	0.49	0.01	0.98	0.13	0.51	−0.19	0.34	0.03	0.88	−0.34	0.09	0.36	0.06
C17:0	−0.22	0.27	−0.19	0.34	0.22	0.28	0.18	0.38	−0.09	0.65	−0.01	0.96	0.37	0.06
TOBCFA	0.05	0.81	0.03	0.89	0.38	0.049	−0.03	0.89	0.17	0.40	−0.18	0.38	0.57	0.002
C15:0 + C17:0	0.16	0.42	0.06	0.77	0.16	0.44	−0.04	0.83	−0.03	0.90	0.24	0.23	0.32	0.098
Iso-C15:0 + *iso*-C17:0	−0.06	0.77	0.13	0.53	0.28	0.16	0.01	0.95	0.11	0.59	0.09	0.67	0.38	0.051
Anteiso-C15:0 + *anteiso*-C17:0	−0.005	0.98	0.09	0.66	0.17	0.39	−0.22	0.27	0.06	0.78	−0.30	0.13	0.27	0.18
Total *iso*-fatty acids	0.02	0.93	−0.03	0.89	0.23	0.25	0.13	0.52	0.06	0.76	−0.10	0.61	0.31	0.12
Total *anteiso*-fatty acids	−0.01	0.96	0.08	0.69	0.16	0.41	−0.22	0.27	0.05	0.79	−0.30	0.13	0.26	0.18
Total odd-chain fatty acids	0.18	0.38	0.11	0.59	0.19	0.35	−0.04	0.84	0.01	0.97	0.24	0.22	0.31	0.12
Total branched-chian fatty acids	0.04	0.86	0.11	0.59	0.18	0.37	−0.15	0.45	0.08	0.69	−0.28	0.16	0.39	0.04
C15:0/C17:0	0.32	0.104	0.33	0.09	−0.05	0.81	−0.32	0.11	0.10	0.63	0.17	0.41	−0.10	0.61
Iso-C15:0/*iso*-C17:0	0.47	0.01	−0.05	0.82	0.13	0.51	−0.10	0.63	−0.04	0.86	0.07	0.75	−0.21	0.28
Anteiso-C15:0/*anteiso*-C17:0	0.23	0.24	0.29	0.15	−0.12	0.54	0.21	0.28	0.17	0.39	0.48	0.01	−0.31	0.12

OBCFA, odd- and branched-chain fatty acids; TOBCFA, the total odd- and branched-chain fatty acids. *r* = correlation coefficient.

Table 8. Correlation between microbial protein marked by nucleic acids, marker:N ratio in rumen (when using nucleic acids as the microbial marker), and milk OBCFA pattern (n = 27).

Item	Cytosine		Uracil		Guanine		Adenine		Cytosine/N		Uracil/N		Guanine/N		Adenine/N	
	r	*p*	*r*	*p*	*r*	*p*	*r*	*p*	*r*	*p*	*r*	*p*	*r*	*p*	*r*	*p*
C11:0	0.05	0.81	0.20	0.33	0.07	0.74	0.16	0.42	0.02	0.91	0.10	0.62	0.19	0.35	0.08	0.68
C13:0	0.12	0.54	−0.09	0.65	0.13	0.51	0.21	0.28	0.27	0.17	0.00	0.99	−0.23	0.24	0.13	0.50
Iso-C15:0	−0.11	0.58	−0.18	0.36	−0.06	0.78	0.18	0.36	−0.13	0.52	−0.05	0.79	−0.14	0.48	0.20	0.32
Anteiso-C15:0	−0.06	0.76	0.12	0.56	0.07	0.74	0.49	0.01	−0.07	0.72	−0.17	0.39	0.16	0.44	0.08	0.68
C15:0	−0.19	0.35	−0.32	0.11	−0.28	0.16	0.30	0.13	−0.25	0.21	0.04	0.86	−0.14	0.48	0.10	0.63
Iso-C16:0	0.09	0.67	0.38	0.051	0.31	0.12	−0.09	0.66	0.35	0.07	−0.07	0.72	0.26	0.19	0.20	0.31
Iso-C17:0	−0.39	0.04	−0.51	0.006	−0.54	0.004	−0.21	0.29	−0.17	0.40	−0.24	0.22	−0.49	0.01	−0.59	0.001
Anteiso-C17:0	−0.18	0.36	−0.12	0.55	−0.25	0.20	−0.13	0.52	−0.22	0.27	−0.09	0.65	−0.17	0.40	−0.47	0.01
C17:0	−0.30	0.13	0.11	0.59	−0.23	0.24	−0.39	0.04	−0.12	0.54	−0.13	0.51	0.19	0.35	0.20	0.32
TOBCFA	−0.50	0.007	−0.28	0.16	−0.41	0.03	0.03	0.90	−0.40	0.04	−0.33	0.09	−0.17	0.40	−0.37	0.06
C15:0 + C17:0	−0.28	0.16	−0.04	0.84	−0.19	0.34	0.00	0.99	−0.24	0.23	−0.09	0.65	0.08	0.69	0.22	0.28
Iso-C15:0 + *iso*-C17:0	−0.35	0.08	−0.54	0.003	−0.53	0.005	−0.09	0.66	−0.21	0.28	−0.27	0.17	−0.55	0.003	−0.55	0.003
Anteiso-C15:0 + *anteiso*-C17:0	−0.18	0.38	−0.04	0.83	−0.21	0.28	−0.001	1.00	−0.26	0.19	−0.16	0.42	−0.13	0.52	−0.49	0.01
Total *iso*-fatty acids	−0.28	0.16	−0.33	0.09	−0.31	0.11	−0.17	0.39	0.05	0.81	−0.34	0.08	−0.37	0.06	−0.42	0.03
Total *anteiso*-fatty acids	−0.18	0.37	−0.05	0.82	−0.22	0.27	−0.01	0.97	−0.26	0.19	−0.16	0.42	−0.13	0.52	−0.49	0.01
Total odd-chain fatty acids	−0.24	0.23	−0.03	0.89	−0.15	0.45	0.04	0.83	−0.23	0.24	−0.08	0.67	0.09	0.66	0.23	0.24
Total branched-chian fatty acids	−0.23	0.26	−0.11	0.60	−0.26	0.19	−0.04	0.86	−0.21	0.30	−0.26	0.19	−0.24	0.22	−0.56	0.002
C15:0/C17:0	0.16	0.42	0.00	0.99	−0.05	0.81	0.44	0.02	0.01	0.96	0.23	0.24	−0.03	0.87	0.06	0.78
Iso-C15:0/*iso*-C17:0	0.10	0.60	0.23	0.25	0.23	0.25	0.21	0.28	0.04	0.83	0.10	0.61	0.25	0.21	0.50	0.01
Anteiso-C15:0/*anteiso*-C17:0	0.06	0.77	0.22	0.27	0.17	0.39	0.38	0.053	0.02	0.91	−0.05	0.79	0.24	0.22	0.34	0.08

OBCFA, odd- and branched-chain fatty acids; TOBCFA, the total odd- and branched-chain fatty acids. r = correlation coefficient.

The quadratic, ratio and reciprocal of milk OBCFAs were used in the predicted models shown in Tables 9–11, respectively. The equations of fermentation parameters, with the exception of that for valerate, were all significant. According to the R^2 value, the isobutyrate and butyrate models were more accurate than those for other VFAs. The regression equations showed that the ruminal acetate, propionate, and isobutyrate molar proportions and pH had inverse proportional relationships to milk C11:0 concentrations. In addition, inverse proportional relationships existed between the ruminal isobutyrate proportion and milk C17:0 concentrations, between the ruminal butyrate proportion and milk *anteiso*-C15:0 concentrations, and between ruminal NH_3-N and milk *anteiso*-C15:0 concentrations. The accuracy of the prediction equations for different bacterial populations in the rumen was variable. Regression analysis showed that milk C11:0 concentrations had inverse proportional relationships with *Eubacterium ruminantium*, *Ruminococcus albus*, and *Streptococcus bovis* populations in the rumen. Inverse proportional relationships were found between milk *iso*-C15:0 content and ruminal *Fibrobacter succinogenes* populations as well as between milk C17:0 content and ruminal *Selenomonas ruminantium* populations. Significant regression equations for ruminal microbial markers were obtained using milk OBCFAs as an independent data set. Additionally, the compositions of these equations were complex.

Table 9. Predicted equations of ruminal fermentation parameters from milk OBCFA.

Variables	Predicted Equation	RMSE	R^2	*P*
Acetate (mmol/mol)	$Y = 55.73 - 1.60/(C11:0)$	2.38	0.15	0.049
Propionate (mmol/mol)	$Y = 26.99 + 1.15/(C11:0)$	1.75	0.12	0.0496
Isobutyrate (mmol/mol)	$Y = -0.32 + 0.48/(C17:0) + 20.23 \times C11:0 \times anteiso - C15:0 + 1.04/(C11:0)$	0.17	0.46	0.002
Butyrate (mmol/mol)	$Y = 14.37 - 0.37/(anteiso - C15:0) - 4.08 \times C11:0 \times C17:0 - 102.26 \times C13:0 \times iso - C16:0$	0.41	0.35	0.02
Isovalerate (mmol/mol)	$Y = 1.13 - 0.88 \times C15:0 - 0.68 \times anteiso - C17:0 \times C17:0$	0.21	0.28	0.02
Valerate (mmol/mol)	$Y = 3.21 - 1.36 \times C11:0 \times C17:0 - 0.38 \times C15:0 \times anteiso - C17:0$	0.21	0.27	0.08
TVFA (mmol/L)	$Y = 24.56 + 155.95 \times C11:0 \times C15:0 + 38.78 \times anteiso - C17:0 \times C17:0$	12.03	0.27	0.02
NH3-N (mmol/L)	$Y = 5.46 + 0.11/(iso - C16:0) + 3.71 \times C11:0 \times iso - C16:0$	2.95	0.40	0.002
pH	$Y = 6.03 + 0.15/(C11:0)$	0.16	0.22	0.04

R^2 = coefficient of determination; RMSE = root mean square error. TVFA, the total contents of acetate, propionate, isobutyrate, butyrate, isovalerate, and valerate in rumen.

Table 10. Predicted equations of ruminal bacterial populations ($\log_{10}$ copies/mL) from milk OBCFA.

Variables	Predicted Equation	RMSE	R^2	*P*
Fibrobacter succinogenes	$Y = 8.48 - 0.15 \times (iso - C17:0)/(iso - C15:0)$	0.20	0.29	0.02
Ruminococcus albus	$Y = 6.87 - 0.007/(C11:0)$	0.26	0.18	0.03
Ruminococcus flavafaciens	$Y = 7.04 - 0.01/(C11:0) + 35.41 \times C11:0 \times anteiso - C15:0$	0.24	0.41	0.002
Butyrivibro flarisolvens	$Y = 7.30 + 73.80 \times C11:0 \times iso - C15:0 + 0.53 \times anteiso - C17:0 \times C17:0$	0.12	0.38	0.09
Eubacterium ruminantium	$Y = 5.86 - 0.22/(C11:0)$	0.28	0.15	0.04
Streptococcus bovis	$Y = 7.08 - 0.22/C17:0 + 1.20 \times C11:0 \times anteiso - C17:0 + 0.65 \times C15:0 \times anteiso - C17:0$	0.14	0.53	0.0005

R^2 = coefficient of determination; RMSE = root mean square error.

Table 11. Predicted equations of ruminal bacterial bases from milk OBCFA.

Variables	Predicted Equation	R^2	RMSE	*P*
Ruminal bases (g/kg DM)				
Cytosine	$Y = 1.54 - 0.79 \times C11:0 \times C17:0 - 0.93 \times iso - C17:0 \times anteiso - C17:0$	0.32	0.09	0.01
Uracil	$Y = 1.18 - 4.11 \times iso - C15:0 \times iso - C17:0 + 1.59 \times anteiso - C15:0 \times iso - C16:0$	0.50	0.10	0.0002
Guanine	$Y = 2.23 + 0.01/(C11:0) + 1.17 \times C11:0 \times anteiso - C17:0 + 23.00 \times C13:0 \times anteiso - C17:0 - 1.48 \times iso - C17:0 \times anteiso - C17:0$	0.58	0.20	0.001
Adenine	$Y = 1.96 + 2.33 \times iso - C16:0 \times iso - C16:0 + 52.67 \times C11:0 \times anteiso - C15:0 - 91.61 \times C13:0 \times iso - C16:0 - 0.5 \times anteiso - C17:0 \times C17:0$	0.49	0.12	0.004
Ruminal bases /N (g/100 g N)				
Cytosine:N	$Y = 1.72 + 0.01/(C13:0) + 0.20/(iso - C16:0) - 1.90 \times iso - C16:0 \times anteiso - C17:0$	0.58	0.18	0.002
Uracil:N	$Y = 2.93 + 0.01/(C13:0) - 2.88 \times iso - C17:0 \times anteiso - C17:0 + 273.13 \times C11:0 \times C13:0 - 1.44 \times (iso - C15:0)/(iso - C17:0)$	0.60	0.23	0.0004
Guanine:N	$Y = 7.17 - 2.12 \times iso - C17:0 \times anteiso - C17:0$	0.37	0.50	0.001
Adenine:N	$Y = 3.27 - 2.53 \times iso - C17:0 \times C17:0 + 3.81 \times C15:0 \times iso - C16:0 + 0.73 \times (anteiso - C15:0)/(iso - C16:0)$	0.62	0.28	<0.0001

N, the total contents of nitrogen in rumen. R^2 = coefficient of determination; RMSE = root mean square error.

4. Discussion

The OBCFA in milk are mainly derived from the ruminal bacterial cell membrane, which was already recognized half a century ago [7]. For reasons previously described by Fievez et al. [33], the milk OBCFA profile was not completely similar to the rumen OBCFA profile. Different proportions of OBCFAs in the rumen and milk were also found in this research. In this study, the fatty acids with 15-carbon atoms were more plentiful than other components in the rumen, but the contents of these fatty acids and *iso*-C16:0 were much lower than those in milk. The summary statistics by Vlaeminck et al. [34] suggested that the linear odd-chain fatty acids accounted for the major proportion of milk OBCFAs. The linear odd-chain fatty acid contents were also much higher in the rumen than in milk in our study. Previous studies have indicated that the mammary gland tissue can de novo synthesize odd-chain fatty acids and their *anteiso*-isomers, through the incorporation of propionyl-CoA instead of acetyl-CoA [35–37]. Milk secretion of linear odd-chain fatty acids is higher than that of these fatty acids in duodenal flow, which further demonstrates endogenous chain elongation in the mammary gland [34]. This research also found that the contents of odd-chain fatty acids, except those of C15:0, were higher in milk than in the rumen. Scheerlinck et al. [38] summarized a representative subset of 92 milk samples adopted from a sample database of different dietary treatments, and they reported that the highest concentrations of *iso*-C17:0, *anteiso*-C17:0, and some mono-unsaturated fatty acids were observed in milk. Additionally, Vlaeminck et al. [39] reported that two-carbon elongation of branched-chain FAs occurs postruminal. Increased contents of *anteiso*-C17:0 in milk were found in this research.

At the beginning of lactation, cows are in a negative energy balance, causing the mobilization of adipose fatty acids and the synthesis of these long-chain fatty acids in milk fat [40]. The OBCFAs with 14- and 15-carbon atoms, such as *iso*-C14:0, *iso*-C15:0, C15:0, and *anteiso*-C15:0, were increased in early lactation, which changed with the lactation curves of the short- and medium-chain fatty acids in milk [41]. Increased levels of ruminal C15:0 contents were also found in early lactation in this study. In contrast, levels of OBCFAs with 17-carbon atoms decreased during early lactation, which showed a similar pattern to that of long-chain fatty acids in milk [41]. In the present study, however, the concentrations of *iso*-C17:0 in milk were more abundant in early lactation, which may be caused by the synthesis of these fatty acids. A previous study found that long-chain fatty acids are de novo synthesized from some fatty acids, which causes rumen acidosis in milk at an early lactation stage [40].

In the mammary gland, fatty acids with 4- to 14-carbon atoms mainly come from endogenous synthesis, and the long-chain fatty acids greater than 16-carbon atoms are mostly derived from exogenous transformation [42]. In addition, Verbeke et al. [43] found that the mammary gland had the ability to synthesize fatty acids from 2-methylbutyryl-CoA, isovaleryl-CoA, and isobutyryl-CoA. The same types of fatty acids with less than 16-carbon atoms collected from rumen fluid and milk were significantly or nearly related to each other in this research. These results may be due to the minor amount of fatty acids synthesized in the mammary gland. Odd-chain fatty acids are further metabolized under the action of dehydrogenase in the mammary gland, but the transformation of C17:0 to C17:1 was statistically significant [44]. Hence, negative correlations existed between the milk fatty acids with 17-carbon atoms and other ruminal fatty acids in this study. The $C15{:}0_{milk}/C15{:}0_{duodenum}$ and $C17{:}0 + C17{:}1_{milk}/C17{:}0_{duodenum}$ values changed from 1.5 to 2.25 [34,45,46], which suggested that the metabolic and transport processes of the two types of fatty acids were different in the mammary gland.

Increasing the acetate supply has a positive effect on milk yield and milk fat content [47]. In this study, a positive correlation existed between acetate content in the rumen and the total concentrations of OBCFAs in milk. Vlaeminck et al. [10] found that milk *iso*-C14:0 and *iso*-C15:0 contents are positively correlated with ruminal proportions of acetate. We also found a positive relationship between the concentrations of *iso*-C17:0 in milk and the molar proportions of ruminal acetate, but this relationship was not significant. Moreover, milk *iso*-C14:0 and *iso*-C15:0 contents are negatively related to ruminal propionate proportions according to Vlaeminck et al. [10]. In this work, the contents of ruminal propionate were negatively related to *iso*-fatty acids, especially milk *iso*-C17:0. Negative relationships were found between ruminal *iso*-fatty acids and the contents of total OBCFAs in milk. Isoacids are also used as primers for branched-chain fatty acids [34], and their contents are low in the rumen [48,49]. Compared to linear and *iso*-fatty acids, *anteiso*-fatty acids are more susceptible to effects on bacterial membrane properties [50]. Cabrita et al. [51] found a negative correlation between the concentrations of ruminal NH_3 and milk *anteiso*-C17:0. In this study, there was a negative relationship between NH_3-N concentrations in the rumen and *anteiso*-C15:0 concentrations in milk. However, milk *anteiso*-C17:0 concentrations were significantly positively related to ruminal NH_3 concentrations. In addition, the ruminal NH_3-N concentrations were positively related to *iso*-C15:0/*iso*-C17:0 or *anteiso*-C15:0/*anteiso*-C17:0 values.

Variation in the OBCFA profile leaving the rumen has been suggested to reflect changes in the relative abundance of specific ruminal bacterial populations [33]. Any deviation in the rumen's inner circumstance may influence the microbial population and its fermentation products [52]. Vlaeminck et al. [34] summarized that *Fibrobacter succinogenes* abundantly produces linear odd-chain fatty acids and fatty acids with 15-carbon atoms. In the present study, *Fibrobacter succinogenes* populations in the rumen were not significantly related to separate OBCFA concentrations, but they were positively correlated with the *iso*-C15:0/*iso*-C17:0 ratio in milk. Hence, the different results from the cultivable ruminal bacteria might be due to de novo fatty acid synthesis and elongation in the mammary gland. Some previous studies have reported that cellulolytic bacteria contain a high amount of *iso*-fatty acids, with differences between different species [19,53–55]. Specifically, *Ruminococcus flavefaciens* has a high level of odd-chain *iso*-fatty acids, and *Ruminococcus albus* has enriched even-chain *iso*-fatty acids [34]. Although the contribution of OBCFA endogenous synthesis in milk from dairy cows was negligible [56], linear odd-chain fatty acids, or their *anteiso*-isomers, can be de novo synthesized in the mammary gland [35]. Ifkovits and Ragheb [53] found that a pure culture of *Eubacterium ruminantium* contains high levels of C15:0 and *iso*-C15:0, but Minato et al. [54] reported that *Eubacterium ruminantium* is enriched in *anteiso*-C13:0. These studies suggest that different isolated strains of *Eubacterium ruminantium* have variable OBCFA profiles. In the present work, *Ruminococcus albus*, *Ruminococcus flavefaciens*, and *Eubacterium ruminantium* abundance were correlated with C13:0 contents in milk, which suggested that milk C13:0 might reflect the population of cellulolytic bacteria. The copy number of *Fibrobacter succinogenes* was positively correlated with the ratio of *iso*-C15:0 to *iso*-C17:0, which indicated that the transformation of *iso*-C15:0 to *iso*-C17:0 may occur in *Fibrobacter succinogenes*. Various bacterial strains of *Butyrivibrio fibrisolvens* form a heterogeneous group of bacteria that ferment a wide array of substrates [19], including fiber, starch, and fatty acids [52], and produce large amounts of lactate and acetate [57], which can increase both the growth rate [58] and butyrate production rate [59]. These previous studies showed that *Butyrivibrio fibrisolvens* is a key link between long-chain and short-chain fatty acids, which may be attributed to the relationship between the ruminal microbial population and milk OBCFA concentrations. Previous studies have suggested that amylolytic bacteria are relatively enriched in linear odd-chain fatty acids [34]. The populations of *Selenomonas ruminantium* in the rumen are highly correlated with total OBCFAs and branched-chain fatty acid concentrations in milk, suggesting that OBCFA concentrations in milk are involved in a complex interaction with ruminal microorganisms.

Vlaeminck et al. [9] first suggested that milk OBCFAs can be used as markers for the duodenal flow of microbial matter and found that changes in OBCFAs are closely associated with changes in uracil and purine bases in the rumen, thus confirming the potential of OBCFAs to predict microbial matter in the rumen. Several significant correlations existed between ruminal microbial base concentrations and milk OBCFA contents in the present study and in previous studies. Cabrita et al. [51] found that milk C17:0, *iso*-C17:0, and *anteiso*-C17:0 contents are significantly and negatively associated with dietary crude protein content, and they suggested that C17:0 is a marker of protein deficiency. This study implied that fatty acids with 17-carbon atoms in milk are associated with protein degradability in the rumen. Additionally, protein degradability is one of the factors that may affect rumen microbial growth [60]. The odd-chain fatty acid synthesis process is different from that of branched-chain fatty acids. [34]. Wongtangtintharn et al. [61] showed that all branched-chain fatty acids equally inhibit fatty acid synthesis. In this study, changes in the relationships between milk branched-chain fatty acids and ruminal microbial base contents were found. Moreover, the elongation of *iso*-C15:0 and *anteiso*-C15:0 adds to the existing *iso*-C17:0 and *anteiso*-C17:0 content [33], which may explain the lower *iso*-C15:0/*iso*-C17:0 and *anteiso*-C15:0/*anteiso*-C17:0 ratios in milk compared to the ratios found in rumen bacteria or duodenal content. The *iso*-C15:0/*iso*-C17:0 ratio in milk was positively correlated with the adenine/N ratio in the rumen, which suggested that the *iso*-C15:0/*iso*-C17:0 ratio in milk reflects the abundance of ruminal microbial.

Vlaeminck et al. [10] developed equations based on milk OBCFAs to predict molar proportions of individual VFAs in the rumen. Bhagwat et al. [12] improved the prediction accuracy of VFA proportions from measured milk OBCFA concentrations through the development and application of quadratic terms and interactions as well as the ratio of linear expressions. Bhagwat et al. [12] indicated that the more complex methods provide better predictions with fewer OBCFAs. Hence, we established an equation with mixed variables containing the first and second degree of dependent data as well as the ratio of any two datasets and the reciprocal of each dependent datasets in this research. Using this method, we generated a prediction model for major bacterial populations and microbial protein markers. In this model, there were some inversely proportional and quadratic relationships from the predicted equations, which suggested that there were some nonlinear relationships between milk OBCFAs and ruminal fermentation parameters, ruminal bacterial populations, and base contents.

5. Conclusions

OBCFAs originating from milk were significantly correlated with ruminal fermentation parameters, ruminal bacterial populations, and base contents. The results suggested that milk odd-chain fatty acids have the potential to be used as a noninvasive technique to assess rumen function in terms of microbial populations, substrates, and interactions. To increase the accuracy of the predicted equations for ruminal parameters, ruminal bacterial populations, and base contents established based on milk OBCFAs, a large number of experiments are required.

Author Contributions: Conceptualization, K.L.; Data curation, K.L., Y.L. and G.L.; Funding acquisition, Y.L., H.X., Y.Z., and G.L.; Writing—original draft, K.L.; Writing—review and editing, K.L., Y.Z.

References

1. Bauman, D.E.; Griinari, J.M. Nutritional regulation of milk fat synthesis. *Annu. Rev. Nutr.* **2003**, *23*, 203–227. [CrossRef] [PubMed]
2. Bainbridge, M.L.; Cersosimo, L.M.; Wright, A.D.G.; Kraft, J. Content and composition of branched-chain fatty acids in bovine milk are affected by lactation stage and breed of dairy cow. *PLoS ONE* **2016**, *11*, e0150386. [CrossRef]
3. Gross, J.; Van Dorland, H.A.; Bruckmaier, R.M.; Schwarz, F.J. Milk fatty acid profile related to energy balance in dairy cows. *J. Dairy Res.* **2011**, *78*, 479–488. [CrossRef] [PubMed]
4. Stoop, W.M.; Bovenhuis, H.; Heck, J.M.L.; van Arendonk, J.A.M. Effect of lactation stage and energy status on milk fat composition of Holstein-Friesian cows. *J. Dairy Sci.* **2009**, *92*, 1469–1478. [CrossRef] [PubMed]
5. Garnsworthy, P.C.; Masson, L.L.; Lock, A.L.; Mottram, T.T. Variation of milk citrate with stage of lactation and de novo fatty acid synthesis in dairy cows. *J. Dairy Sci.* **2006**, *89*, 1604–1612. [CrossRef]
6. Dewhurst, R.J.; Davies, D.R.; Merry, R.J. Microbial protein supply from the rumen. *Anim. Feed Sci. Technol.* **2000**, *85*, 1–21. [CrossRef]
7. Keeney, M.; Katz, I.; Allison, M.J. On the probable origin of some milk fat acids in rumen microbial lipids. *J. Am. Oil Chem. Soc.* **1962**, *39*, 198–201. [CrossRef]
8. Kaneda, T. Iso- and anteiso-fatty acids in bacteria: Biosynthesis, function, and taxonomic significance. *Microbiol. Mol. Biol. Rev.* **1991**, *55*, 288–302.
9. Vlaeminck, B.; Dufour, C.; van Vuuren, A.M.; Cabrita, A.R.J.; Dewhurst, R.J.; Demeyer, D.; Fievez, V. Use of odd and branched-chain fatty acids in rumen contents and milk as a potential microbial marker. *J. Dairy Sci.* **2005**, *88*, 1031–1042. [CrossRef]
10. Vlaeminck, B.; Fievez, V.; Tamminga, S.; Dewhurst, R.J.; Van Vuuren, A.; De Brabander, D.; Demeyer, D. Milk odd- and branched-chain fatty acids in relation to the rumen fermentation pattern. *J. Dairy Sci.* **2006**, *89*, 3954–3964. [CrossRef]
11. Vlaeminck, B.; Fievez, V.; Van Laar, H.; Demeyer, D. Rumen odd and branched chain fatty acids in relation to in vitro rumen volatile fatty acid productions and dietary characteristics of incubated substrates. *J. Anim. Physiol. Anim. Nutr.* **2004**, *88*, 401–411. [CrossRef] [PubMed]
12. Bhagwat, A.M.; De Baets, B.; Steen, A.; Vlaeminck, B.; Fievez, V. Prediction of ruminal volatile fatty acid proportions of lactating dairy cows based on milk odd- and branched-chain fatty acid profiles: New models, better predictions. *J. Dairy Sci.* **2012**, *95*, 3926–3937. [CrossRef] [PubMed]
13. Dijkstra, J.; Ellis, J.L.; Kebreab, E.; Strathe, A.B.; López, S.; France, J.; Bannink, A. Ruminal pH regulation and nutritional consequences of low pH. *Anim. Feed Sci. Technol.* **2012**, *172*, 22–33. [CrossRef]
14. Sutton, J.D. Digestion and Absorption of Energy Substrates in the Lactating Cow. *J. Dairy Sci.* **1985**, *68*, 3376–3393. [CrossRef]
15. Fujihara, T.; Shem, M.N. Metabolism of microbial nitrogen in ruminants with special reference to nucleic acids. *Anim. Sci. J.* **2011**, *82*, 198–208. [CrossRef]
16. Bryant, M.P.; Robinson, I.M. Effects of diet, time after feeding, and position sampled on numbers of viable bacteria in the bovine rumen. *J. Dairy Sci.* **1968**, *51*, 1950–1955. [CrossRef]
17. Shi, Y.; Odt, C.L.; Weimer, P.J. Competition for cellulose among three predominant ruminal cellulolytic bacteria under substrate-excess and substrate-limited conditions. *Appl. Environ. Microb.* **1997**, *63*, 734–742.
18. Bryant, M.P. Bacterial species of the rumen. *Microbiol. Mol. Biol. Rew.* **1959**, *23*, 125–153.
19. Miyagawa, E. Cellular fatty-acid and fatty alaehyde composition rumen bacteria. *J. Gen. Appl. Microbiol.* **1982**, *28*, 389–408. [CrossRef]
20. Fernando, S.C.; Purvis, H.T., II; Najar, F.Z.; Sukharnikov, L.O.; Krehbiel, C.R.; Nagaraja, T.G.; Roe, B.A.; DeSilva, U. Rumen microbial population dynamics during adaptation to a high-grain diet. *Appl. Environ. Microb.* **2010**, *76*, 7482–7490. [CrossRef]
21. Klieve, A.V.; Hennessy, D.; Ouwerkerk, D.; Forster, R.J.; Mackie, R.I.; Attwood, G.T. Establishing populations of *Megasphaera elsdenii* YE 34 and *Butyrivibrio fibrisolvens* YE 44 in the rumen of cattle fed high grain diets. *J. Appl. Microbiol.* **2003**, *95*, 621–630. [CrossRef] [PubMed]

22. NRC. *Nutrient Requirements of Dairy Cattle*, 7th ed.; National Academy of Science: Washington, DC, USA, 2001.
23. AOAC. *Officical Methods of Analysis of the AOAC*; Association of Official Analytical Chemists Inc.: Arlington, VA, USA, 1990; Volume 2.
24. Van Soest, P.J.; Roberson, J.B.; Lewis, B.A. Methods for dietary fiber, neutral detergent fiber, and nonstarch polysaccharides in relation to animal nutrition. *J. Dairy Sci.* **1991**, *74*, 3583–3597. [CrossRef]
25. Li, Y.; Meng, Q. Effect of different types of fibre supplemented with sunflower oil on ruminal fermentation and production of conjugated linoleic acids in vitro. *Arch. Anim. Nutr.* **2006**, *60*, 402–411. [CrossRef] [PubMed]
26. Tajima, K.; Aminov, R.I.; Nagamine, T.; Matsui, H.; Nakamura, M.; Benno, Y. Diet-dependent shifts in the bacterial population of the rumen revealed with real-time PCR. *Appl. Environ. Microb.* **2001**, *67*, 2766–2774. [CrossRef] [PubMed]
27. Zhang, Y.; Liu, K.; Hao, X.; Xin, H. The relationships between odd- and branched-chain fatty acids to ruminal fermentation parameters and bacterial populations with different dietary ratios of forage and concentrate. *J. Anim. Physiol. Anim. Nutr.* **2016**, *101*, 1103–1114. [CrossRef]
28. Reilly, K.; Attwood, G.T. Detection of Clostridium proteoclasticum and closely related strains in the rumen by competitive PCR. *Appl. Environ. Microb.* **1998**, *64*, 907–913.
29. Singha, K.M.; Pandyab, P.R.; Tripathia, A.K.; Patelb, G.R.; Parnerkarb, S.; Kotharic, R.K.; Joshia, C.G. Study of rumen metagenome community using qPCR under different diets. *Meta Gene* **2014**, *2*, 191–199. [CrossRef]
30. Yang, S. Effect of Soybean Oil and Linseed Oil Supplementation on Population of Ruminal Bacteria and Fermentation Parameters in Dairy Cows. Ph.D. Thesis, Chinese Academy of Agricultural Sciences, Beijing, China, 2007.
31. Stevenson, D.M.; Weimer, P.J. Dominance of *Prevolla* and low abundance of classical ruminal bacterial species in the bovine rumen revealed by relative quantification real-time PCR. *Appl. Microb. Biotechnol.* **2007**, *75*, 165–174. [CrossRef]
32. St-Pierre, N.R. Invited review integrating quantitative findings from multiple studies using mixed model methodology. *J. Dairy Sci.* **2001**, *84*, 741–755. [CrossRef]
33. Fievez, V.; Colman, E.; Castro-Montoya, J.M.; Stefanov, I.; Vlaeminck, B. Milk odd- and branched-chain fatty acids as biomarkers of rumen function. *Anim. Feed Sci. Technol.* **2012**, *172*, 51–65. [CrossRef]
34. Vlaeminck, B.; Fievez, V.; Cabrita, A.R.J.; Fonseca, A.J.M.; Dewhurst, R.J. Factors affecting odd- and branched-chain fatty acids in milk: A review. *Anim. Feed Sci. Technol.* **2006**, *131*, 389–417. [CrossRef]
35. Smith, S. The animal fatty acid synthase: One gene, one polypeptide, seven enzymes. *FASEB J.* **1994**, *8*, 1248–1259. [CrossRef] [PubMed]
36. Dodds, P.F.; Guzman, M.G.; Chalberg, S.C.; Anderson, G.J.; Kumar, S. Acetoacetyl-CoA reductase activity of lactating bovine mammary fatty acid synthase. *J. Biol. Chem.* **1981**, *256*, 6282–6290. [PubMed]
37. Massart-Leèn, A.M.; Roets, E.; Peeters, G.; Verbeke, R. Propionate for fatty acid synthesis by the mammary gland of the lactating goat. *J. Dairy Sci.* **1983**, *7*, 1445–1454. [CrossRef]
38. Scheerlinck, K.; De Baets, B.; Stefanov, I.; Fievez, V. Subset selection from multi-experiment data sets with application to milk fatty acid profiles. *Comput. Electron. Agric.* **2010**, *73*, 200–212. [CrossRef]
39. Vlaeminck, B.; Gervais, R.; Rahman, M.M.; Gadeyne, F.; Gorniak, M.; Doreau, M.; Fievez, V. Postruminal synthesis modifies the odd- and branched-chain fatty acid profile from the duodenum to milk. *J. Dairy Sci.* **2015**, *98*, 4829–4840. [CrossRef]
40. Belyea, R.L.; Adams, M.W. Energy and nitrogen utilization of high versus low producing dairy cows. *J. Dairy Sci.* **1990**, *73*, 1023–1030. [CrossRef]
41. Craninx, M.; Steen, A.; Laar, H.V.; Nespen, T.V.; Martín-Tereso, J.; Baets, B.D.; Fievez, V. Effect of Lactation Stage on the Odd- and Branched-Chain Milk Fatty Acids of Dairy Cattle Under Grazing and Indoor Conditions. *J. Dairy Sci.* **2008**, *91*, 2662–2677. [CrossRef]
42. Bauman, D.E.; Mellenberger, R.W.; Ingle, D.L. Metabolic adaptations in fatty acid and lactose biosynthesis by sheep mammary tissue during cessation of lactation. *J. Dairy Sci.* **1974**, *57*, 719–723. [CrossRef]
43. Verbeke, R.; Lauryssens, M.; Peeters, G.; James, A.T. Incorporation of DL-[1-^{14}C] leucine and [1-^{14}C] isovaleric acid into milk constituents by the perfused cow's udder. *Biochem. J.* **1959**, *73*, 24–29.
44. Fievez, V.; Vlaeminck, B.; Dhanoa, M.S.; Dewhurst, R.J. Use of principal component analysis to investigate the origin of heptadecenoic and conjugated linoleic acids in milk. *J. Dairy Sci.* **2003**, *86*, 4047–4053. [CrossRef]

45. Dewhurst, R.J.; Moorby, J.M.; Vlaeminck, B.; Fievez, V. Apparent recovery of duodenal odd- and branched-chain fatty acids in milk of dairy cows. *J. Dairy Sci.* **2007**, *90*, 1775–1780. [CrossRef] [PubMed]
46. Gervais, R.; Vlaeminck, B.; Fanchone, A.; Nozière, P.; Doreau, M.; Fievez, V. Odd-and branched-chain fatty acids duedenal flows and milk yield in response to N underfeeding and enegy source in dairy cows. *J. Dairy Sci.* **2011**, *94*, 125–126.
47. Dijkstra, J. Production and absorption of volatile fatty acids in the rumen. *Livest. Prod. Sci.* **1994**, *39*, 61–69. [CrossRef]
48. Misra, A.K.; Thakur, S.S. Influence of isobutyric acid supplementation on nutrient intake, its utilization and growth performance of crossbred calves fed low protein and urea containing rations. *Indian J. Anim. Sci.* **2002**, *72*, 476–479.
49. Felix, A.; Cook, R.M.; Huber, J.T. Effect of feeding isoacids with urea on growth and nutrient utilization by lactating cows. *J. Dairy Sci.* **1980**, *63*, 1943–1946. [CrossRef]
50. Kaneda, T. Fatty acids of the genus Bacillus: An example of branched-chain preference. *Microbiol. Mol. Biol. Rev.* **1977**, *41*, 391–418.
51. Cabrita, A.R.J.; Fonseca, A.J.M.; Dewhurst, R.J.; Gomes, E. Nitrogen supplementation of corn silages. 2. Assessing rumen function using fatty acid profiles of bovine milk. *J. Dairy Sci.* **2003**, *86*, 4020–4032. [CrossRef]
52. Buccioni, A.; Decandia, M.; Minieri, S.; Molle, G.; Cabiddu, A. Lipid metabolism in the rumen: New insights on lipolysis and biohydrogenation with an emphasis on the role of endogenous plant factors. *Anim. Feed Sci. Technol.* **2012**, *174*, 1–25. [CrossRef]
53. Ifkovits, R.W.; Ragheb, H.S. Cellular Fatty Acid Composition and Identification of Rumen Bacteria. *Appl. Environ. Microb.* **1968**, *16*, 1406–1413.
54. Minato, H.; Ishibashi, S.; Hamaoka, T. Cellular fatty-acid and suger composition of reppesentative strains of rumen bacteria. *J. Gen. Appl. Microbiol.* **1988**, *34*, 303–319. [CrossRef]
55. Saluzzi, L.; Smith, A.; Stewart, C.S. Analysis of bacterial phospholipid markers and plant monosaccharides during forage degradation by Ruminococcus flavefaciens and Fibrobacter succinogenes in co-culture. *J. Gen. Microbiol.* **1993**, *139*, 2865–2873. [CrossRef] [PubMed]
56. Croom, W.J.; Bauman, D.E.; Davis, C.L. Methylmalonic acid in low-fat milk syndrome. *J. Dairy Sci.* **1981**, *64*, 649–654. [CrossRef]
57. Diez-Gonzalez, F.; Bond, D.R.; Jennings, E.; Russell, J.B. Alternative schemes of butyrate production in Butyrivibrio fibrisolvens and their relationship to acetate utilization, lactate production, and phylogeny. *Arch. Microbiol.* **1999**, *171*, 324–330. [CrossRef]
58. Roché, C.; Albertyn, H.; Van Gylswyk, N.O.; Kistner, A. The growth response of cellulolytic acetate-utilizing and acetate-producing *Butyrivibrios* to volatile fatty acids and other nutrients. *J. Gen. Microbiol.* **1973**, *78*, 253–260. [CrossRef]
59. Latham, M.J.; Legakis, N.J. Cultural factors influencing the utilization or production of acetate by Butyrivibrio fibrisolvens. *J. Gen. Microbiol.* **1976**, *94*, 380–388. [CrossRef]
60. Stern, M.D.; Hoover, W.H. Methods for determining and factors affecting rumen microbial protein synthesis: A review. *J. Anim. Sci.* **1979**, *49*, 1590–1603. [CrossRef]
61. Wongtangtintharn, S.; Oku, H.; Iwasaki, H.; Toda, T. Effect of branched-chain fatty acids on fatty acid biosynthesis of human breast cancer cells. *J. Nutr. Sci. Vitaminol.* **2004**, *50*, 137–143. [CrossRef]

12

Autochtonous Strain *Enterococcus faecium* EF2019(CCM7420), its Bacteriocin and their Beneficial Effects in Broiler Rabbits

Monika Pogány Simonová [1,*], Ľubica Chrastinová [2] and Andrea Lauková [1,*]

[1] Institute of Animal Physiology, Centre of Biosciences of Slovak Academy of Sciences, Šoltésovej 4-6, 04001 Kosice, Slovakia

[2] Institute for Nutrition, National Agricultural and Food Centre, Hlohovecká 2, 951 41 Nitra-Lužianky, Slovakia; lubica.chrastinova@nppc.sk

* Correspondence: simonova@saske.sk (M.P.S.); laukova@saske.sk (A.L.)

Simple Summary: Weaning is the most important and critical period in rabbits breeding; the cecal digestion is very complex and only small dietary and/or environmental changes can disturb the stable microbial population/fermentation and gut health, leading to digestive dysbiosis and increased morbidity, often with fatal outcome and big economic losses. Control of the microbiota, prevention of digestive disturbances and improving gut health and immunity can be achieved through the natural substances application in rabbit nutrition. While probiotics are frequently used in rabbit farms, the in vivo administration of bacteriocins (antimicrobial substances produced by bacteria, which usually also possess probiotic properties) in these animals is often limited and has become an area of research activity. Moreover, the most of probiotic strains used in rabbits are non-autochthonous (have a different origin than the rabbits ecosystem). Therefore, our study focused on improving rabbits' health using the autochthonous strain *Enterococcus faecium* EF2019 (CCM7420) and its enterocin (Ent7420) in broiler rabbits. The antibacterial and anticoccidial effect of additives was observed, with good colonization ability of the CCM7420 strain. Both additives showed a tendency to modulate the serum biochemistry parameters and to improve the immunity, jejunal morphology, weight gains, feed conversion ratio and meat quality (physicochemical traits and mineral content).

Abstract: The present review evaluates and compares the effects achieved after application of rabbit-derived bacteriocin-producing strain *Enterococcus faecium* CCM7420 with probiotic properties and its bacteriocin Ent7420. The experiments included varying duration of application (14 and 21 days), form of application (fresh culture and lyophilized form), combination with herbal extract and application of the partially purified enterocin—Ent7420, produced by this strain. Results from these studies showed that *E. faecium* CCM7420 strain was able to colonize the gastrointestinal tract (caecum) of rabbits (in the range < 1.0–6.7 log cycle, respectively 3.66 log cycle on average), to change the composition of intestinal microbiota (increased lactic acid bacteria, reduced counts of coliforms, clostridia and staphylococci), to modulate the immunity (significant increase of phagocytic activity), morphometry (enlargement absorption surface in jejunum, higher villi height:crypt depth (VH:CD) ratio), physiological (serum biochemistry; altered total proteins, glucose and triglycerides levels) and parasitological (*Eimeria* sp. oocysts) parameters and to improve weight gains (in the range 4.8–22.0%, respectively 11.2% on average), feed conversion ratio and meat quality (physicochemical traits and mineral content).

Keywords: rabbit; *Enterococcus faecium*; enterocin; microbiota; intestinal morphology; phagocytic activity; serum biochemistry; meat quality; weight gain

---

1. Introduction

Rabbit breeding has a great potential because of the small body size, short generation interval, rapid growth rate, high productive capacity and healthy, easily digestible meat of rabbits [1,2]. Moreover, rabbits can convert a higher amount (20%) of the protein they eat into edible meat, compared with pigs (16–18%) and cattle (8–12%; [3]). In several European countries, in which the rabbit breeding has a long history and high production efficiency, presently is regressing, whereas in the developing countries of the world rabbit farming has become to an important emerging enterprise. In growing rabbits, the most critical period is the weaning, when the kits are separated from mother and the milk is substituted with solid feed [4]. During these environmental and physiological changes/stresses, the rabbits are very sensitive to digestive disturbances, also called non specific enteritis (usually caused by dietary stresses, parasites—*Coccidia* and bacteria—*Clostridia* sp. and enteropathogenic *Escherichia coli*) and gastrointestinal infections (epizootic rabbit enteropathy—ERE, a multifactorial gastrointestinal syndrome, [5]). These dietary and bacterial changes are the main reason of morbidity and mortality and have negative effects on feed consumption, growth performance and health status of animals in this period and also on the economic aspects of rabbit farming.

To overcome this period, to reduce economic losses and to improve and stabilize the health status and gastrointestinal tract development, antibiotic growth promoters (AGPs) have been widely used for years. Although, these synthetic drugs showed good effects on production indicators, on the other hand, there was a risk of increasing resistance to antibiotics and transferring of antibiotic resistance genes from animal to human, which also threatened the human health and quality of meat and food [6]. For this reason, AGPs have been banned by the European Union (began in 1986 in Sweden and completely banned in January 2006, when the last four antibiotics have been permitted as feed additives was no longer allowed to be marketed or used from this date; IP/03/1058; [7,8]). As a result of the ban, researchers had to substitute AGPs and to find new feed additives that were supposed to be safer, without leaving residues and spreading resistance to themselves, but also improving health and productivity of rabbits. Therefore, the antibiotics have been replaced with new, naturally based supplements: probiotics, prebiotics, synbiotics, enzymes, bacteriocins, organic acids, herbs and their extracts, which are well-tried tools for disease prevention and therapy in various animal species, including rabbits [9].

The last two decades have seen a substantial increase in the use natural supplements and/or additives in animal nutrition, in which their antimicrobial activity has been highlighted many times. The EU in its Regulation EC 1831/2003 defined the terms "feed additives" as "substances, microorganisms or preparations, other than feed material and premixtures, which are intentionally added to feed or water" and the "antimicrobials" as "substances produced either synthetically or naturally, used to kill or inhibit the growth of microorganisms, including bacteria, viruses or fungi, or of parasites, in particular protozoa" [10]. According to the World Health Organization (WHO) probiotics are defined as "live microorganisms which, when administered in adequate amount, confer a health benefit on the host". Although this definition is widely accepted, a 2007 guidance document from the European Commission on Regulation EC 1924/2006 on nutrition and health claims (NHCR) categorizes the term 'probiotic' as a health claim on the basis that it implies a health benefit. For this reason, the term "beneficial microbes" is used more often instead of a "probiotic microorganism" [11].

The use of several natural feed additives has already been reviewed in rabbit breeding [10,12–15]. The positive effects of probiotics and their antibacterial products—bacteriocins on health, growth performance, nutrient utilization and metabolism changes, microbial composition [14,16–35], blood serum biochemistry, oxidative stress, immune response, intestinal morphology [21,24,28–31,34,36–38] and meat quality of rabbits [32,34,39–43] was described. However, in spite of the achieved results, there are still few declared probiotic preparations based (Lactina, Toyocerin; [44,45]; Prorabbit—declared in Slovakia; [25]) and detailed studies of microorganisms with beneficial properties that have also the ability to produce antimicrobial substances, enterocins.

The aim of this review was to summarize all achieved properties and physiological effects of the bacteriocin-producing strain with probiotic properties *Enterococcus faecium* CCM7420 (EF2019 previous working labeling, [46]) isolated in 2003 from rabbit feces in the Laboratory of Animal Microbiology of the Institute of Animal Physiology, Centre of Biosciences of the Slovak Academy of Sciences (Košice, Slovakia) and tested to date in 180 rabbits. These experiments included varying duration of application (2 and 3 weeks), form of application (fresh culture in water; the concentration of cells was $\times 10^9$ CFU/mL in a dose 500 µL/animal/day; lyophilized (freeze-dried) form rehydrated in water ($\times 10^9$ CFU/mL; dose 500 µL/animal/day) as well as mixed in feed and pelleted (15 g/100 kg feed), application of its partially purified bacteriocin (PPB)—enterocin (Ent) EF2019 (applied into water) and fresh culture in combination with natural substance (*Eleutherococcus senticosus*).

2. *Enterococcus faecium* CCM7420 (EF2019) and Its Bacteriocin-Enterocin (Ent7420)

Enterococcus faecium EF2019 (CCM7420) is a bacteriocin-producing strain [47], which was isolated from the rabbit feces and genetically confirmed by the PCR method and subsequently by MALDI-TOF mass spectrophotometry as well as the sequencing procedure of this strain was provided (Dr. Kopčáková, IAP CBs SAS). This strain produces lactic acid, tolerates low pH (3.0; 63% surviving of cells) and is able to grow even in 5% oxgall—bile (80% surviving of cells), shows sensitivity to antibiotics, including vancomycin [25,48] and possess lipolytic activity [49]. Other unpublished data suggests that the CCM7420 does not produce biogenic amines and enzymes such as β-glucuronidase, β-galactosidase or N-acetyl-β-glucosaminidase (enzymes produced by unfriendly gut bacteria; their increased levels are usually the indicators of colon cancer), and it does not show any gelatinase (absence of the *gelE* gene) or hemolytic activities with low ability to form biofilm (0.092). The strain was deponed into Czech Collection of Microorganisms in Brno, Czech Republic to have number CCM7420. This strain showed the broadest inhibitory activity from all tested rabbits enterococcal strains against the indicators *E. avium* EA5, *Listeria innocua* LMG13568 and *L. monocytogenes* CCM4699 and against other tested enterococci and staphylococci tested such as clostridia, pseudomonads, enterobacteria and coliform bacteria [48]. The presence of the structural genes for enterocins (ent) A, P and L50B was detected; however, the CCM7420 did not possessed gene for ent B [47]. The molecular mass of its bacteriocin-like substance ranged from 3 to 10 kDa. Proteinaceous substance produced by CCM7420 strain was partially purified (partially purified bacteriocin (PPB) or enterocin (Ent) 2019 =7420). It is thermostable substance as well as stable at pH 4.0, 7.0 and 9.0. Its production starts in early logarithmic growth phase and it culminates in the late logarithmic phase of CCM7420 strain growth. By its properties, it can probably be included in the II. classification group of bacteriocins. Ent2019 or Ent7420 added to the growing strain *L. innocua* LMG13568 (after 4 h) inhibited its growth already at 1 h after enterocin addition with a difference of 1.5 log cycles (5 h of cultivation). This effect was prolonged up to 24 h. The Ent7420 was tested against more than 300 strains of enterococci, staphylococci, clostridia, pseudomonads, enterobacteria and coliforms [48]. The inhibitory activity of this substance was preserved after 24 months of storage at −20 °C (6400 AU/mL; [48]) and also after lyophilization (freeze dried) and redissolution in PBS buffer (25600 AU/mL; not published data). The CCM7420 strain is currently available in the ProRabbit, probiotic product for rabbits and other rodents, made by the International Probiotic Company Košice (Slovakia).

3. Application Effects of *E. faecium* CCM7420 and Its Enterocin Ent7420 Observed in Experiments

3.1. Effect on Growth Performance

The results from all these experiments indicated that the CCM7420 strain could improve the average daily weight gain (ADWG; between 4.8 and 22.0%; Table 1) regardless of the form of bacteriocin-producing strain (fresh, $p < 0.001$ compared to control data; or freeze dried-lyophilized—numerical increase) and its application time (2 or 3 weeks). The feed conversion ratio (FCR) was influenced only through the Ent7420 administration and the combinative application of the CCM7420 strain with *Eleutherococcus senticosus* extract. The effect of probiotics, including registered probiotic preparations and new beneficial microorganisms on the growth performance of rabbits have been already described/reviewed; they usually confirmed the increased body weight [14,15,17,19,24,25,31,32]. The most of these studies described faster growth and higher weight gain of rabbits using probiotic preparations and feed additives based on the following bacterial strains and yeasts alone or in their combinations: *Saccharomyces cerevisiae*, *S. boulardii*, *Bacillus licheniformis*, *B. cereus*, *B. cereus* var. *toyoi*, *Pediococcus acidilactici*, *Lactococcus lactis*, *Lactobacillus acidophilus*, *L. plantarum*, *L. helveticus*, *L. delbrueckii* and *L. sporogenes* [14]. Up to now, only several commercial products recommended especially for rabbits contain especially the species strain *Enterococcus faecium* as a component of the probiotic bacterial mix (Lactina—*E. faecium* NBIMCC 8270 [44]; Prorabbit—*E. faecium* CCM7420 [25]) or for companion animals with diarrhea (Pro-enteric Triplex—*E. faecium* DSM 10663/NCIMB 10415 [50]). In all experiments, higher ADWG was noted during CCM7420 administration. Surprisingly, the highest increase of ADWG (by 22%, Table 1) was noted after 2 weeks addition (model experiment, only seven animals were used in the group). The three weeks long CCM7420 dietary inclusion also improved the growth performance; however, only by 6.7% on average (from results of fresh culture application in two experiments; [25,26]). Similarly to our results, Lauková et al. [21] and Szabóová et al. [35] also reported higher ADWG through bacteriocinogenic and probiotic *E. faecium* AL41 (CCM8558) and CCM4231 strains application in rabbits. Improved body weight in rabbits was also noted after probiotics administration by Bovera et al. [30], Bhatt et al. [32] and Kalma et al. [51]. On the other hand, after PPB CCM7420 addition increased the body weight only slightly (by 2.2%, Table 2), but in this group, better FCR was achieved, compared to strain application. The bacteriocin dietary inclusion showed better feed conversion also in the case of other bacteriocins: EntM, EntCCM4231, EntEF55 and gallidermin [21,52–54], applied in rabbit husbandry.

Table 1. The effect of *Enterococcus faecium* CCM7420 and its enterocin Ent7420 on the growth performance of rabbits.

Tested Zootechnical Parameters	*E. faecium* CCM7420 Strain					Enterocin (Ent) 7420
	Fresh Culture	Fresh Culture	Lyophilized Form Resolved in Water	Lyophilized Form Mixed into Pellets	Fresh Culture + *E. senticosus*	
Reference number of publication	[24]	[25]	[55]	[55]	[26]	[25]
Length of application Number of rabbits	14 days (n = 7)	21 days (n = 24)	21 days (n = 24)	21 days (n = 24)	21 days (n = 24)	21 days (n = 24)
Initial live weight (35 d of age; 0 d of experiment), g	1136.0 ± 100.0	977.0 ± 97.0	1002.3 ± 162.3	1042.5 ± 315.7	1077.5 ± 102.2	963.0 ± 101.0
Intermediate live weight (49/56 d of age; 14/21 d of experiment), g	1507.0 ± 120.0	1850.0 ± 152.0	1664.4 ± 170.0	1856.9 ± 361.4	1943.3 ± 222.2	1788.0 ± 199.0
Final weight (70/77 d of age; 35/42 d of experiment), g	2325.0 ± 260.0	2622.0 ± 104.0	2206.7 ± 164.6	2319.2 ± 164.6	2723.3 ± 204.7	2431.0 ± 142.0
Average daily weight gain (g/d; increase compare to control data%)	28.00 (22.0%)	39.17 (4.8%)	38.35 (10.0%)	38.51 (10.5%)	39.19 (8.5%)	34.95 (2.2%)
Feed conversion ratio between 35 and 56 days of age (g/g)	Not tested	2.71	2.23	2.28	2.59	2.68
Feed conversion ratio between 56 and 77 days of age (g/g)	Not tested	4.67	3.41	3.42	3.97	3.87
Feed conversion ratio per kg gain	Not tested	3.47	2.82	2.85	3.28	3.22
Mortality (n(%) in experimental group/n(%) in control group)	0(0.0%)/0(0.0%)	3(12.5%)/5(20.8%)	3(12.5%)/7(29.2%)	3(12.5%)/7(29.2%)	0(0.0%)/4(16.7%)	1(0.04%)/4(16.7%)

The fresh culture of the CCM7420 strain was applied into water (at concentration of cells $\times 10^9$ CFU/mL; dose 500 µL/animal/day); lyophilized (freeze-dried) form rehydrated in water ($\times 10^9$ CFU/mL; dose 500 µL/animal/day) as well as mixed in feed and pelleted (15 g/100 kg feed).

Table 2. The effect of *Enterococcus faecium* CCM7420 and its enterocin Ent7420 on the fecal microbiota of rabbits.

	E. faecium CCM7420 Strain					Enterocin (Ent)7420
Tested Microorganisms	**Fresh Culture**	**Fresh Culture**	**Lyophilized form Resolved in Water**	**Lyophilized Form Mixed into Pellets**	**Fresh Culture + *E. senticosus***	
Reference number of publication	[24]	[25]	[55]	[55]	[26]	[25]
Length of application	14 days	21 days	21 days	21 days	21 days	21 days
Enterococci	Increased	Increased ($p < 0.01$)	Increased	Increased	Unchanged	Increased ($p < 0.05$)
Lactic acid bacteria (LAB)	Increased ($p < 0.01$)	Unchanged	Increased	Increased	Unchanged	Decreased
Clostridia	Decreased	Decreased	Unchanged	Unchanged	Decreased ($p < 0.05$)	Unchanged
Coagulase-positive staphylococci	Decreased ($p < 0.01$)	Unchanged	Unchanged	Unchanged	Decreased	Decreased
Staphylococcus aureus	Unchanged	Decreased	Unchanged	Unchanged	Decreased ($p < 0.001$)	Decreased
Coliforms	Unchanged	Decreased	Unchanged	Unchanged	Decreased ($p < 0.001$)	Decreased
CCM7420	6.7 log cycle	6.7 log cycle	<1.0 log cycle	2.8 log cycle	1.1 log cycle	-

The fresh culture of the CCM7420 strain was applied into water (at concentration of cells $\times 10^9$ CFU/mL; dose 500 μL/animal /day); lyophilized (freeze-dried) form rehydrated in water ($\times 10^9$ CFU/mL; dose 500 μL/animal/day) as well as mixed in feed and pelleted (15 g/100 kg feed). Statistical analysis was performed using one-way analysis of variance (ANOVA) with the post hoc Tukey test with the level of significance set at ($p < 0.05$), within experimental groups during each individual experiment.

3.2. Effect on Fecal Microbiota

In all experiments, administration of *E. faecium* CCM7420 was associated with increased enterococci ($p < 0.01$) and lactic acid bacteria (LAB) counts during the treatment period by a 1.3 log cycle on average (Table 2), while the application form or time length had no impact on the bacterial counts increase. The numerical increase of enterococci and LAB was recorded also in previous experiments with non-autochthonous *E. faecium* probiotic strains inclusion in rabbits [21,29,56]. Other authors also noted abundance of microflora in caecum and higher lactobacilli counts after *Lactobacillus* strains application [57]. Outgoing from results of several studies focusing on the molecular profiling of rabbits gut [58–60] and colonization ability of applied enterococcal probiotic strains [33], enterococci were found as predominant microbiota from the phyla *Firmicutes* and they were able also to colonize the caecum and the small intestine (ileum and jejunum) in sufficient counts. The microbiological examinations in fecal samples confirmed the presence of *E. faecium* CCM7420 strain. This strain was able to colonize the digestive tract of rabbits, reaching counts in the range 2.8–6.7 log cycle during the 2 or 3 weeks treatment. These numbers are comparable with the level of autochthonous probiotic strains of several *Enterococcus* spp. after their application in rabbits [35]. Obviously, decreased CCM7420 counts (1.0–3.3 log cycle) were noted in the post-treatment period (3 weeks after strain cessation), but it was still able to persist in the rabbit's intestine. However, the lowest counts in fecal samples were achieved during the experimental application of the lyophilized CCM7420 strain [60], this level is comparable to that one achieved through fresh culture addition of non-autochthonous, bacteriocinogenic and the probiotic *E. faecium* AL41 (CCM8558) strain in rabbits [21]. Important results concerning the spoilage microbiota, including clostridia, coliforms and staphylococci were described in our experiments. Significant reduction of coagulase-positive staphylococci was noted in fecal samples of rabbits, administering fresh culture of CCM7420 ($p < 0.01$; [24]); the *S. aureus* counts were also reduced by 0.5–1.2 log; $p < 0.001$ [25,26]. The CCM7420 strain seems to be useful in rabbits suffering from diarrhea disturbances involving coliforms (reduced by 1.6 log; $p < 0.001$ [26]) and/or clostridia (reduced by 0.5–1.5 log; $p < 0.05$ [25,26]). Similarly to our results, decreased counts of coliforms, *S. aureus* and clostridia were presented in probiotic-treated rabbits [21,23,29]. In rabbits administering the enterocin Ent7420, reduction in coliforms, coagulase-positive staphylococci, including *S. aureus* was noted [25]. Our results are in accordance to those presented by Lauková et al. [21], who tested the effect of nisin on the rabbits gut microbiota and noted reduced counts of the most bacterial species. The gut microbiota is an important constituent in the intestine's mucosal barrier; the increase of the host defense had been already demonstrated by the application of potentially beneficial microorganisms and other natural antimicrobials (bacteriocins, organic acids and plant extracts; [61]).

Changes in bacterial composition—decrease of fecal coliforms, *Pseudomonas*-like sp., *Clostridium*-like sp. and *S. aureus*—during the additives supplementation confirm the antibacterial effect of CCM7420 strain. The inhibitory effect of the bacteriocin-producing and probiotic enterococci and their enterocins on rabbits' intestinal microbiota was already reported in our previous studies [24,25,28,29,52]. Kritas et al. [23] also described the lower frequency of *E. coli* and *C. perfringens* in rabbits treated by probiotic. The dominancy in antimicrobial activity of CCM7420 strain combined with *E. senticosus* can be confirmed by the fact that the *E. senticosus* extract possesses slight or no antimicrobial activity [26]. We supposed that the dietary modulation of the gastrointestinal microbiota by natural antimicrobial substances could result in an enhancement of colonization resistance against potentially pathogenic bacteria.

3.3. *Effect on Eimeria sp. Oocysts*

Coccidiosis is one of the most frequent and prevalent parasitic diseases in rabbit farms. The most markedly and typical symptoms are weight loss, diarrhea (from mild intermittent to severe) with the presence of blood and/or mucus in feces, which leads, through dehydration, to the mortality of animals. The high morbidity and mortality rates among all ages, especially in the young rabbits may be responsible for important economic losses [62]. The oocysts are always present in the intestines of rabbits and they cannot be completely eliminated even by the use of coccidiostat because of the caecotrophy and the symptomless, but potential source of infections of adults. These oocysts are able to cause not only pure eimeriosis after their multiplication and massive infection, but they may also be the cause of multifactorial diseases, when associated with other bacterial or viral infections. Eimeria infections can cause severe disease depending on *Eimeria* species, especially in young animals and the highest incidence of oocysts was usually found around the weaning period [63]. The EU has banned the use of antibiotics as feed additives for growth promotion in animals since 2005 [64]. Today, these antibiotics are replaced with alternative anticoccidials, including prebiotics and probiotics, based on their bactericidal and/or bacteriostatic activities, with immunostimulation and improved growth performance and productivity of the host organism. The experimental application of the CCM7420 strain and its Ent7420 was associated with the reduction of fecal *Eimeria* sp. oocysts. The different treatment period had no impact on oocysts reduction. When the fresh culture of the CCM7420 strain at concentration $\times 10^9$ CFU/mL/g was applied only for 2 weeks to compare its effect with the commonly used probiotic strain *Lactobacillus rhamnosus* GG, the decrease in oocysts counts was observed after probiotic application (8.3×10^1 OPG; oocysts per 1 g of feces) compared to control data (1.5×10^4 OPG), but also compared to the initial counts in experimental group (7.5×10^2 OPG). This reduction effect was maintained until the end of the experiment, also after cessation of the CCM7420 strain. Moreover, at the end of the experiment (at day 42), the difference one order of magnitude in *Eimeria* sp. oocysts was found comparing the control (1.5×10^4 OPG) and the experimental groups (7.2×10^3 OPG; [26]). On the other hand, when CCM7420 was applied with its bacteriocin Ent7420 in rabbits through 3 weeks, after 1 week of their addition, oocysts showed a trend towards a numerical reduction (not significant; [25]) in both experimental groups compared to the control group (Table 1). Surprisingly, at the end of CCM7420 strain addition, increased oocyst counts was observed in this group, while Ent7420 administration was more effective due to a significant reduction in oocysts ($p < 0.05$) at the end of probiotic application (day 21). This finding could be explained by the irregular excretion of oocysts. In the group with Ent7420 addition, a decreasing tendency of oocysts occurrence (not significant) was observed up to the end of enterocin substance application. This could lead to consideration that longer application of CCM7420 strain did not influence more the *Eimeria* sp. oocysts counts in rabbits.

While probiotics are widely used in animals because they improve the growth performance, productivity, health status and stimulate the immunity, studies concerning the protective effect against *Eimeria* sp. are still limited and focused on mainly the avian coccidiosis [65,66]. In rabbits, natural alternatives—prebiotics and herbal extracts—to coccidiostats have been studied [52,67–69]. To the best of our knowledge, only several works demonstrated the anticoccidial effect of beneficial microbes and/or probiotics as well as their antimicrobial products bacteriocins in rabbits [24,25,28,52,53,55,70]. The in vitro effect of four probiotic/bacteriocin-producing strains towards poultry *Eimeria* sp. oocysts was also documented by Strompfová et al. [71] and no differences in the reductive effect of bacteriocin-producing and non-producing strains ($p < 0.05$) were found in this experiment. The in vivo administration of bacteriocin-producing and probiotic strains decreased *Eimeria* sp. oocysts in rabbits. Outgoing from these results, we supposed that anticoccidial effect could be done due to the lactic acid production or by the effect of bacteriocins produced by the mentioned strains. Moreover, the potential protective effect of the CCM7420 strain against zoonotic *Trichinella spiralis* infection was also investigated in the framework of a new therapeutic strategy aimed at using probiotics to control parasitic zoonoses [72], when the authors demonstrated the reduced intensity of *T. spiralis* infection and female fecundity ex vivo and in vitro (about 60%) throughout the CCM7420 administration.

3.4. Effect on Serum Biochemistry

The measurement of biochemical parameters is data mostly used for diagnostic investigations and presents a useful way for controlling the health of animals. However, in some cases, several "components" of the host biochemistry are less specific because of the reparation/compensation ability of healthy tissues/organs and/or metabolic processes.

The tested serum parameters were in the range of normal values defined for these parameters in previous studies with rabbits [73–75], although there are differences in physiological or reference ranges in rabbit serum. During the *E. faecium* CCM7420 application, increased (even though not significantly) concentration of the total protein (TP) was noted in most experiments, and remained stable or higher also three weeks after the strain ceasing (cessation). The highest increase in TP was measured through the fresh CCM7420 culture administration in rabbits (Table 3) [24–26], while the lyophilized (freeze dried form resolved in water and mixed in pellets did not affect the TP level [60]. Application of enterocin Ent7420 as well as the combinative application of CCM7420 with the *E. senticosus* extract also improved the TP concentration [25,26]; similarly as was reported by Fathi et al. [33] and Kalma et al. [51] who observed a slight increase in serum TP after probiotic supplementation in rabbits. The increased level of TP could be explained by better resorption and utilization from the gut; this finding could be also confirmed by higher ADWG. On the other hand, application of non-autochthonous probiotic strains *E. faecium* CCM4231 and AL41 as well as their enterocins did not affect the TP in blood serum [21,29,38]. Blood glucose is an important source of energy for many cells and this is a parameter of the balance between glucose source/availability and utilization. Similarly to TP, higher glucose content was observed during fresh (both 2 and 3 weeks) and lyophilized CCM7420 culture mixed in feed as well as during Ent7420 application. The increased glucose level can be explained by conversion of lactic acid to pyruvate through the gluconeogenesis in the liver. Oppositely, reduced glucose level was observed in the case of the rehydrated–lyophilized CCM7420 strain and its combination with *E. senticosus* ($p < 0.01$). It could be that the glucose concentration was reduced by increased H^+ concentration due to higher organic acid values in the cecum content, which inhibited gluconeogenesis [76]. The increased H^+ (lactate accumulation) in the organism first stimulates physicochemical mineral dissolution by increasing the osteoclast and osteoblast activity (bone resorption) and mostly the Ca^{2+} and Mg^{2+} reabsorption in renal tubules for pH neutralization; it is usually confirmed also with higher serum calcium levels. This hypothesis was confirmed also by us, when the slight increase of calcium content was noted during three weeks CCM7420 fresh culture and its Ent7420 application. The rabbits' blood calcium levels fluctuate widely, dependent upon the level of calcium in their diet and the intestinal absorption as well [77]; this is a difference in the calcium metabolism from other mammals. While Lauková et al. [21] and Szabóová et al. [35] described no influence of probiotic strains on serum glucose, triglycerides and calcium levels, Fathi et al. [33] presented numerical increase of triglycerides associated with dietary probiotic treatment in rabbits, similarly to our achievements [25]. The hypocholesterolemic effect of probiotics in rabbits has been already presented [30,34,51]; surprisingly, our results did not confirm those findings. Moreover, increased/higher cholesterol levels (however, without significant changes) were measured during fresh culture CCM7420 and its combinative administration with *E. senticosus* [26,55] in rabbits; the detected levels were still within the physiological norm.

Table 3. The effect of *Enterococcus faecium* CCM7420 and its enterocin Ent7420 on the serum biochemistry of rabbits.

	E. faecium CCM7420 Strain					Enterocin (Ent)7420
Tested Blood Parameters	**Fresh Culture**	**Fresh Culture**	**Lyophilized Form Resolved in Water**	**Lyophilized Form Mixed into Pellets**	**Fresh Culture + *E. senticosus***	
Reference number of publication	[24]	[25]	[55]	[55]	[26]	[25]
Length of application	14 days	21 days	21 days	21 days	21 days	21 days
Total proteins (g/L)	Increased ($p < 0.05$)	Increased	Unchanged	Unchanged	Increased	Increased ($p < 0.05$)
Total lipids (g/L)	Increased	Increased	Unchanged	Unchanged	Unchanged	Increased
Cholesterol (mmol/L)	Not tested	Not tested	Unchanged	Unchanged	Increased	Not tested
Glucose (mmol/L)	Unchanged	Increased	Increased	Decreased	Decreased	Increased
Calcium (mmol/L)	Unchanged	Increased	Unchanged	Unchanged	Not tested	Increased
Glutathione-peroxidase (GSH-Px; U/mL)	Increased	Decreased	Increased	Increased	Decreased	Decreased
Phagocytic activity (%)	Not tested	Not tested	Increased	Increased	Increased ($p < 0.0001$)	Not tested

The fresh culture of CCM7420 strain was applied into water (at concentration of cells 1×10^9 CFU/mL; dose 500 μL/animal /day); lyophilized (freeze-dried) from rehydrated in water (1×10^9 CFU/mL; dose 500 μL/animal /day) as well as mixed in feed and pelleted (15 g/100 kg feed). Statistical analysis was performed using one-way analysis of variance (ANOVA) with the post hoc Tukey test with the level of significance set at ($p < 0.05$), within experimental groups during each individual experiments.

Exogenous factors such as manipulation, nutritional, weather and temperature changes (mainly hot environmental conditions) often induce physiological oxidative stress, which is avoided by the host's defense system. The host's reaction to stress can be marked mostly by the glutathione-peroxidase (GPx) enzyme activity in blood. In addition, there were no significant differences in GPx activity in blood among experimental groups whereas they were differently affected by CCM7420 (Table 3). Application of CCM7420 strain did not disturb the GPx level; while Ent7420 has a reducing effect on GPx during its application (day 21; $p < 0.05$; [25]). Comparing our previous experimental applications of probiotic strains and their enterocins in rabbits, we suppose that enterocins were more active to protect the host organism; our assumptions were confirmed by reduced GPx levels during Ent7420 and EntM application [25,36]. Outgoing from results we suppose that application of CCM7420 and its bacteriocin Ent7420 did not evoke oxidative stress in the rabbits, similarly to other probiotic strains or enterocins administration [21,29,38].

3.5. *Effect on Organic Acids*

In rabbits, approximately 40% of digested organic matter of the feed is digested in the caeco-colic segment; so the caecum and the proximal colon are the primary fermenters [78]. The digestion process of nutrients continues in the small intestine by the digestive enzymes of the host, but some components, e.g., plant cell walls and fibers (mainly lignins, cellulose, hemicellulose, pectins) are hydrolyzed by bacterial enzymes into soluble smaller compounds and fermented into the end products: volatile fatty acids (VFA: acetic, propionic and butyric acid), ammonia, intermediary metabolites (lactic, succinic and formic acid) and gas (CO_2, CH_4 and H_2; [79]). The stable microbial fermentation is essential for rabbit health, and only small dietary and environmental changes can lead to increased morbidity and/or mortality via microbial dysbiosis and digestive disturbances. Natural substances applications could prevent those disturbances [12–14]. The concentrations of VFA are usually measured in the cecal content of rabbits [21,28–30,38,78]; in our first experiment we decided to follow the VFA and organic acids concentrations in feces (it was a model experiment with a low number of rabbits in the experimental groups). Application of *E. faecium* CCM7420 to rabbits led to an increase of fecal levels of acetic acid ($p < 0.001$) compared to control animals [24], while other tested organic acids (butyric, succinic and lactic acids) were unaffected with the CCM7420 treatment. Similarly to our results, application of other probiotics (*E. faecium* CCM4231; [28] or bacteriocins (EntM and nisin; [21,38]) in rabbits did not influence the molar proportion of VFA in caecum, while the total VFA production and cecal fermentative activity was increased after *L. plantarum* spray application [29]. Concluded from these results—enhanced enzymatic activity and organic/fatty acid production, better feed conversion ratio and improved jejunal morphology (data shown below) during the CCM7420 strain application, we hypothesized a positive correlation between weight gain and cecal fermentation, improved gut functionality (jejunal morphology) and nutrient absorption. Despite many reports presenting the beneficial results during natural substance application in rabbits, there is a need to extend the existing knowledge and find new possibilities to improve the cecal fermentation and rabbit gut health.

3.6. *Effect on Immunity and Jejunal Morphometry*

Knowledge of the immune response and homeostasis in farm animals represents important information to protect animals from especially bacteria-derived diseases, to improve their health and productivity. The overall organization of the rabbit's digestive immune—lymphoid system is mostly similar to other species, but at the same time it is very special. There are two additional structures identified only in this species, the sacculus rotundus and the vermiform appendix (a place of lymphoid cells differentiation and maturation); they generally act synergistically. The gut microbiota contribute to intestinal homeostasis via inducing the intestinal immune cells and also influence the systemic host immunity. The probiotic consumption shows a beneficial effect in several ways, including intestinal microbiota balance and ability to modulate host innate and specific immune response. Their effect on non-specific immunity was reported as enhanced phagocytosis of pathogenic bacteria and modified cytokine production [80], while the specific way is usually followed through immunoglobulins testing. Nevertheless, the effect of probiotic administration on the immune system of rabbits has been reported on a limited scale [61,81]. Our studies with *E. faecium* CCM7420 alone and in combination with *Eleutherococcus senticosus* demonstrated significant increases in total phagocytic activity (PA) of leukocytes and PA of neutrophils at the end of the treatment period (21 days) and also after three weeks of the ost-treatment period (42 days; $p < 0.0001$; Table 3; [26]). The freeze-dried CCM7420 strain has not influenced the PA in rabbits during its application, either resolved in water or composed in pellets. However, the prolonged effect of CCM7420 strain rehydrated in water was observed at the end of the experiment (42 days; $p < 0.0001$; [55]). Another rabbit studies with non-autochthonous strains *E. faecium* CCM4231 (ruminal isolate) and AL41 (CCM8558; isolate from animal waste) also showed stimulation of non-specific immune reaction in rabbits; the significant increase of PA ($p < 0.001$) was noted in both experiments during the treatment and increased several weeks after the strain's cessation. Fathi et al. [33] also represented improved cell-mediated immunity adding 400 g probiotic/t feed in rabbits´ diet. Contrary to results reported above, Wang et al. [57] noted no probiotic influence on the number of mast cells in duodenum and jejunum, but increased the number of mast cells in caecum and also, increased IgG and IgM in serum. During enterocin Ent7420 administration in rabbits, the prolonged immuno-stimulative effect was observed, which was demonstrated by a significant increase of PA in the experimental group ($p < 0.05$) compared also to the control data [82]. The same immunomoderate influence was noted during experimental applications of enterocins produced by strains CCM4231 and AL41 (Ent4231 and EntM/EntAL41; [21,28]) as well as in the case of nisin feed inclusion in rabbits [38]. We expected that enterocins are able to stimulate the immune system through the gut microbiota modulation on behalf of lactic acid bacteria (LAB) and via supporting/improving the GALT by the stimulation of the IgA system. This fact could explain the "timeshift" of the enterocins influence compared to the probiotic strain application, since probiotic bacteria begins to act and multiply immediately, while bacteriocins firstly modulate the environment on behalf of beneficial microbes (LAB) inhibiting other bacterial species in the gut. On the other hand, the prolonged or maintaining immunostimulative effect (even at 21 days after ceasing the administration of enterocins) should be also explained by the adopting of animals on them. More studies testing the effect of the CCM7420 strain as well as its enterocin Ent7420 on the immune response, containing PA and other parameters of innate immunity in rabbits are necessary. Moreover, we would like to focus also on the intestinal immunity, mainly on the IgA level.

It is well known, that probiotic supplementation improves not only the growth rate and enhances the efficiency of feed conversion but also may positively influence the health status via enhancing gut health in rabbits [61]. Gut health, including microbial and immunological stability, is often influenced by exogenous factors (dietary changes, stress from manipulation, transfer, climate changes, etc.), mainly around the weaning period. Therefore, the alternative strategies are required to improve the animal's health. The stable or improved intestinal environment (gut microflora, mucosal immunity, epithelial morphology and function) directly influences the health status and growth performance of animals due to better nutrient absorption in the gut [61,83]. The effect of natural feed additives, including probiotics on the intestinal histomorphology in animals is often presented, mostly in chicken and pigs. However, studies reported changes of morphometric parameters in the small intestine of rabbits during probiotic application are limited [31,82–84]. Our results showed that the surface area and villi height: crypt depth (VH:CD) ratio also increased throughout the *E. faecium* CCM7420 strain administration [83]. These results could support the hypothesis about higher weight gain, better feed conversion and nutrient utilization due to enlargement of the absorption surface and improved morphometry parameters. Similarly to these results, increased surface area, VH, VH:CD and decreased CD was observed after Ent2019 addition in rabbits. To the best of our current knowledge, our team is the first that deals with experiments regarding the morphological changes in rabbits during enterocins administration (Ent2019 and EntM; [80,85]). This knowledge leads us to a more detailed study of these physiological changes in the rabbit's digestive tract, which will not only allow us to extend our knowledge, but also gain new information to understand the complexity of physiological, microbiological and immunological processes in the host organism.

3.7. Effect on Meat—Nutrient Content and Physicochemical Properties

Rabbit meat is greatly valued for its high nutritional and dietary quality, especially its low amounts of cholesterol, fat and sodium and high content of polyunsaturated fatty acids (PUFA), potassium, phosphorus and magnesium [85,86]; for these reasons it is recommended mainly for children, pregnant women and patients with cardiovascular illnesses. Previously there have been many studies/reviews concerning rabbit meat, including its production, quality, physicochemical properties and composition [1,2]. A lot of them dealing with the effects of dietary supplementation with functional compounds, mainly probiotics, prebiotics, fatty and organic acids, vitamins, selenium and antioxidants and their combinations on rabbit carcass quality [12–14,31,33,40,87–89]. In general, there is a lack of studies testing the beneficial microbiota and/or their antimicrobial substances (bacteriocins) on rabbit meat quality and composition, including fatty acids, amino acids and minerals [31,41–45]. Moreover, to the best of our knowledge we presented the first reports about the effect of beneficial/probiotic strains as well as bacteriocins on the minerals and amino acids of rabbit meat [39,40,42] and also the effect of bacteriocins on the fatty acid content of rabbit meat [41–43]. All presented results concerning the physicochemical properties—pH, protein, fat, ash and water content, energy value, lightness and color of rabbit meat—showed that probiotic administration in rabbits had no negative effect on the rabbit meat quality. On the other hand, Fathi et al. [33] presented a significant effect of probiotic supplementation on moisture, dry matter, organic matter and ash content. Only limited information about rabbit minerals are available, in response to natural feed additives. Our result showed increased iron content ($p = 0.0011$) in treated groups with freeze-dried CCM7420 strain (both, rehydrated in water and enriched in feed pellets), in contrary to other findings with reduced iron content during microbial fermented feed utilization with *Lactobacillus plantarum* and *Pediococcus acidilactici* [90] and after enterocin administration [42]. Although the pH of the luminal content was not measured, we hypothesized a more acidic environment in the gut due to the previous results of higher lactic acid production and lower pH in the caecum during CCM7420 strain administration in rabbits [48]. This acidic environment can enhance the ionization of minerals, which in turn results in passive diffusion [91] and could be one alternative explanation of the higher iron absorption from the gut. Another hypothesis could be the larger absorption surface due to enterocyte proliferation, which is confirmed by improved morphometry parameters—villus height, crypt depth and villus height:crypt depth ratio—also recorded during our previous experiments with CCM7420 administration [82] and lantibiotic–nisin application to rabbits [38]. The enlargement of the luminal surface could ensure better mineral absorption and their inclusion to rabbit meat. On the other hand, significantly decreased concentrations of copper ($p = 0.0004$) and calcium ($p < 0.0001$) were noted. Similarly to these results, the copper concentrations in rabbit meat also reduced during the enterocin M addition to rabbits, however, not significant but only numerically changes were recorded. Similarly to us, lower concentrations of copper, zinc and manganese was found by Shah et al. [90] after probiotic supplementation. Copper is an essential trace element, performing important biochemical functions; its level in rabbit meat varies widely [92,93]. Regarding the higher iron concentration, we hypothesized an iron competitive influence on the copper intestinal absorption and its lower deposition to meat. The calcium metabolism in rabbits is very unique, widely fluctuates and its intestinal absorption is very vitamin D independent, in contrast to most mammals [77]. Despite the generally known fact that probiotics can increase the organic and short chain fatty acids in caecum, which also stimulate the minerals ionization and diffusion through the intestine, we noted decreased calcium concentration, although, still in the range presented in the literature [86,92]. Inferring from achieved results, we assumed no adverse effect of the CCM7420 strain on the meat characteristics; in addition, it could enhance the mineral quality of meat and also increased its value to the functional food level.

4. Conclusions

The strain *E. faecium* CCM7420 in different application forms (fresh culture at concentration 1×10^9 CFU/mL of cells in a dose 500 μL/animal /day applied into drinking water) was lyophilized (freeze-dried) from rehydrated water (1×10^9 CFU/mL; dose 500 μL/animal/day) as well as mixed in feed and pelleted (15 g/100 kg feed), either alone and in combination with *Eleutherococcus senticosus* and its enterocin Ent7420 (50 μL/animal/day applied into drinking water) were tested in rabbits. During these experiments, the following effects of the strain and its enterocin were observed: improved average daily weight gain and feed conversion ratio, good colonization ability of the tested strain with maximum counts in the first 2–3 weeks of application, increased lactic acid bacteria and reduced coagulase-positive staphylococci including *S. aureus*, coliforms and clostridia population as well as the *Eimeria* sp. oocysts counts in the rabbit's gut. Improved biochemical blood parameters (total proteins, glucose and triglycerides) have been noted during the CCM7420 strain application; however, the glutathione-peroxidase level was not disturbed and oxidative stress was not evoked through the additives application. Another interesting finding was the significant stimulation of blood phagocytic activity and also the improved morphometry parameters (enlargement of the absorption surface in jejunum and higher villi height: crypt depth (VH:CD) ratio). The physicochemical properties of rabbit meat were not negatively affected by the CCM7420 strain, while the meat iron content significantly increased during its application, which improved the rabbit meat quality. It could be also emphasized that knowing the probiotic properties and the ability of *Enterococcus faecium* CCM7420 to produce enterocin Ent7420 with an antimicrobial effect is of great interest mainly in the case of several disease/pathologies, such as epizootic rabbit enteropathy, which are difficult to prevent and combat because their etiology is not known and there is no vaccine. This strain is the main component of the Prorabbit probiotic preparation, which is often used in Slovak rabbit farms (at dosage 1–2 g/animal/day for 21 days as prevention and 3 g/animal/day with a therapeutic effect; resolved in water or mixed into feed). Moreover, to the best of our current knowledge, our team is the first that deals with experiments regarding the morphological changes in the jejunum of rabbits during enterocins administration; the first reports regarding the effect of beneficial/probiotic strains and bacteriocins on the mineral and amino acid concentrations as well as the effect of bacteriocins on fatty acid content of rabbit meat were also published by our team.

Author Contributions: Conceptualization, M.P.S. and A.L.; data curation, M.P.S. and Ľ.C.; funding acquisition, A.L.; investigation, M.P.S., A.L. and Ľ.C.; project administration, A.L.; resources, A.L.; validation, A.L.; visualization, M.P.S.; writing—original draft preparation, M.P.S.; writing—review and editing, A.L., M.P.S. All authors have read and agreed to the published version of the manuscript.

Acknowledgments: This work was supported by the project VEGA 2/0006/17. All care and experimental procedures involving animals followed the guidelines stated in the Guide for the Care and Use of Laboratory Animals approved by the State Slovak Veterinary and Food Administration and the Ethics Committees of both institutions (permission code: SK CH 17016 and SK U 18016). None of the authors of this paper has a financial or personal relationship with other people or organizations that could inappropriately influence or bias the content of the paper. We would like to thank Mrs. M. Bodnárová and Mr. P. Jerga for their skillful technical assistance with sample collection and processing. We are grateful to A. Kopčáková for the *E. faecium* CCM7420 strain sequencing, Ľ. Ondruška, V. Párkányi and R. Jurčík, from the National Agricultural and Food Centre (NAFC) in Nitra for blood sampling and Mr. J. Pecho for animals slaughtering, to Z. Formelová and M. Chrenková (NAFC) for meat samples processing and meat quality properties analyses. We also thank to V. Strompfová for blood biochemistry analyses, K. Čobanová and Š. Faix for glutathione-peroxidase activity measuring, I. Plachá for phagocytic activity determination, S. Gancarčíková (University of Veterinary Medicine and Pharmacy, Laboratory of Gnotobiology, Košice) for organic acids measurement, Z. Vasilková (Parasitological Institute of SAS) for *Eimeria* sp. oocysts detection, R. Žitňan for intestinal morphometry testing.

References

1. Dalle Zotte, A.; Szendrő, Z. The role of rabbit meat as functional food. *Meat Sci.* **2011**, *88*, 319–331. [CrossRef]
2. Cullere, M.; Dalle Zotte, A. Rabbit meat production and consumption: State of knowledge and future perspectives. *Meat Sci.* **2018**, *143*, 137–146. [CrossRef]
3. Basavaraj, M.; Nagabhushana, V.; Prakash, N.; Appannavar, M.M.; Prashanth, W.; Mallikarjunappa, S. Effect of dietary supplementation of curcuma longaon the biochemical profile and meat characteristics of broiler rabbits under summer stress. *Vet. World* **2011**, *4*, 15–18. [CrossRef]
4. Carabaño, R.; Badiola, I.; Licois, D.; Gidenne, T. The digestive ecosystem and its control through nutritional or feeding strategies. In *Recent Advances in Rabbit Sciences*, 1st ed.; Maertens, L., Coudert, P., Eds.; ILVO: Merehleke, Belgium, 2006; pp. 221–227.
5. Marlier, D.; Dewrée, R.; Lassence, C.; Licois, D.; Mainil, J.; Coudert, P.; Meulemans, L.; Ducatelle, R.; Vindevogel, H. Infectious agents associated with epizootic rabbit enteropathy: Isolation and attempts to reproduce the syndrome. *Vet. J.* **2006**, *172*, 493–500. [CrossRef] [PubMed]
6. Salyers, A.A.; Gupta, A.; Wang, Y.P. Human intestinal bacteria as reservoirs for antibiotic resistance genes. *Trends Microbiol.* **2004**, *12*, 412–416. [CrossRef] [PubMed]
7. European Commission. *IP/03/1058 Council and Parliament Prohibit Antibiotics as Growth Promoters: Commissioner Byrne Welcomes Adoption of Regulation on Feed Additives*; European Commission: Brussels, Belgium, 2003.
8. Barug, D.; de Jong, J.; Kies, A.K.; Verstegen, M.W.A. *Antimicrobial Growth Promoters. Where Do We Go from Here?* 1st ed.; Wageningen Academic Publishers: Wageningen, The Netherlands, 2006; pp. 1–196.

9. Falcão-e-Cunha, L.; Castro-Solla, L.; Maertens, L.; Marounek, M.; Pinheiro, V.; Freire, J.; Mourão, J.L. Alternatives to antibiotic growth promoters in rabbit feeding: A review. *World Rabbit Sci.* **2007**, *15*, 127–140. [CrossRef]
10. European Commission. *Regulation EC/1831/2003 of the European Parliament and of the Council on Additives for Use in Animal Nutrition*; European Commission: Brussels, Belgium, 2003.
11. European Commission. *Regulation EC 1924/2006 of the European Parliament and of the Council on Nutrition and Health Claims Made on Foods*; European Commission: Brussels, Belgium, 2006.
12. Papatsiros, V.G.; Christodoulopoulos, G. The use of organic acids in rabbit farming. *Online J. Anim. Feed Res.* **2011**, *1*, 434–438.
13. Dalle Zotte, A.; Celia, C.; Szendrő, Z. Herbs and spices inclusion as feedstuff or additive in growing rabbit diet and as additive in rabbit meat: A review. *Livest. Sci.* **2016**, *189*, 82–90. [CrossRef]
14. Kalma, R.P.; Patel, V.K.; Joshi, A.; Umatiya, R.V.; Parmar, K.N.; Damor, S.V.; Chauhan, H.D.; Srivastava, A.K.; Sharma, H.A. Probiotic supplementation in rabbit: A review. *Int. J. Agric. Sci.* **2016**, *8*, 2811–2815.
15. Sharma, K.G.; Vidyarthi, K.V.; Archana, K.; Zuyie, R. Probiotic supplementation in the diet of rabbits—A Review. *Livest. Res. Int.* **2016**, *4*, 1–10.
16. Canzi, E.; Zanchi, R.; Camaschella, P.; Cresci, A.; Greppi, G.F.; Orpianesi, C.; Serrantoni, M.; Ferrari, A. Modulation by lactic-acid bacteria of the intestinal ecosystem and plasma cholesterol in rabbits fed a casein diet. *Nutr. Res.* **2000**, *20*, 1329–1340. [CrossRef]
17. Amber, K.H.; Yakout, H.M.; Hamed Rawya, S. Effect of feedings diets containing Yucca extract or probiotic on growth, digestibility, nitrogen balance and caecal microbial activity of growing New Zealand White rabbits. In Proceedings of the 8th World Rabbit Congress, Puebla, Mexico, 7–10 September 2004; pp. 737–745.
18. Trocino, A.; Xiccato, G.; Carraro, L.; Jiménez, G. Effect of diet supplementation with Toyocerin® (*Bacillus cereus* var. *Toyoi*) on performance and health of growing rabbits. *World Rabbit Sci.* **2005**, *13*, 17–28. [CrossRef]
19. Matusevičius, P.; Ašmenskaitė, L.; Žilinskienė, A.; Gugolek, A.; Lorek, M.O.; Hartman, A. Effect of probiotic Bioplus 2B® on performance of growing rabbits. *Vet. Zootech* **2006**, *36*, 54–59.
20. Lauková, A.; Strompfová, V.; Skřivanová, V.; Volek, Z.; Jindřichová, E.; Marounek, M. Bacteriocin producing strain of *Enterococcus faecium* EK13 with probiotic character and its application in the digestive tract of rabbits. *Biol. Bratisl.* **2006**, *61*, 779–782. [CrossRef]
21. Lauková, A.; Chrastinová, L'.; Pogány Simonová, M.; Strompfová, V.; Plachá, I.; Čobanová, K.; Formelová, Z.; Chrenková, M.; Ondruška, L'. *Enterococcus faecium* AL 41: Its enterocin M and their beneficial use in rabbits husbandry. *Probiotics Antimicrob. Proteins* **2012**, *4*, 243–249. [CrossRef]
22. Lauková, A.; Pogány Simonová, M.; Chrastinová, L'.; Kandričáková, A.; Ščerbová, J.; Plachá, I.; Čobanová, K.; Formelová, Z.; Ondruška, L'.; Štrkolcová, G.; et al. Beneficial effect of bacteriocin-producing strain *Enterococcus durans* ED 26E/7 in model experiment using broiler rabbits. *Czech J. Anim. Sci.* **2017**, *62* (Suppl. 4), 168–177.
23. Kritas, S.K.; Petridou, E.I.; Fortomaris, P.; Tzika, E.; Arsenos, G.; Koptopoulos, G. The effect of probiotics on microbiology, health and performance of fattening rabbits. *Asian Australas. J. Anim. Sci.* **2008**, *21*, 1312–1317. [CrossRef]
24. Simonová, M.; Marciňáková, M.; Strompfová, V.; Čobanová, K.; Gancarčíková, S.; Vasilková, Z.; Lauková, A. Effect of probiotics *Lactobacillus rhamnosus* GG and new isolate *Enterococcus faecium* EF2019 (CCM 7420) on growth, blood parameters, microbiota and coccidia oocysts excretion in rabbits. *Int. J. Prob. Preb.* **2008**, *3*, 7–14.
25. Pogány Simonová, M.; Lauková, A.; Chrastinová, L'.; Strompfová, V.; Faix, Š.; Vasilková, Z.; Ondruška, L'.; Jurčík, R.; Rafay, J. *Enterococcus faecium* CCM7420, bacteriocin PPB CCM7420 and their effect in thedigestive tract of rabbits. *Czech J. Anim. Sci.* **2009**, *54*, 376–386. [CrossRef]
26. Pogány Simonová, M.; Lauková, A.; Chrastinová, L'.; Plachá, I.; Čobanová, K.; Strompfová, V.; Formelová, Z.; Chrenková, M. Combined administration of bacteriocin-producing, probiotic strain *Enterococcus faecium* CCM7420 with *Eleutherococcus senticosus* and their effect in rabbits. *Pol. J. Vet. Sci.* **2013**, *8*, 730–734.
27. Pogány Simonová, M.; Chrastinová, L'.; Kandričáková, A.; Kubašová, I.; Formelová, Z.; Chrenková, M.; Miltko, R.; Belzecki, G.; Strompfová, V.; Lauková, A. Enterocin M and sage supplementation in post-weaning rabbits: Effects on growth performance, caecal microbiota, fermentation and enzymatic activity. *Probiotics Antimicrob. Proteins* **2020**, *12*, 732–739. [CrossRef] [PubMed]

28. Szabóová, R.; Lauková, A.; Chrastinová, Ľ.; Pogány Simonová, M.; Strompfová, V.; Plachá, I.; Čobanová, K.; Vasilková, Z.; Chrenková, M. Enterocin 4231 produced by *Enterococcus faecium* CCM 4231 and its use in rabbits. *Acta Vet. Beogr.* **2011**, *61*, 523–529. [CrossRef]
29. Bovera, F.; Iannaccone, F.; Mastellone, V.; Nizza, S.; Lestingi, A.; De Martino, L.; Lombardi, P.; Mallardo, K.; Ferrara, M.; Nizza, A. Effect of spray application of *Lactobacillus plantarum* on in vivo performance, caecal fermentations and haematological traits of suckling rabbits. *Ital. J. Anim. Sci.* **2012**, *11*, 145–149. [CrossRef]
30. Oso, A.O.; Idowu, O.M.O.; Haastrup, A.S.; Ajibade, A.J.; Olowonefa, K.O.; Aluko, A.O.; Ogunade, I.M.; Osho, S.O.; Bamgbose, A.E. Growth performance, apparent nutrient digestibility, caecal fermentation, ileal morphology and caecal microflora of growing rabbits fed diets containing probiotics and prebiotics. *Livest. Sci.* **2013**, *157*, 184–190. [CrossRef]
31. Bhatt, R.S.; Agrawal, A.R.; Sahoo, A. Effect of probiotic supplementation on growth performance, nutrient utilization and carcass characteristics of growing Chinchilla rabbits. *J. Appl. Anim. Res.* **2017**, *45*, 304–309. [CrossRef]
32. Cunha, S.; Mendes, Â.; Rego, D.; Meireles, D.; Fernandes, R.; Carvalho, A.; Martins da Costa, P. Effect of competitive exclusion in rabbits using an autochtonous probiotic. *World Rabbit Sci.* **2017**, *25*, 123–134. [CrossRef]
33. Fathi, M.; Abdelsalam, M.; Al-Homidan, I.; Ebeid, T.; El-Zarei, M.; Abou-Emera, O. Effect of probiotic supplementation and genotype on growth performance, carcass traits, hematological parameters and imunity of growing rabbits under hot environmental conditions. *Anim. Sci. J.* **2017**, *88*, 1644–1650. [CrossRef]
34. Phuoc, T.L.; Jamikorn, U. Effects of probiotic supplement (*Bacillus subtilis* and *Lactobacillus acidophilus*) on feed efficiency, growth performance, and microbial population of weaning rabbits. *Asian Austral. J. Anim. Sci.* **2017**, *30*, 198–205. [CrossRef]
35. Szabóová, R.; Chrastinová, Ľ.; Lauková, A.; Haviarová, M.; Simonová, M.; Strompfová, V.; Faix, Š.; Vasilková, Z.; Chrenková, M.; Plachá, I.; et al. Bacteriocin - producing strain *Enterococcus faecium* CCM4231 and its use in rabbits. *Int. J. Prob. Preb.* **2008**, *3*, 77–82.
36. Pogány Simonová, M.; Lauková, A.; Plachá, I.; Čobanová, K.; Strompfová, V.; Szabóová, R.; Chrastinová, Ľ. Can enterocins affect phagocytosis and glutathione peroxidase in rabbits? *Cent. Eur. J. Biol.* **2013**, *8*, 730–734.
37. Pogány Simonová, M.; Chrastinová, Ľ.; Kandričáková, A.; Gancarčíková, S.; Bino, E.; Plachá, I.; Ščerbová, J.; Strompfová, V.; Žitňan, R.; Lauková, A. Can have enterocin M in combination with sage beneficial effect on microbiota, blood biochemistry, phagocytic activity and jejunal morphometry in broiler rabbits? *Anim.* **2020**, *10*, 115. [CrossRef]
38. Lauková, A.; Chrastinová, Ľ.; Plachá, I.; Kandričáková, A.; Szabóová, R.; Strompfová, V.; Chrenková, M.; Čobanová, K.; Žitňan, R. Beneficial effect of lantibiotic nisin in rabbit husbandry. *Probiotics Antimicrob. Proteins* **2014**, *6*, 41–46. [CrossRef]
39. Pogány Simonová, M.; Szabóová, R.; Haviarová, M.; Chrastinová, Ľ.; Mojto, J.; Strompfová, V.; Lauková, A.; Rafay, J. Quality of rabbit meat after application of bacteriocinogenic and probiotic strain *Enterococcus faecium* CCM 4231 in rabbits. *Int. J. Prob. Preb.* **2009**, *4* (Suppl. 1), 1–6.
40. Pogány Simonová, M.; Chrastinová, Ľ.; Lauková, A. Dietary supplementation of a bacteriocinogenic and probiotic strain of *Enterococcus faecium* CCM7420 and its effect on the mineral content and quality of *Musculus longissimus dorsi* in rabbits. *Anim. Prod. Sci.* **2016**, *56* (Suppl. 12), 2140–2145.
41. Pogány Simonová, M.; Chrastinová, Ľ.; Lauková, A. Lantibiotic nisin applied in broiler rabbits and its effect on the growth performance and carcass quality. *Probiotics Antimicrob. Proteins* **2019**, *11*, 1414–1417. [CrossRef]
42. Pogány Simonová, M.; Chrastinová, Ľ.; Chrenková, M.; Formelová, Z.; Kandričáková, A.; Bino, E.; Lauková, A. Benefits of enterocin M and sage combination on the physico-chemical traits, fatty acid, amino acid, and mineral content of rabbit meat. *Probiotics Antimicrob. Proteins* **2019**. [CrossRef]
43. Chrastinová, Ľ.; Lauková, A.; Chrenková, M.; Poláčiková, M.; Formelová, Z.; Kandričáková, A.; Glatzová, E.; Ščerbová, J.; Bučko, O.; Ondruška, Ľ.; et al. Effects of enterocin M and durancin ED26E/7 substances applied in drinking water on the selected carcass characteristics and meat quality of broiler rabbits. *Arch. Zootech.* **2018**, *21*, 5–17.
44. EFSA Panel on Additives and Products or Substances used in Animal Feed (EFSA FEEDAP PANEL). Safety and efficacy of Probiotic Lactina® (*Enterococcus faecium* NBIMCC8270, *Lactobacillus acidophilus* NBIMCC 8242, *Lactobacillus helveticus* NBIMCC 8269, *Lactobacillus delrueckei* ssp. *lactis* NBIMCC 8250, *Lactobacillus delbrueckei* ssp. *bulgaricus* NBIMCC 8244 and *Streptococcus thermophilus* NBIMCC 8253) as a feed additive for chickens for fattening and suckling and weaned rabbits. *EFSA J.* **2019**, *17*, 5646.

45. Matusevičius, P.; Bartkeviciute, Z.; Cernauskenie, J.; Kozłowski, K.; Jeroch, H. Effect of probiotic preparation "ToyoCerin (R)" and phytobiotic preparation "Cuxarom Spicemaster" on growing rabbits. *Arch. Geflügelkunde* **2011**, *75*, 67–71.
46. Simonová, M.; Lauková, A. Isolation of faecal *Enterococcus faecium* strains from rabbits and their sensitivity to antibiotics and ability to bacteriocin production. *Bull. Vet. Inst. Pulawy* **2004**, *48*, 383–386.
47. Simonová, M.; Lauková, A. Bacteriocin activity of enterococci from rabbits. *Vet. Res. Comm.* **2007**, *31*, 143–152. [CrossRef]
48. Simonová, M. Probiotic and Bacteriocin-Producing Bacteria and Their Effect on Physiology of Digestion in Rabbits. Ph.D. Thesis, Institute of Animal Physiology, Slovak Academy of Sciences, Košice, Poland, 2006.
49. Simonová, M.; Lauková, A.; Sirotek, K. (2008b): Lipolytic activity of potential probiotic enterococci and additive staphylococci. *Acta Vet. Brno* **2008**, *77*, 575–580. [CrossRef]
50. El Dimerdash, M.Z.; Dalia, M.H.; Hanan, F.A.; Doaa, S.A. Studies on the effect of some probiotics in rabbits. *Suez Canal Vet. Med. J.* **2011**, *16*, 151–168.
51. Kalma, R.P.; Chauhan, H.D.; Shrivastava, A.K.; Pawar, M.M. Growth and blood profile of broiler rabbits on probiotics supplementation. *Ind. J. Small Rum.* **2018**, *24*, 66–69. [CrossRef]
52. Szabóová, R.; Chrastinová, L.; Lauková, A.; Strompfová, V.; Pogány Simonová, M.; Vasilková, Z.; Plachá, I.; Čobanová, K.; Chrenková, M. Effect of combinative administration of bacteriocin-producing and probiotic strain *Enterococcus faecium* CCM 4231 and sage plant extract on rabbits. *Afr. J. Microbiol. Res.* **2012**, *6*, 4868–4873. [CrossRef]
53. Lauková, A.; Chrastinová, Ľ.; Plachá, I.; Szabóová, R.; Kandričáková, A.; Pogány Simonová, M.; Formelová, Z.; Ondruška, Ľ.; Goldová, M.; Chrenková, M.; et al. Enterocin 55 produced by non rabbit-derived strain *Enterococcus faecium* EF55 in relation with microbiota and selected parameters in broiler rabbits. *Int. J. Agric. Environ. Res.* **2017**, *3*, 45–52.
54. Lauková, A.; Pogány Simonová, M.; Chrastinová, Ľ.; Gancarčíková, S.; Kandričáková, A.; Plachá, I.; Chrenková, M.; Formelová, Z.; Ondruška, Ľ.; Ščerbová, J.; et al. Assessment of Lantibiotic type bacteriocin—Gallidermin application in model experiment with broiler rabbits. *Int. J. Anim. Sci.* **2018**, *2*, 1028.
55. Pogány Simonová, M.; Lauková, A.; Chrastinová, Ľ.; Strompfová, V.; Plachá, I.; Szabóová, R.; Chrenková, M.; Čobanová, K.; Goldová, M.; Žitňan, R. Benefity of probiotic and bacteriocin-producing strain Enterococcus faecium EF2019 (CCM7420) for rabbits health (In Slovak). In Proceedings of the 12th Conference of New Ways in Intensive and Hobbies Rabbits Husbandries, Praha, Czech Republic, 6 November 2013; pp. 37–43.
56. Benato, L.; Hastie, P.; O'Shaughnessy, P.; Murray, J.A.; Meredith, A. Effects of probiotic Enterococcus faecium and Saccharomyces cerevisiae on the faecal microflora of pet rabbits. *J. Small Anim. Pr.* **2014**, *55*, 442–446. [CrossRef]
57. Wang, C.; Zhu, Y.; Li, F.; Huang, L. The effect of *Lactobacillus* isolates on growth performance, immune response, intestinal bacterial community composition of growing Rex rabbits. *Anim. Physiol. Anim. Nutr.* **2017**, *101*, e1–e13. [CrossRef]
58. Linaje, R.; Coloma, M.D.; Pérez-Martínez, G.; Zúñiga, M. Characterization of faecal enterococci from rabbits for the selection of probiotic strains. *J. Appl. Microbiol.* **2014**, *96*, 761–771. [CrossRef]
59. Abecia, L.; Fondevila, M.; Balcells, J.; Edwards, J.E.; Newbold, J.C.; McEwan, N.R. Molecular profiling of bacterial species in the rabbit cecum. *FEMS Microbiol. Lett.* **2005**, *244*, 111–115. [CrossRef] [PubMed]
60. Simonová, M.; Lauková, A.; Štyriak, I. Enterococci from rabbits—potential feed additives. *Czech J. Anim. Sci.* **2005**, *50*, 416–421. [CrossRef]
61. Fortune-Lamothe, L.; Boullier, S. A review on the interactions between gutmicroflora and digestive mucosal imunity. Possible ways to improve the health of rabbits. *Livest. Sci.* **2007**, *107*, 1–18. [CrossRef]
62. Pakandl, M. Coccidia of rabbit: A review. *Folia Parasitol.* **2009**, *56*, 153–166. [CrossRef]
63. Zita, L.; Tůmová, E.; Skřivanová, V.; Ledvinka, Z. The effect of weaning age on performance and nutrient digestibility of broiler rabbits. *Czech J. Anim. Sci.* **2007**, *52*, 341–347. [CrossRef]
64. EC (2005) European Union Commission. *Ban on Antibiotics as Growth Promoters in Animal Feed Enters into Effect: Regulation 1831/2003/EC on Additives for Use in Animal Nutrition, Replacing Directive 700/524//333c on Additives in Feedstuffs*; European Union Commission: Brussels, Belgium, 2005.
65. Travers, M.-A.; Florent, I.; Kohl, L.; Grellier, P. Probiotics for the control of parasites: An overview. *J. Parasitol. Res.* **2011**, *2011*, 610769. [CrossRef]

66. Markowiak, P.; Śliżewska, K. The role of probiotics, prebiotics and synbiotics in animal nutrition. *Gut Pathog.* **2018**, *10*, 21. [CrossRef]
67. Kowalska, D.; Bielański, P.; Nosal, P.; Kowal, J. Natural alternatives to coccidiostats in rabbit nutrition. *Ann. Anim. Sci.* **2012**, *12*, 561–574. [CrossRef]
68. Nosal, P.; Kowalska, D.; Bielanski, P.; Kowal, J.; Kornas, S. Herbal formulations as feed additives in the course of rabbit subclinical coccidiosis. *Ann. Parasitol.* **2014**, *60*, 65–69.
69. El-Ashram, S.A.; Aboelhadid, S.M.; Abdel-Kafy, E.-S.M.; Hashem, S.A.; Mahrous, L.N.; Farghly, E.M.; Moawad, U.K.; Kamel, A.A. Prophylactic and therapeutic efficacy of prebiotic supplementation against intestinal coccidiosis in rabbits. *Animals* **2019**, *9*, 965. [CrossRef]
70. Szabóová, R.; Lauková, A.; Chrastinová, Ľ.; Strompfová, V.; Pogány Simonová, M.; Vasilková, Z.; Čobanová, K.; Plachá, I.; Chrenková, M. Effect of combined administration of enterocin 4231 and sage in rabbits. *Pol. J. Vet. Sci.* **2011**, *14*, 359–366. [CrossRef] [PubMed]
71. Strompfová, V.; Lauková, A.; Marciňáková, M.; Vasilková, Z. Testing of probiotic and bacteriocin-producing lactic acid bacteria towards *Eimeria* sp. *Pol. J. Vet. Sci.* **2010**, *13*, 389–391. [PubMed]
72. Bucková, B.; Hurníková, Z.; Lauková, A.; Revajová, V.; Dvorožňáková, E. The anti-parasitic effect of probiotic bacteria via limiting the fecundity of *Trichinella spiralis* female adults. *Helminthol.* **2018**, *55*, 102–111. [CrossRef] [PubMed]
73. Jenkins, J.F. Clinical Pathology. In *Manual of Rabbits Medicine and Surgery*, 2nd ed.; Meredith, A., Flecknell, P., Eds.; BSAVA: Gloucester, UK, 2006.
74. Martinec, M.; Härtlová, H.; Chodová, D.; Tůmová, E.; Fučíková, A. Selected haematological and biochemical indicators in different breeds of rabbits. *Acta Vet. Brno* **2012**, *81*, 371–375. [CrossRef]
75. Özkan, C.; Kaya, A.; Akgül, Y. Normal values of haematological and some biochemical parameters in serum and urine of New Zealand rabbits. *World Rabbit Sci.* **2012**, *20*, 253–259. [CrossRef]
76. Illes, R.A.; Cohen, R.D.; Rist, A.H.; Baron, P.G. The mechanism of inhibition by acidosis of gluconeogenesis from lactate in rat liver. *Biochem. J.* **1977**, *164*, 185–191. [CrossRef] [PubMed]
77. Redrobe, S. Calcium metabolism in rabbits. In *Seminars in Avian and Exotic Pet Medicine*; WB Saunders: Philadelphia, PA, USA, 2002; Volume 11, pp. 94–101.
78. Gidenne, T. Estimation of volattile fatty acids and of their energetic supply in the rabbit caecum: Effect of the dietary fibre level. In Proceedings of the Conference Vleme Journees de la Recherche Cunicole, Paris, France, 6–7 December 1994; pp. 293–299.
79. Combes, S.; Fortune-Lamothe, L.; Cauqui, L.; Gidenne, T. Engineering the rabbit digestive ecosystem to improve digestive health and efficacy. *Animal* **2013**, *7*, 1429–1439. [CrossRef]
80. Maldonado, G.C.; Lemme-Dumit, J.M.; Thieblemont, N.; Carmuega, E.; Weill, R.; Perdigón, G. Stimulation of innate immune cells induced by probiotics: Participation of toll-like receptors. *J. Clin. Cell. Immunol.* **2015**, *6*, 1000283.
81. Deptuła, W.; Niedźwiedzka-Rystwej, P.; Śliwa, J.; Kaczmarczyk, M.; Tokarz-Deptuła, B.; Hukowska-Szematowicz, B.; Pawlikowska, M. Values of selected immune indices in healthy rabbits. *Centr. Eur. J. Immunol.* **2008**, *33*, 190–192.
82. Pogány Simonová, M.; Lauková, A.; Žitňan, R.; Chrastinová, Ľ. Effect of rabbit origin enterocin-producing strain *Enterococcus faecium* CCM7420 application on growth performance and gut morphometry in rabbits. *Czech J. Anim. Sci.* **2015**, *60*, 509–512. [CrossRef]
83. Liu, L.; Zeng, D.; Yang, M.; Wen, B.; Lai, J.; Zhou, Y.; Sun, H.; Xiong, L.; Wang, J.; Lin, Y.; et al. Probotic *Clostridium butyricum* improves the growth perfomance, immune function, and gut microbiota of weaning rex rabbits. *Probiotics Antimicrob. Proteins* **2019**, *11*, 1278–1292. [CrossRef]
84. Matur, E.; Eraslan, E. The impact of probiotics on the gastrointestinal physiology. In *New Advances in the Basic and Clinical Gastroenterology*; Brzozowski, T., Ed.; InTech: London, UK, 2013; pp. 51–74.
85. Dalle Zotte, A. Rabbit farming for meat purposes. *Anim. Front.* **2014**, *4*, 62–67. [CrossRef]
86. Dalle Zotte, A. Perception of rabbit meat quality and major factors influencing the rabbit carcass and meat quality. *Livest. Prod. Sci.* **2002**, *75*, 11–32. [CrossRef]
87. Li, R.G.; Wang, X.P.; Wang, C.Y.; Ma, M.W.; Li, F.C. Growth performance, meat quality and fatty acid metabolism response of growing meat rabbits to dietary linoleic acid. *Asian Astr. J. Anim. Sci.* **2012**, *25*, 1169–1177. [CrossRef] [PubMed]

88. Mattioli, S.; Cardinali, R.; Balzano, M.; Pacetti, D.; Castellini, C.; Dal Bosco, A.; Mattioli, N.G.F. Influence of dietary supplementation with prebiotic, oregano extract and vitamin E on fatty acid profile and oxidative status of rabbit meat. *J. Food Qual.* **2017**, *2017*, 3015120. [CrossRef]
89. Ayyat, M.S.; Al-Sagheer, A.A.; Abd El-Latif, K.M.; Khalil, B.A. Organic selenium, probiotics, and prebiotics effects on growth, blood biochemistry, and carcass traits of growing rabbits during summer and winter seasons. *Biol. Trace Elem. Res.* **2018**, *186*, 162–173. [CrossRef] [PubMed]
90. Shah, A.A.; Liu, Z.; Qian, C.; Wu, J.; Sultana, N.; Zhong, X. Potential effect of the microbial fermented feed utilization on physicochemical traits, antioxidant enzyme and trace mineral analysis in rabbit meat. *J. Anim. Physiol. Anim. Nutr.* **2020**, *104*, 767–775. [CrossRef]
91. Cummings, J.H.; Macfarlane, G.T.; Englyst, H.N. Prebiotic digestion and fermentation. *Am. J. Clin. Nutr.* **2001**, *73*, 415S–420S. [CrossRef]
92. Hermida, M.; Gonzalez, M.; Miranda, M.; Rodríguez-Otero, J.L. Mineral analysis in rabbit meat from Galicia (NW Spain). *Meat Sci.* **2006**, *73*, 35–639. [CrossRef]
93. Valenzuela, C.; Lopez de Romaña, D.; Schmiede, C.; Sol Morales, M.; Olivares, M.; Pizzaro, F. Total iron, heme iron, zinc and copper content in rabbit meat and viscera. *Biol. Trace Elem. Res.* **2011**, *143*, 1489–1496. [CrossRef] [PubMed]

13

Methane Production of Fresh Sainfoin, with or without PEG, and Fresh Alfalfa at Different Stages of Maturity is Similar but the Fermentation End Products Vary

Pablo José Rufino-Moya, Mireia Blanco, Juan Ramón Bertolín and Margalida Joy *

Centro de Investigación y Tecnología Agroalimentaria de Aragón (CITA), Instituto Agroalimentario de Aragón–IA2 (CITA-Universidad de Zaragoza), Avda, Montañana 930, 50059 Zaragoza, Spain; pjrufino@cita-aragon.es (P.J.R.-M.); mblanco@aragon.es (M.B.); jrbertolin@cita-aragon.es (J.R.B.)

* Correspondence: mjoy@aragon.es;

Simple Summary: In the last years, there has been increasing interest in the use of forages containing condensed tannins (CT) in ruminant nutrition. Condensed tannins can reduce the methane emissions and the ruminal degradation of protein, improving the animal performances to different extents depending on the source and dose of CT. In vitro fermentation of sainfoin has not been studied in fresh forage. The effect of CT can be studied in comparison with a similar CT-free forage or using polyethylene glycol (PEG), which is a tannin-blocking agent. The maturity stage influences the chemical composition to a different degree depending on the legume species, and can affect the content and fractions of CT. The aims of this trial were to compare the fermentation parameters of sainfoin with or without PEG, to detect the differences due to CT, at different maturity stages (vegetative, start-flowering, and end-flowering) and compare them with the fermentation parameters of alfalfa. The main results were that sainfoin had greater in vitro organic matter degradability (IVOMD) and lower ammonia and acetic:propionic ratio than alfalfa. Sainfoin CT had effect on ammonia and individual fatty acid proportions. In conclusion, fermentation end products were affected both by the chemical composition and CT contents.

Abstract: Alfalfa and sainfoin are high-quality forages with different condensed tannins (CT) content, which can be affected by the stage of maturity. To study the effects of CT on fermentation parameters, three substrates (alfalfa, sainfoin, and sainfoin+PEG) at three stages of maturity were in vitro incubated for 72 h. Sainfoin had greater total polyphenol and CT contents than alfalfa. As maturity advanced, CT contents in sainfoin decreased ($p < 0.05$), except for the protein-bound CT fraction ($p > 0.05$). The total gas and methane production was affected neither by the substrate nor by the stage of maturity ($p > 0.05$). Overall, sainfoin and sainfoin+PEG had greater in vitro organic matter degradability (IVOMD) than alfalfa ($p < 0.05$). Alfalfa and sainfoin+PEG presented higher ammonia content than sainfoin ($p < 0.001$). Total volatile fatty acid (VFA) production was only affected by the stage of maturity ($p < 0.05$), and the individual VFA proportions were affected by the substrate and the stage of maturity ($p < 0.001$). In conclusion, alfalfa and sainfoin only differed in the IVOMD and the fermentation end products. Moreover, CT reduced ammonia production and the ratio methane: VFA, but the IVOMD was reduced only in the vegetative stage.

Keywords: polyethylene glycol; gas production; in vitro organic matter degradability; condensed tannins; ammonia; volatile fatty acid; in vitro assay

---

1. Introduction

There is increasing interest in legume-based forage production systems because of their low reliance on fertilizer nitrogen, potential environmental benefits, and high protein content that contribute to low-input and sustainable livestock production systems [1]. Alfalfa (*Medicago sativa L.*) and sainfoin (*Onobrychis viciifolia Scop.*) are two pluriannual legumes widely grown in the Mediterranean area, presenting high forage productive capacity, high nutritional value, and restorative action for soil fertility [2]. However, alfalfa has a low content of polyphenols and is considered virtually free of condensed tannins (CTs), whereas sainfoin presents a high content of polyphenols and a medium to high content of CTs [3–5].

Alfalfa is usually fed as hay to ruminants mainly to avoid bloat, although continuous grazing is possible without bloat incidence both in sheep [6] and growing cattle [7]. Thus, the majority of studies that compared the ruminal fermentation of alfalfa and sainfoin have been performed using hays [3,4,8]. The differences between alfalfa and sainfoin have been ascribed to differences in the chemical compositions [4], but also to the presence of CTs [8]. To clarify whether the differences between alfalfa and sainfoin are only due to CTs, polyethylene glycol (PEG) must be added as a tannin-blocking agent [8]. To the best of our knowledge, the ruminal fermentation of both species was compared in fresh forages only by McMahon et al. [9] and Chung et al. [10]. Depending on their content, characteristics, and properties [11], CTs from sainfoin alter both the breakdown of protein in the rumen to ammonia and methane, gas and the production of total volatile fatty acids (VFAs) [3,4,12]. These, in turn, are associated with the variety, stage of maturity, and environmental factors [5,13,14].

The objectives of this trial were to compare the in vitro fermentation of alfalfa and sainfoin at three stages of maturity and to clarify whether the differences between both legumes were due to the CTs of sainfoin using PEG.

2. Materials and Methods

2.1. Experimental Design

Three substrates (alfalfa, sainfoin, and sainfoin+PEG) at three stages of maturity (vegetative, start-flowering, and end-flowering) were used to evaluate the effect of sainfoin CTs on in vitro fermentation. Alfalfa was used as a tannin-free legume.

2.2. Animal and Diets

2.2.1. Forages, Crop Management, and Harvest

The experiment was performed at the CITA Research Institute at Zaragoza (41°42′ N, 0°49′ W), altitude 216 m a.s.l., located in Ebro Valley, north-eastern Spain. The silt loam soil at the site had pH 8.1 and 1.81% organic matter and contained 16% clay, 53.5% silt, and 30.5% sand. Alfalfa and sainfoin were cultivated and managed as described by Lobón et al. [15]. Samples of forage were collected fortnightly from 14 April to 22 September 2015. The stage of maturity of the sainfoin and alfalfa was classified into vegetative, start-flowering, and end-flowering according to Borreani et al. [16] and Kalu and Fick [17], respectively. In each sampling, 10 samples per legume were obtained from 0.25 m^2 areas randomly allocated in the plot. These samples were mixed homogenously, and a part of the sample was separated manually into stems, leaves, and flowers to study their respective percentages. Another part of the samples was dried at 60 °C for 48 h for chemical analyses, and the rest of the sample was freeze-dried in a Genesis Freeze Dryer 25 (Hucoa Erlöss, SA/Thermo Fisher Scientific, Madrid, Spain) for polyphenol and CT analyses and in vitro fermentation assays. Samples for the chemical analyses were ground and sieved through a 1-mm screen (Rotary Mill, ZM200 Retsch, Hann, Germany), and a part of the samples was ground and sieved through a 0.2-mm screen for crude protein (CP), polyphenol, and CT determination. All the samples were stored at −20 °C in total darkness.

The number of samples of vegetative, start-flowering and end-flowering stages of each forage to study the in vitro fermentation were chosen in concordance with the representativeness of the plant development. The vegetative stage was the most frequent stage (55%), followed by end-flowering (27%) and start-flowering (18%) stages. Three samples of the vegetative stage and two samples of the start- and end-flowering of each legume species were studied.

2.2.2. Animals and Sampling of Ruminal Digesta

The procedures used in the trial followed the Spanish guidelines for experimental animal protection (RD 53/2013) and were approved by the Institutional Animal Care and Use Committee of the Research Centre (Procedure number 2011-05). Rasa Aragonesa wethers (n = 4; Live weight: 65 ± 2.1 kg) were used as donors of ruminal content. The management of the rumen-cannulated wethers and the sampling of the ruminal digesta was made as reported in Rufino-Moya et al. [18]. Briefly, wethers were fed a diet composed by alfalfa hay (70%) and barley grain (30%) at energy maintenance level. Before morning feeding, ruminal digesta from each wether was collected into a prewarmed insulated thermo, individually strained through four layers of cheesecloth and homogenized. Rumen fluid from the four wethers was mixed (pH: 6.76 ± 0.099), and a buffer solution was added in a proportion of 1:2 (*v*/*v*) based on the protocol of Menke and Steingass [19]

2.2.3. In Vitro Gas Production Technique and Sampling

The production of gas was measured with the Ankom system (Ankom Technology Corporation, Fairport, NY, USA), which had 310 mL capacity bottles fitted with pressure and temperature sensors. The valve open time was one second, the threshold for gas release was 5 PSI and the bottles were not shaken. Five hundred mg of freeze-dried substrate (alfalfa, sainfoin, or sainfoin+PEG) were incubated with 60 mL of buffered solution:rumen fluid (2:1 *v*/*v*) in a water bath (at 39 °C) for 72 h. To make the sainfoin+PEG samples, PEG-4000 (Merck, Darmstadt, Germany) was added to the buffered rumen fluid at a concentration of 2.3 g/L [12]. Four runs were conducted on four separate days, and each sample was incubated in duplicate in each run. Gas production and corrected with the blanks (two bottles without substrate were included in each run).

After 72 h of incubation, the bottles were placed in ice to stop fermentation (5–10 min), and then tempered (at room temperature for 10–15 min). A sample of the gas produced was transferred (at atmospheric pressure) with a syringe attached to a manometer into a Vacutainer®tube to determine CH_4 and conserved at 4 °C until analysis. At the end of gas sampling, the pH was measured immediately with a microPH 2002 (Crison Instruments S.A., Barcelona, Spain). The sampling to determine ammonia (NH_3-N) content and VFA were carried out as reported in Rufino-Moya et al. [18]. Briefly, to determine the ammonia content, 2.5 mL of liquid was mixed with 2.5 mL HCl 0.1 N. For VFA determination, 0.5 mL of the liquid was added to 0.5 mL of deproteinizing solution and 1 mL of distilled water. Tubes with samples for determination of ammonia and VFAs were stored at −20 °C until future analyses. The entire incubated sample was filtered through a preweighed bag (50 μm; Ankom) to estimate the in vitro organic matter degradability (IVOMD). Briefly, the bags were sealed, washed, dried at 103 °C for 48 h, and finally, placed in a muffle at 550 °C to obtain the ashes. The organic matter of the bag content was obtained as DM-ashes, and the IVOMD was calculated.

2.3. Analytical Methods

2.3.1. Chemical Composition

All the analyses of the chemical composition were analyzed as reported in Rufino-Moya et al. [18] according to official methods. Briefly, AOAC methods were used to determine the contents of dry matter (index no. 934.01), ash (index no. 942.05), and CP (index no. 968.06) [20]. Neutral detergent fiber (NDFom), acid detergent fiber (ADFom), and lignin (sa) contents were determined according to the method described by Van Soest et al. [21] using the Ankom 200/220 fiber analyzer (Ankom Technology

Corporation). The NDFom was assayed with a heat stable amylase. The lignin (sa) was analyzed in ADF residues by the solubilization of cellulose with sulfuric acid. All the values were corrected for ash-free content. The ether extract (EE) was determined following the approved procedure Am 5-04 [22] using an XT10 Ankom extractor (Ankom Technology Corporation). The nonstructural carbohydrates (NSC) were calculated as $NSC = 1000 - CP - EE - NDF - ash$, according to Guglielmelli et al. [3].

The content of total polyphenol (TP) and the fractions of CT were determined as described in Rufino-Moya et al. [18]. Briefly, the TP were extracted using the method of Makkar [23] and were quantified following the method of Julkunen-Tiitto [24]. The extractable CT (ECT), protein-bound CT (PBCT), and fiber-bound CT (FBCT) were extracted and fractioned following the method of Terrill et al. [25] and quantified by the colorimetric HCl-butanol method described by Grabber et al. [26]. The standard used for the quantification was extracted and purified from sainfoin using the method described by Wolfe et al. [27]. An Heλios β spectrophotometer was used to measure the samples and standard calibration at 550 nm.

2.3.2. Determination of Parameters of the In Vitro Gas Production Technique

Gas production, recorded hourly for 72 h, was used to estimate the parameters of the kinetics of fermentation adjusting the gas produced to the model described by France et al. [28]:

$$P = A \times \left(1 - e^{-ct}\right)$$

where P is the cumulative gas production (mL) at time t (h), A is the potential gas production (mL), and c is the rate of gas production (h^{-1}).

An HP-4890 (Agilent, St. Clara, CA, USA) gas chromatograph (GC) equipped with a flame ionization detector (FID) and a TG-BOND Q+ capillary column (30 m × 0.32 mm id × 10 μm film thickness, Thermo Scientific, Waltham, MA, USA) was used to determine CH_4. Helium was used as the carrier gas at a flow rate of 1 mL/min. The oven temperature was maintained at 100 °C (isothermal program). The splitless injection volume was 200 μL. Methane identification was based on the retention time compared with the standard. The methane concentration was calculated from the peak concentration:area ratio using the peak area generated from standard gas as the reference (CH_4; 99.995% purity [C45], Carburos Metálicos, Barcelona, Spain).

The content of ammonia was measured in Epoch microplate Spectrophotometer (BioTek Instruments, Inc., Winooski, VT, USA) using a colorimetric method described by Chaney and Marbach [29].

A Bruker Scion 460 GC (Bruker, Billerica, MA, USA) equipped with CP-8400 autosampler, FID and a BR-SWax capillary column (30 m × 0.25 mm i.d. × 0.25 μm film thickness, Bruker) was used to determine the concentration of acetic, propionic, iso-butyric, butyric, iso-valeric, and valeric acids. Helium was the carrier gas (flow rate of 1 mL/min). The oven temperature program was 100 °C, followed by a 6 °C/min increase to 160 °C. The injection volume was 1 μL at a split ratio of 1:50. The individual VFAs were identified based on retention time comparisons with commercially available standards of acetic, propionic, iso-butyric, butyric, iso-valeric, valeric, and 4-methyl-valeric acids at ≥99% purity (Sigma-Aldrich, St. Louis, MO, USA).

2.4. Statistical Analyses

Data were analyzed using SAS v. 9.3 (SAS Inst. Inc., Cary, NC, USA). The fermentation kinetic parameters (A and c) were estimated using a nonlinear regression model (NLIN procedure). The contents of secondary compounds were analyzed using the GLM procedure with the substrate, the stage of maturity and its interaction as fixed effects. Total gas and CH_4 production, A, c, IVOMD and the fermentation end products were analyzed using mixed models considering the substrate, the stage of maturity and its interaction as fixed effects and the run as random effect. If the interaction was not significant, it was deleted from the model and the analysis was repeated. The least square means, their associated standard errors and the differences were obtained. Pearson correlation coefficients between variables were calculated using the CORR procedure. For the entire test, the effects were considered a significant probability at a value of $p < 0.05$ or a trend at $p = 0.05$.

3. Results

3.1. Chemical Composition

The chemical composition and the percentage of stems, leaves, and flowers of both legume species at the three stages of maturity are shown in Table 1. On average, alfalfa and sainfoin had similar ADFom (231 g/kg DM), CP (198 g/kg DM), EE (15 g/kg DM), and NSC (335 g/kg DM) contents. Alfalfa, however, had higher ash and NDFom contents and a lower lignin (sa) content than sainfoin. For both forages, NDFom and ADFom content increased as the stage of maturity progressed, whereas the CP content decreased. As the forage matured, the lignin (sa) content increased only in alfalfa whereas the contents of EE and NSC decreased only in sainfoin.

Table 1. Chemical composition and plant components of alfalfa and sainfoin at three stages of maturity.

	Alfalfa			Sainfoin		
Item	Vegetative	Start-Flowering	End-Flowering	Vegetative	Start-Flowering	End-Flowering
Chemical composition						
Dry Matter (DM) (g/kg)	249	261	333	241	262	224
Ash (g/kg DM)	103	111	98	83	113	82
CP [1] (g/kg DM)	227	207	166	204	201	181
NDFom [2] (g/kg DM)	336	352	405	313	324	394
ADFom [3] (g/kg DM)	201	230	276	199	213	264
Lignin (sa), (g/kg DM)	45	53	66	78	76	76
Ether extract (g/kg DM)	18	10	16	22	15	11
NSC [4] (g/kg DM)	317	320	316	379	347	332
Plant components (%)						
Leaves	59.3	53.7	42.7	66.3	56.0	38.5
Stems	40.7	43.7	51.3	33.7	40.8	44.6
Flowers	0.0	2.5	5.9	0.0	3.2	16.8

[1] crude protein; [2] Neutral detergent fiber; [3] acid detergent fiber; [4] nonstructural carbohydrates.

Alfalfa and sainfoin had similar proportions of leaves (51.9 vs. 53.6%). However, alfalfa had a greater proportion of stems (45.2 vs. 39.7%) and a lower proportion of flowers (2.8 vs. 6.7%) than sainfoin. Regarding the stage of maturity, the proportion of stems and flowers increased, whereas the proportion of leaves decreased as the stage of maturity advanced.

3.2. Contents of Total Polyphenols and Condensed Tannins

The content of total polyphenols and the total (TCT) and fractions of CT were affected by the interaction between the legume species and the stage of maturity ($p < 0.05$) (Figure 1). Alfalfa presented steady contents of total polyphenols, TCTs, ECTs, PBCTs, and FBCTs, which were lower than those of sainfoin ($p < 0.001$) regardless of the stage of maturity. The contents of polyphenols, TCTs, ECTs, and FBCTs decreased as maturity advanced ($p < 0.05$).

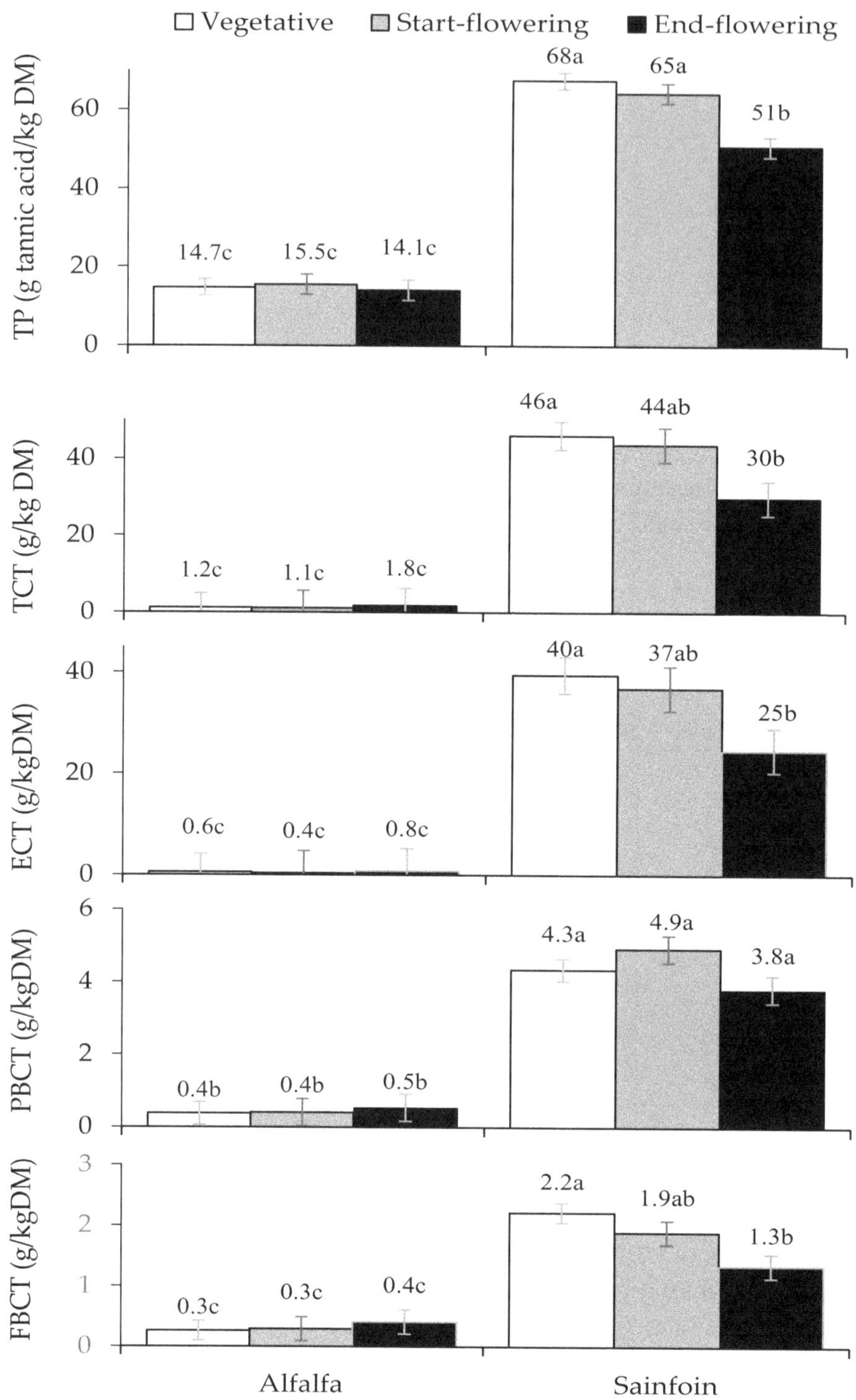

Figure 1. Effect of the species and the stage of maturity on the contents of total polyphenols (TP), total condensed tannins (TCT), extractable CT (ECT), protein-bound CT (PBCT), and fiber-bound (FBCT). Within a parameter, means with different letter differ at $p < 0.05$. (Within each species: n = 3 for the vegetative stage, n = 2 for the start-flowering, and n = 2 for the end-flowering stages.)

3.3. *In Vitro Fermentation*

The pH was affected by the interaction between the substrate and the stage of maturity ($p < 0.01$; Table 2). Alfalfa had greater pH than sainfoin and sainfoin+PEG ($p < 0.05$). Sainfoin and sainfoin+PEG were affected similarly by the stage of maturity ($p < 0.001$), with the greatest pH value in the vegetative stage ($p < 0.05$) Total gas and CH_4 production were affected neither by the substrate nor by the stage of maturity ($p > 0.05$; Table 2). However, the interaction between the substrate and the stage of maturity affected A ($p < 0.001$) and c ($p = 0.05$). Alfalfa showed lower A values at start-and end-flowering stages ($p < 0.001$) and greater c in the vegetative and start-flowering stages ($p < 0.05$) than sainfoin and sainfoin+PEG ($p < 0.05$). Regarding the effect stage of maturity, sainfoin and sainfoin+PEG had the lowest A values in the vegetative stage ($p < 0.05$). As the stage of maturity progressed, c increased in sainfoin and sainfoin+PEG substrates ($p < 0.05$).

The IVOMD was also affected by the interaction between the substrate and the stage of maturity ($p < 0.001$; Table 2). Alfalfa had lower IVOMD than both sainfoin substrates in the start-flowering and end-flowering stages ($p < 0.05$), whereas sainfoin+PEG had greater IVOMD than alfalfa and sainfoin in the vegetative stage ($p < 0.05$). In alfalfa, the IVOMD decreased as the stage of maturity advanced ($p < 0.05$). The sainfoin and sainfoin+PEG showed the greatest IVOMD in the start-flowering stage ($p < 0.05$). The IVOMD was correlated with A ($r = 0.60$; $P < 0.01$) and with the total VFA production ($r = 0.51$; $p < 0.05$).

The NH_3-N content was only affected by the substrate ($p < 0.001$); sainfoin produced a lower NH_3-N concentration than alfalfa and sainfoin+PEG (Table 2). In contrast, total VFA production was only affected by the stage of maturity ($p < 0.05$), the start-flowering stage presented greater VFA production than the rest of the stages (105, 99, and 100 mmol/L for start-flowering, vegetative, and end-flowering, respectively). Regarding the individual VFAs, alfalfa had a lower acetic acid proportion and greater proportions of the rest of the individual VFAs than sainfoin ($p < 0.001$). When comparing both sainfoin substrates, sainfoin had a greater acetic acid proportion and lower proportions of the rest of the VFAs than sainfoin+PEG ($p < 0.001$). Sainfoin presented the greatest C_2:C_3 ratio, followed by sainfoin+PEG and alfalfa, which had the lowest ratio ($p < 0.001$). Regarding the effect of the stage of maturity, the vegetative stage had a lower proportion of acetic acid and greater proportions of propionic, iso-butyric, and iso-valeric acid than the rest of stages of maturity ($p < 0.001$). The vegetative stage had a lower C_2:C_3 ratio than the other stages ($p < 0.001$). The CH_4:VFA ratio was only affected by the substrate ($p = 0.01$); sainfoin+PEG presented a greater CH_4:VFA ratio than the other substrates ($p < 0.05$).

Table 2. Effect of the substrate (S) and the stage of maturity [1] (SM) on gas and methane production (CH_4), potential gas production (*A*), rate of gas production (*c*), in vitro organic matter degradability (IVOMD), ammonia (NH_3-N), and volatile fatty acids (VFAs).

Item	Alfalfa			Sainfoin			Sainfoin+PEG				*p*-Values		
	VEG	Start-F	End-F	VEG	Start-F	End-F	VEG	Start-F	End-F	r.s.d. [2]	S	SM	SxSM
pH	6.42 ^{a}	6.44 ^{a}	6.41 ^{a}	6.35 ^{bx}	6.32 ^{by}	6.33 ^{by}	6.37 ^{bx}	6.31 ^{bz}	6.34 ^{by}	0.032	<0.001	0.002	0.01
Gas production (mL/g dOM [3])	179	184	188	183	163	181	180	199	173	26.5	0.43	0.97	0.12
A (mL)	68 ^{x}	67 ^{bxy}	62 ^{y}	63 ^{y}	74 ^{bx}	70 ^{x}	65 ^{y}	78 ^{ax}	67 ^{y}	5.7	0.01	<0.001	<0.001
c (h^{-1})	0.2 ^{a}	0.22 ^{a}	0.19	0.14 ^{by}	0.16 ^{bxy}	0.19 ^{y}	0.15 ^{bx}	0.15 ^{bx}	0.19 ^{y}	0.037	<0.001	0.02	0.05
CH_4 production (mL/g dOM [3])	37	38	39	37	33	38	37	39	36	5.0	0.29	0.73	0.17
CH_4: gas (mL/L)	145	142	141	141 ^{xy}	134 ^{y}	148 ^{x}	145	144	144	8.8	0.35	0.13	0.11
IVOMD (g/kg)	851 ^{bx}	837 ^{bx}	770 ^{by}	850 ^{by}	926 ^{ax}	856 ^{ay}	868 ^{ay}	936 ^{ax}	864 ^{ay}	16.9	<0.001	<0.001	<0.001
NH_3-N (mg/L)	240 ^{a}	247 ^{a}	210	206 ^{b}	184 ^{b}	201	242 ^{a}	234 ^{a}	215	31.7	<0.001	0.06	0.20
Total VFAs (mmol/L)	102 ^{xy}	105 ^{x}	97 ^{y}	96 ^{y}	103 ^{x}	101 ^{xy}	99 ^{y}	105 ^{x}	102 ^{xy}	7.1	0.61	0.02	0.35
Acetic acid (C_2) (mol/100 mol)	63.1 ^{cy}	63.9 ^{bx}	64.0 ^{bx}	65.4 ^{ay}	66.5 ^{ax}	66.3 ^{ax}	63.9 ^{by}	64.6 ^{bx}	64.5 ^{bx}	0.76	<0.001	<0.001	0.95
Propionic acid (C_3) (mol/100 mol)	15.5 ^{a}	15.3 ^{a}	15.3 ^{a}	15 ^{bx}	14.5 ^{cy}	14.5 ^{by}	15.3 ^{ax}	14.9 ^{by}	15.2 ^{ax}	0.28	<0.001	<0.001	0.09
Butyric acid (mol/100 mol)	13.0 ^{a}	12.6 ^{a}	12.6 ^{a}	12.1 ^{c}	12.2 ^{b}	12 ^{b}	12.5 ^{b}	12.8 ^{a}	12.4 ^{a}	0.45	<0.001	0.20	0.16
Iso-butyric acid (mol/100 mol)	1.97 ^{ax}	1.91 ^{axy}	1.87 ^{ay}	1.82 ^{bx}	1.68 ^{by}	1.72 ^{by}	1.97 ^{ax}	1.83 ^{ay}	1.84 ^{ay}	0.100	<0.001	<0.001	0.63
Valeric acid (mol/100 mol)	2.14 ^{axy}	2.09 ^{ay}	2.22 ^{ax}	1.87 ^{bx}	1.7 ^{cy}	1.83 ^{cx}	2.12 ^{ay}	1.95 ^{bx}	2.04 ^{by}	0.107	<0.001	<0.001	0.11
Iso-valeric acid (mol/100 mol)	4.27 ^{ax}	4.09 ^{axy}	4.04 ^{ay}	3.87 ^{bx}	3.5 ^{by}	3.65 ^{by}	4.19 ^{ax}	3.91 ^{ay}	3.93 ^{ay}	0.223	<0.001	<0.001	0.69
C_2:C_3 (mol/mol)	4.1 ^{cy}	4.18 ^{cxy}	4.2 ^{bx}	4.39 ^{ay}	4.62 ^{ax}	4.59 ^{ax}	4.18 ^{by}	4.36 ^{bx}	4.25 ^{by}	0.109	<0.001	<0.001	0.13
CH_4:VFA (mL/mol)	2.41	2.24 ^{b}	2.35	2.53	2.22 ^{b}	2.48	2.62	2.73 ^{a}	2.4	0.313	0.01	0.26	0.10

[1] VEG: vegetative; Start-F: start of flowering; End-F: end of flowering; [2] residual standard deviation; [3] degraded organic matter. Within a parameter, means with different superscript (a,b,c) differ at $p < 0.05$ for the substrate effect in each stage of maturity; with different superscripts (x,y,z) at $p < 0.05$ for the stage of maturity effect in each substrate.

4. Discussion

In Mediterranean areas, there is an increasing interest to reintroduce forage-based systems in ovine production to ensure the viability and sustainability of the farms. Legumes are especially advisable due to their nutritional quality for ruminants and environmental benefits [1]. Moreover, the presence of CT in some legumes may decrease CH_4 production and improve the performance of ovine to different extents depending on the source and dose of CT. As these legumes are usually fed conserved, either as silage or as hay, there is scarce information on the fermentation parameters when alfalfa and sainfoin are offered fresh. Previous experiments showed differences between the fermentation parameters of alfalfa and sainfoin hay and silage [4,30], however, it is not clear if the differences were due to the chemical composition, the presence of CT in sainfoin or both. With the present study, the fermentation parameters of alfalfa and sainfoin, with or without PEG, in fresh at different stages of maturity were compared to try to clarify the origin of the differences in fermentation. The use of gas production technique is a good tool to evaluate the effect of CT on fermentation parameters, but the fermentation is influenced by the time of incubation, species of the animal donor and the diet [31,32]. Furthermore, the effects of sainfoin CT vary according the variety, harvest time, and cultivation site [13,14], making it difficult to compare the results with other studies. In the present study, the content of total CT and their fractions were analyzed, however, the chemical characteristics of CT (molecular weight, degree of polymerization, prodelphinidin/procyanidin ratio, cis/trans ratio, etc.) were not evaluated.

4.1. Effect of the Substrate

References showed noticeable variability in the chemical composition of alfalfa and sainfoin among studies, which is related to the stage at harvest, leaf:stem ratio, soil characteristics, weather conditions, and the cultivars utilized [5,13,14]. In the present study, the similar CP and ADF contents and different NDF contents of alfalfa and sainfoin agree partially with the results reported by Chung et al. [10], who observed similar NDF, ADF and CP contents in fresh alfalfa and sainfoin at the late vegetative stage. However, at the early vegetative stage, the same authors observed greater NDF and ADF contents in alfalfa than sainfoin and similar CP contents.

Alfalfa has a low content of total polyphenols and is considered a CT-free legume [5], although it may present very low CT content in the seed coats. Therefore, the present results related to the presence of CTs agree with previous studies that analyzed both legumes offered fresh [33,34] or as preserved forages [4,30].

The pH did not negatively affect the fermentation environment because the values were within the range of 6.2 to 6.8 and these values ensure a favorable environment for the activity of cellulolytic bacteria [35]. The inclusion of PEG in the current experiment did not affect pH, as reported in fresh sainfoin [12] and sainfoin hay [8]. Regarding gas production, the similar production of alfalfa and sainfoin agrees with the similar gas production observed in alfalfa and sainfoin leaves incubated in Rusitec units [9] and alfalfa and sainfoin silages [30]. However, when alfalfa and sainfoin hays were studied, differences in gas production were reported [3,4,8]. The inconsistency of the results of the type of substrate on gas production might be related to the differences in chemical composition, the different characteristics of CTs and of type and settings of the in vitro assay [14]. In that sense, the similar gas production between sainfoin and sainfoin+PEG was unexpected because previous experiments reported increases in gas production when PEG was added to fresh sainfoin [12,14]. According to Azuhnwi et al. [13], the inclusion of PEG increased gas production by 2.7 to 9.6%, depending on the sainfoin variety, site, and stage at harvest. In the current experiment, the inclusion of PEG slightly increased the gas production, although not statistically significantly.

The similar CH_4 production of alfalfa and sainfoin recorded in the present study agrees with results reported using fresh forages [10] and silages [30]. However, the inclusion of extracts from sainfoin accessions in alfalfa decreased CH_4 production, but the effect was greatly dependent on the accession and the dose of inclusion [11]. Moderate CT content may have beneficial effects reducing rumen CH_4 emission production [36]. The action of CT on methanogenesis can be attributed to indirect effects via reduced hydrogen production (and presumably reduced forage digestibility) and via direct inhibitory effects on methanogens [34]. Regarding the effect of PEG, the inclusion of PEG in sainfoin did not affect CH_4 production in previous studies [12,37] as in the current experiment. The structural features of condensed tannins affect in vitro CH_4 production, which may be linked to the interaction of CTs with dietary substrate or microbial cells [11,38]. Therefore, the type of CT and dose present in the current experiment might not be sufficient to modify CH_4 production.

The reduced *A* in alfalfa compared with sainfoin and sainfoin+PEG in the start- and end-flowering stages can be related to the higher fiber fraction, as reported by Guglielmelli et al. [3]. In the current experiment, the presence of CTs in sainfoin had no effect on *A*, as reported by Calabrò et al. [8]. The higher *c* in alfalfa when compared to sainfoin agrees with the results reported by Hatew et al. [11], although the effect on *c* depends on the types and concentrations of sainfoin CTs.

The lower IVOMD of alfalfa, when compared to sainfoin and sainfoin+PEG, was also reported using fresh forage estimated in situ [10] and in vitro [39] and could be related to the greater fiber fraction. The increased IVOMD in sainfoin+PEG with respect to sainfoin at the vegetative stage could be related to the blockage of CTs by the PEG. However, Theodoridou et al. [12] reported no effect of the inclusion of PEG in sainfoin on IVOMD studied at 24 h, regardless of the stage of maturity. The discrepancy between studies could be related to the content, characteristics and structures of CTs, which depends on the botanical species and variety of the source [13,14].

The NH_3-N contents recorded in the present study are in line with most similar studies that compared alfalfa and sainfoin [3,4,12]. The reduced NH_3-N concentration in sainfoin with respect to alfalfa and sainfoin+PEG confirmed the inhibition elicited by CTs in the ruminal degradation of dietary proteins due to the formation of complexes CT-protein at ruminal pH [40]. In contrast, the effect of CTs on total VFA production is not clear. Some studies reported a lower total VFA production in sainfoin silage than in alfalfa silage [30], and the inclusion of different doses of extracted accession of sainfoin in alfalfa decreased or maintained the total VFA production, depending on the accession [11]. In the current experiment, the total VFA production was not affected by the legume species, as observed by other authors [4,10]. The inclusion of PEG did not affect total VFA production in the current experiment as reported for sainfoin hay [8,37], which is contrary to the increase of total VFA production observed by Hatew et al. [14]. The differences between the studies could be due to the time of incubation, species of the animal donor, chemical structure, and biological activity of CTs [14,31].

Generally, the presence of CT from sainfoin leads to an increase in the propionic acid proportion and a reduction in the C_2:C_3 ratio [10,11,38]. However, the effect on each individual VFA proportion is variable due to the type of substrate, types, and contents of CTs and length of incubation period or the donor animal [11,31]. In the present study, sainfoin had a greater acetic acid proportion than alfalfa, which was similar to results from Guglielmelli et al. [3] using hays and Grosse-Brinkhaus et al. [30] using silages. The C_2:C_3 ratio recorded in the present study was greater in sainfoin than in alfalfa and sainfoin+PEG, which is in contrast with other studies that did not observe effects of the type of substrate or the addition of PEG [3,12,37]. Sainfoin had lower valeric acid and branched-chain VFA proportions than alfalfa and sainfoin+PEG because of the presence of CTs [4,30,38]. Condensed tannins reduce the proportions of branched-chain VFAs due to reduce protein degradation in the rumen because these VFA are products of the breakdown of the carbon skeleton of amino acids during rumen fermentation [41].

4.2. Effect of the Stage of Maturity

The decrease in CP content and the concomitant increase in the cell wall (NDFom and ADFom) content as the stage of maturity progressed in both forages is a result of the decrease of the proportion of leaves to stems and the increase of lignified tissues [10]. The steady lignin (sa) content in sainfoin during the development of plants can be due to some interference between this compound and CTs during analysis, as reported by Guglielmelli et al. [3].

In the current experiment, the TP and TCT contents were affected by the stage of maturity, as reported in previous studies [3,5,42], although the magnitude of the effect varied among the studies. Regarding the CT fractions, there is little information about the influence of the stage of maturity. As in the present study, Chung et al. [10] observed a reduction of the ECT fraction in the end-flowering stage with respect to the vegetative stage in sainfoin. However, Jin et al. [43] observed greater TCT, ECT, and PBCT contents in *Dalea purpurea* at flowering than at the vegetative stage due to the high percentage of flowers, which are very rich in CTs [44]. From a physiological point of view, the reduction in the secondary compound concentrations as maturity advances could be due to a sort of dilution as a consequence of the growth and expansion of plant cells [45] and to the decrease in the leaf: stem ratio as a consequence of the reduction of leaves, which are rich in CTs [44].

The lack of an effect of the stage of maturity on pH values in alfalfa agrees with the results reported by Chung et al. [10], but the reduction of pH in sainfoin at both flowering stages disagrees with the abovementioned study. In this sense, the stage of maturity had no effect on pH when sainfoin hay was incubated [3,8]. More studies considering the stage of maturity, the chemical composition and the presence of secondary compounds in alfalfa and sainfoin must be performed to discern the importance of these factors on the ruminal pH.

Previous studies reported a reduction of in vitro gas and CH_4 production as the stage of maturity advanced, associated with the chemical composition and the CT content [12,46]. However, the stage of maturity had no effect on gas and CH_4 production in the current experiment, which is in agreement with in vitro [3] and in vivo [10] experiments. The similar chemical composition observed between stages of maturity, the low biological activity of CT and the interactions between nutritive components and antinutritional factors could be responsible for the similar gas and CH_4 production [3].

As expected, the IVOMD of alfalfa decreased as the fiber fraction increased with maturity [47]. In contrast, sainfoin, with or without PEG, presented considerably high IVOMD at the start-flowering stage with respect to the rest of the stages in agreement with Theodoridou et al. [12]. These results could be due to the different biological activity of CTs at vegetative and start-flowering stages [5], because the chemical composition and CT contents were similar in both stages.

The advancing of the stage of maturity tended to reduce NH_3-N production as a consequence of the decrease of CP content in concordance with several studies carried out in vitro [3,12] and in vivo [10,42]. Studies concerning the effect of the stage of maturity on the total production of VFAs and their proportions show discrepancies. The chemical composition of the substrates, the length of the incubation period, and the inoculum donor animal are determinant factors that can influence VFA production and proportions [31]. In the current experiment, the start-flowering stage presented the highest total VFA production, in concordance with the highest IVOMD observed. However, Theodoridou et al. [12] studied the effect of the stage of maturity of fresh sainfoin with a similar CT content in a 24 h in vitro assay and did not find an effect on the total VFA production.

In relation to the proportion of individual VFAs, the effect of the stage of maturity on these parameters has been reported in vitro and in vivo in previous studies [3,10], but the results are not consistent. In the current experiment, as maturity advances, there is an increase in acetic acid and a decrease in propionic acid proportions, thus increasing the C_2:C_3 ratio due probably to increase of fiber and reduction of CT content [42]. The reduction of the proportion of iso-butyric and iso-valeric acids in the start- and end-flowering stages and valeric acid at the start-flowering stage in comparison with the vegetative stage might be explained by the decrease in CP content, because they are products of the breakdown of the carbon skeleton of amino acids during rumen fermentation, as the maturity of the forage advanced [10,47].

5. Conclusions

In conclusion, sainfoin might be an alternative to alfalfa due to the high IVOMD and the potential protection against ruminal protein degradation, according to the results of ammonia content, branched-chain VFAs, and valeric acid proportion from sainfoin in vitro. The effect of the stage of maturity was less than expected, probably due to the high quality of the forages. It is required to study the effects of the type of substrate and stage of maturity on animal performance to recommend the best stage of maturity to cut sainfoin and alfalfa.

Author Contributions: Conceptualization, M.B.; Data curation, M.B. and M.J.; Formal analysis, P.J.R.-M. and M.B.; Funding acquisition, M.J.; Investigation, P.J.R.-M. and J.R.B.; Methodology, M.B. and M.J.; Project administration, M.J.; Resources, M.B. and M.J.; Supervision, M.B. and M.J.; Visualization, P.J.R.-M.; Writing—original draft, P.J.R.-M.; Writing—review & editing, M.B. and M.J..

Acknowledgments: Appreciation is expressed to the staff of CITA de Aragón for their help in data collection. Special thanks to the lab team and to the crop team for their assistance.

References

1. Phelan, P.; Moloney, A.; McGeough, E.; Humphreys, J.; Bertilsson, J.; O'Riordan, E.; O'Kiely, P. Forage legumes for grazing and conserving in ruminant production systems. *Crit. Rev. Plant Sci.* **2015**, *34*, 281–326. [CrossRef]
2. Delgado, I.; Muñoz, F.; Andueza, D. Evaluación comparativa de la alfalfa y la esparceta, en condiciones de secano y regadío de Aragón. *Pastos Y PAC* **2014**, *2020*, 304–310.
3. Guglielmelli, A.; Calabrò, S.; Primi, R.; Carone, F.; Cutrignelli, M.I.; Tudisco, R.; Piccolo, G.; Ronchi, B.; Danieli, P.P. In vitro fermentation patterns and methane production of sainfoin (Onobrychis viciifolia Scop.) hay with different condensed tannin contents. *Grass Forage Sci.* **2011**, *66*, 488–500. [CrossRef]
4. Toral, P.G.; Hervás, G.; Missaoui, H.; Andrés, S.; Giráldez, F.J.; Jellali, S.; Frutos, P. Effects of a tannin-rich legume (Onobrychis viciifolia) on in vitro ruminal biohydrogenation and fermentation. *Span. J. Agric. Res.* **2016**, *14*, 1–9. [CrossRef]
5. Theodoridou, K.; Aufrère, J.; Andueza, D.; Le Morvan, A.; Picard, F.; Stringano, E.; Pourrat, J.; Mueller-Harvey, I.; Baumont, R. Effect of plant development during first and second growth cycle on chemical composition, condensed tannins and nutritive value of three sainfoin (Onobrychis viciifolia) varieties and lucerne. *Grass Forage Sci.* **2011**, *66*, 402–414. [CrossRef]
6. Alvarez-Rodriguez, J.; Sanz, A.; Ripoll-Bosch, R.; Joy, M. Do alfalfa grazing and lactation length affect the digestive tract fill of light lambs? *Small Rumin. Res.* **2010**, *94*, 109–116. [CrossRef]
7. Blanco, M.; Joy, M.; Ripoll, G.; Sauerwein, H.; Casasús, I. Grazing lucerne as fattening management for young bulls: Technical and economic performance and diet authentication. *Animal* **2011**, *5*, 113–122. [CrossRef]
8. Calabrò, S.; Guglielmelli, A.; Iannaccone, F.; Danieli, P.P.; Tudisco, R.; Ruggiero, C.; Piccolo, G.; Cutrignelli, M.I.; Infascelli, F. Fermentation kinetics of sainfoin hay with and without PEG. *J. Anim. Physiol. Anim. Nutr.* **2012**, *96*, 842–849. [CrossRef]
9. McMahon, L.; Majak, W.; McAllister, T.; Hall, J.; Jones, G.; Popp, J.; Cheng, K.-J. Effect of sainfoin on in vitro digestion of fresh alfalfa and bloat in steers. *Can. J. Anim. Sci.* **1999**, *79*, 203–212. [CrossRef]
10. Chung, Y.H.; Mc Geough, E.J.; Acharya, S.; McAllister, T.A.; McGinn, S.M.; Harstad, O.M.; Beauchemin, K.A. Enteric methane emission, diet digestibility, and nitrogen excretion from beef heifers fed sainfoin or alfalfa. *J. Anim. Sci.* **2013**, *91*, 4861–4874. [CrossRef]
11. Hatew, B.; Stringano, E.; Mueller-Harvey, I.; Hendriks, W.H.; Carbonero, C.H.; Smith, L.M.J.; Pellikaan, W.F. Impact of variation in structure of condensed tannins from sainfoin (Onobrychis viciifolia) on in vitro ruminal methane production and fermentation characteristics. *J. Anim. Physiol. Anim. Nutr.* **2016**, *100*, 348–360. [CrossRef]
12. Theodoridou, K.; Aufrère, J.; Niderkorn, V.; Andueza, D.; Le Morvan, A.; Picard, F.; Baumont, R. In vitro study of the effects of condensed tannins in sainfoin on the digestive process in the rumen at two vegetation cycles. *Anim. Feed Sci. Technol.* **2011**, *170*, 147–159. [CrossRef]

13. Azuhnwi, B.; Boller, B.; Martens, M.; Dohme-Meier, F.; Ampuero, S.; Günter, S.; Kreuzer, M.; Hess, H. Morphology, tannin concentration and forage value of 15 Swiss accessions of sainfoin (Onobrychis viciifolia Scop.) as influenced by harvest time and cultivation site. *Grass Forage Sci.* **2011**, *66*, 474–487. [CrossRef]
14. Hatew, B.; Hayot Carbonero, C.; Stringano, E.; Sales, L.F.; Smith, L.M.J.; Mueller-Harvey, I.; Hendriks, W.H.; Pellikaan, W.F. Diversity of condensed tannin structures affects rumen in vitro methane production in sainfoin (Onobrychis viciifolia) accessions. *Grass Forage Sci.* **2015**, *70*, 474–490. [CrossRef]
15. Lobón, S.; Blanco, M.; Sanz, A.; Ripoll, G.; Joy, M. Effects of feeding strategies during lactation and the inclusion of quebracho in the fattening on performance and carcass traits in light lambs. *J. Sci. Food Agric.* **2019**, *99*, 457–463. [CrossRef]
16. Borreani, G.; Peiretti, P.G.; Tabacco, E. Evolution of yield and quality of sainfoin (Onobrychis viciifolia Scop.) in the spring growth cycle. *Agronomie* **2003**, *23*, 193–201. [CrossRef]
17. Kalu, B.A.; Fick, G.W. Quantifying Morphological Development of Alfalfa for Studies of Herbage Quality. *Crop Sci.* **1981**, *21*, 267–271. [CrossRef]
18. Rufino-Moya, P.J.; Blanco, M.; Bertolín, J.R.; Joy, M. Effect of the method of preservation on the chemical composition and in vitro fermentation characteristics in two legumes rich in condensed tannins. *Anim. Feed Sci. Technol.* **2019**, *251*, 12–20. [CrossRef]
19. Menke, K.H.; Steingass, H. Estimation of the energetic feed value obtained from chemical analysis and in vitro gas production using rumen fluid. *Anim. Res. Dev.* **1988**, *28*, 7–55.
20. AOAC. *Official Methods of Analysis*, 17th ed.; Association of Official Analytical Chemist: Arlington, VA, USA, 2000.
21. Van Soest, P.J.; Robertson, J.B.; Lewis, B.A. Methods for dietary fiber, neutral detergent fiber, and nonstarch polysaccharides in relation to animal nutrition. *J. Dairy Sci.* **1991**, *74*, 3583–3597. [CrossRef]
22. AOCS. *Approved Procedure Am 5-04, Rapid Determination of Oil/Fat Utilizing High. Temperature Solvent Extraction*; AOCS Press: Urbana, IL, USA, 2005.
23. Makkar, H.P.S. *Quantification of Tannins in Tree Foliage*; FAO/IAEA Working Document; IAEA: Vienna, Austria, 2000.
24. Julkunen-Tiitto, R. Phenolic Constituents in the Leaves of Northern Willows: Methods for the Analysis of Certain Phenolics. *J. Agric. Food Chem.* **1985**, *33*, 213–217. [CrossRef]
25. Terrill, T.H.; Windham, W.R.; Hoveland, C.S.; Amos, H.E. Forage preservation method influences on tannin concentration, intake and digestibility of Sericea lespedeza by sheep. *Agron. J.* **1989**, *81*, 435–439. [CrossRef]
26. Grabber, J.H.; Zeller, W.E.; Mueller-Harvey, I. Acetone enhances the direct analysis of procyanidin- and prodelphinidin-based condensed tannins in lotus species by the butanol-HCl-iron assay. *J. Agric. Food Chem.* **2013**, *61*, 2669–2678. [CrossRef] [PubMed]
27. Wolfe, R.M.; Terrill, T.H.; Muir, J.P. Drying method and origin of standard affect condensed tannin (CT) concentrations in perennial herbaceous legumes using simplified butanol-HCl CT analysis. *J. Sci. Food Agric.* **2008**, *88*, 1060–1067. [CrossRef]
28. France, J.; Dijkstra, J.; Dhanoa, M.S.; Lopez, S.; Bannink, A. Estimating the extent of degradation of ruminant feeds from a description of their gas production profiles observed in vitro: Derivation of models and other mathematical considerations. *Br. J. Nutr.* **2000**, *83*, 143–150. [CrossRef] [PubMed]
29. Chaney, A.L.; Marbach, E.P. Modified reagents for determination of urea and ammonia. *Clin. Chem.* **1962**, *8*, 130–132. [PubMed]
30. Grosse-Brinkhaus, A.; Wyss, U.; Arrigo, Y.; Girard, M.; Bee, G.; Zeitz, J.O.; Kreuzer, M.; Dohme-Meier, F. In vitro ruminal fermentation characteristics and utilisable CP supply of sainfoin and birdsfoot trefoil silages and their mixtures with other legumes. *Animal* **2017**, *11*, 580–590. [CrossRef]
31. Frutos, P.; Hervás, G.; Giráldez, F.J.; Mantecón, A.R. An in vitro study on the ability of polyethylene glycol to inhibit the effect of quebracho tannins and tannic acid on rumen fermentation in sheep, goats, cows, and deer. *Aust. J. Agric. Res.* **2004**, *55*, 1125–1132. [CrossRef]
32. Martínez, T.F.; McAllister, T.A.; Wang, Y.; Reuter, T. Effects of tannic acid and quebracho tannins on in vitro ruminal fermentation of wheat and corn grain. *J. Sci. Food Agric.* **2006**, *86*, 1244–1256. [CrossRef]
33. Scharenberg, A.; Arrigo, Y.; Gutzwiller, A.; Soliva, C.R.; Wyss, U.; Kreuzer, M.F.D. Palatability in sheep and in vitro nutritional value of dried and ensiled sainfoin (*Onobrychis viciifolia*) birdsfoot trefoil (*Lotus corniculatus*), and chicory (*Cichorium intybus*). *Arch. Anim. Nutr.* **2007**, *61*, 481–496. [CrossRef] [PubMed]

34. Tavendale, M.H.; Meagher, L.P.; Pacheco, D.; Walker, N.; Attwood, G.T.; Sivakumaran, S. Methane production from in vitro rumen incubations with Lotus pedunculatus and Medicago sativa, and effects of extractable condensed tannin fractions on methanogenesis. *Anim. Feed Sci. Technol.* **2005**, *123–124 Pt 1*, 403–419. [CrossRef]
35. Doane, P.H.; Pell, A.N.; Schofield, P. The Effect of Preservation Method on the Neutral Detergent Soluble Fraction of Forages. *J. Anim. Sci.* **1997**, *75*, 1140–1148. [CrossRef] [PubMed]
36. Bodas, R.; Prieto, N.; García-González, R.; Andrés, S.; Giráldez, F.J.; López, S. Manipulation of rumen fermentation and methane production with plant secondary metabolites. *Anim. Feed Sci. Technol.* **2012**, *176*, 78–93. [CrossRef]
37. Niderkorn, V.; Mueller-Harvey, I.; Le Morvan, A.; Aufrère, J. Synergistic effects of mixing cocksfoot and sainfoin on in vitro rumen fermentation. Role of condensed tannins. *Anim. Feed Sci. Technol.* **2012**, *178*, 48–56. [CrossRef]
38. Huyen, N.; Fryganas, C.; Uittenbogaard, G.; Mueller-Harvey, I.; Verstegen, M.; Hendriks, W.; Pellikaan, W. Structural features of condensed tannins affect in vitro ruminal methane production and fermentation characteristics. *J. Agric. Sci.* **2016**, *154*, 1474–1487. [CrossRef]
39. Lobón, S.; Molino, F.; Legua, A.; Eseverri, M.P.; Céspedes, M.A.; Joy, M. Efecto del Forraje y de La Inclusión de Concentrado En la Dieta Sobre La Producción de Gas y Metano En Ovino. 2015. Available online: http://www.uibcongres.org/imgdb/archivo_dpo19392.pdf (accessed on 1 March 2019).
40. Mueller-Harvey, I. Unravelling the conundrum of tannins in animal nutrition and health. *J. Sci. Food Agric.* **2006**, *86*, 2010–2037. [CrossRef]
41. Hassanat, F.; Benchaar, C. Assessment of the effect of condensed (acacia and quebracho) and hydrolysable (chestnut and valonea) tannins on rumen fermentation and methane production in vitro. *J. Sci. Food Agric.* **2013**, *93*, 332–339. [CrossRef]
42. Theodoridou, K.; Aufrére, J.; Andueza, D.; Le Morvan, A.; Picard, F.; Pourrat, J.; Baumont, R. Effects of condensed tannins in wrapped silage bales of sainfoin (Onobrychis viciifolia) on in vivo and in situ digestion in sheep. *Animal* **2012**, *6*, 245–253. [CrossRef] [PubMed]
43. Jin, L.; Wang, Y.; Iwaasa, A.; Xu, Z.; Schellenberg, M.; Zhang, Y.; Liu, X.; McAllister, T. Effect of condensed tannins on ruminal degradability of purple prairie clover (Dalea purpurea Vent.) harvested at two growth stages. *Anim. Feed Sci. Technol.* **2012**, *176*, 17–25. [CrossRef]
44. Piluzza, G.; Bullitta, S. The dynamics of phenolic concentration in some pasture species and implications for animal husbandry. *J. Sci. Food Agric.* **2010**, *90*, 1452–1459. [CrossRef]
45. Joseph, R.; Tanner, G.; Larkin, P. Proanthocyanidin synthesis in the forage legume Onobrychis viciifolia. A study of chalcone synthase, dihydroflavonol 4-reductase and leucoanthocyanidin 4-reductase in developing leaves. *Funct. Plant Biol.* **1998**, *25*, 271–278. [CrossRef]
46. Bal, M.A.; Ozturk, D.; Aydin, R.; Erol, A.; Ozkan, C.O.; Ata, M.; Karakas, E.; Karabay, P. Nutritive value of sainfoin (Onobrychis viciaefolia) harvested at different maturity stages. *Pak. J. Biol. Sci.* **2006**, *9*, 205–209. [CrossRef]
47. Navarro-Villa, A.; O'Brien, M.; López, S.; Boland, T.M.; O'Kiely, P. In vitro rumen methane output of red clover and perennial ryegrass assayed using the gas production technique (GPT). *Anim. Feed Sci. Technol.* **2011**, *168*, 152–164. [CrossRef]

14

Comparative Grain Chemical Composition, Ruminal Degradation In Vivo, and Intestinal Digestibility In Vitro of *Vicia Sativa* L. Varieties Grown on the Tibetan Plateau

Yafeng Huang [1,2,3], Rui Li [1,2,3], Jeffrey A. Coulter [4], Zhixin Zhang [1,2,3] and Zhibiao Nan [1,2,3,*]

1 State Key Laboratory of Grassland Agro-ecosystems, Lanzhou University, Lanzhou 730020, China; huangyafeng316@163.com (Y.H.); lir15@lzu.edu.cn (R.L.); zhixinzhang.phd@gmail.com (Z.Z.)
2 Key Laboratory of Grassland Livestock Industry Innovation of Ministry of Agriculture and Rural Affairs, Lanzhou University, Lanzhou 730020, China
3 College of Pastoral Agriculture Science and Technology, Lanzhou University, Lanzhou 730020, China
4 Department of Agronomy and Plant Genetics, University of Minnesota, St. Paul, MN 55108, USA; jeffcoulter@umn.edu
* Correspondence: zhibiao@lzu.edu.cn;

Simple Summary: Common vetch (*Vicia sativa* L.) grain is an important source of protein in rations for ruminants, but little information is available on the protein value of common vetch grains, both in terms of chemical composition and protein degradability, and regarding variation between intra-species and year. The objective of this study was to evaluate grain chemical composition, ruminal protein degradability in vivo, and intestinal protein digestibility in vitro of four common vetch varieties over two cropping years on the Tibetan Plateau. This study was also conducted to establish correlations of grain chemical composition with ruminal degradability parameters of grain protein and with intestinal digestibility of grain protein. Results of this study demonstrated that grain quality characteristics varied significantly among varieties and years. The relationship between grain chemical composition and intestinally absorbable digestible protein (IADP) was best described by a linear regression equation, and coefficients of determination remained very high ($R^2 = 0.891$). Overall, the results indicated that in terms of effective crude protein degradability and IADP of grain, common vetch varieties Lanjian No. 2 and Lanjian No. 3 have the greatest potential among varieties examined for supplementing ruminant diets when grown on the Tibetan Plateau.

Abstract: Four varieties of common vetch, three improved varieties and one local variety, were evaluated for grain chemical composition, rumen protein degradability, and intestinal protein digestibility over two cropping years on the Tibetan Plateau. This study also examined correlations of grain chemical composition with rumen degradability parameters of grain protein and with intestinal digestibility of grain protein. Results of this study showed that grain quality attributes varied ($p < 0.05$) among varieties and cropping years. Significant intra-species variation was observed for concentrations (g/kg dry matter) of crude protein (CP; range = 347–374), ether extract (range = 15.8–19.6), neutral detergent fiber (aNDF; range = 201–237), acid detergent fiber (range = 58.2–71.6), ash (range = 27.6–31.0), effective CP degradability (EDCP; range = 732–801 g/kg CP), and intestinally absorbable digestible protein (IADP; range = 136–208 g/kg CP). The relationship between grain chemical composition and IADP was best described by the linear regression equation IADP = −0.828CP + 8.80ash + 0.635aNDF + 70.2 ($R^2 = 0.891$), indicating that chemical analysis offers a quick and reliable method for IADP of common vetch grain. In terms of EDCP and IADP of grain, common vetch varieties, Lanjian No.2 and Lanjian No. 3, have the greatest potential among varieties tested for supplementing ruminant diets when grown on the Tibetan Plateau.

Keywords: common vetch; grain; nutritive value; ruminants

1. Introduction

Livestock sector is the largest land-use system worldwide and is an engine of economic growth that has received significant attention within the last decade [1,2]. The Tibetan Plateau covers an area of 256 million ha^2 and is the largest grassland ecosystem in Eurasia [3]. It supports nearly 41 million Tibetan sheep (*Ovies aries*) and 14.3 million yak (*Bos grunniens*), cattle (*Bos taurus*), and cattle-yak hybrids, which are a predominant part of the local economy for the nomadic population of over 9.8 million [3]. The Tibetan Plateau is also the source of Asia's major river systems, including the Mekong, Yangtze, Yellow, Ganges, and Indus. Therefore, sustainable development and management of the Tibetan Plateau is of vital importance, not only for supporting the livelihood of millions of people, but also for protection of these critical river systems.

Legume grains are important sources of protein in ruminant nutrition [4–6]. Demand for plant-based protein has grown steadily, since the influence of the bovine spongiform encephalopathy which resulted in the banning of meat meal for feeding ruminants in the European Union in 1994 [7]. The Tibetan Plateau has an inherently extreme and unstable climate, and a short growing season, which seriously limits crop production [8], especially for leguminous crops [9]. Nan et al. [9] reported that many annual legume crops [e.g., woolly-pod vetch (*Vicia villosa* ssp. *dasycarpa Roth*) have low grain yield potential in this region. Common vetch was introduced to the Tibetan Plateau of China in 1998 and has shown promising potential for grain production [9]. Recently, new varieties of common vetch designated as Lanjian No.1, Lanjian No.2, and Lanjian No.3, which represent diverse levels of maturity with acceptable grain production on the Tibetan Plateau, were developed by the common vetch (*Vicia sativa* L.) breeding program at Lanzhou University [10].

Common vetch is a cool-season annual legume cultivated in many countries that can satisfy grain demand for feed and food with low requirement for nitrogen input due to biological nitrogen fixation [5,11]. The nutritive value of common vetch grain is relatively high, with an average of 314 g/kg crude protein (CP) and 961 g/kg CP total digestible protein [7,12,13]. This makes common vetch grain a potentially important source of protein in rations for ruminants, such as cattle and sheep, and is of great interest in mixed crop-livestock systems [5,7,13–15], particularly due to the banning of soybean (*Glycine max* (L.) Merr.) meal for feeding organic livestock [16]. Koumas and Economides [14] and Gül et al. [15] reported satisfactory ruminant growth with diets containing common vetch grain as a replacement for soybean (*Glycine max* (L.) Merr.) meal.

For the feed application of common vetch grain, previous studies focused on grain chemical composition [4,11,17] or nutritional values of a single variety [7]. The nutrient value of common vetch for ruminants varies widely in the literature and depends on the common vetch variety and animal species, and chemical composition of common vetch grain. These phenomena indicate the need to analyze the grain of common vetch varieties for nutritive value before using in ration formulations. However, little information is available on protein value of common vetch grains, regarding variation between intra-species and the growing season. Rumen degradability of protein is a key parameter for the evaluation of feed protein value in ruminants [7,18], and the benefits of protein resources are dependent on protein digestion in the small intestine [7,19]. Minerals are a minor ingredient of ruminant feeds, but they are essential for growth and development [20]. Existing data on the mineral composition of common vetch grain is limited. Woods et al. [21,22] reported that chemical analysis of concentrate feedstuffs could be used for accurate prediction of ruminal and intestinal protein degradability. These observations could be more cost and time effective than using in situ nylon bags and in vitro techniques with surgically prepared animals, which are relatively expensive, labor-intensive, and subject to animal welfare problems [21,22]. However, no available information has been reported on the relationship between grain chemical composition with grain rumen degradation parameters of CP and grain intestinal protein digestibility for common vetch.

Understanding the varietal differences in nutritive value of common vetch grain grown on the Tibetan Plateau could increase its use as a protein supplement in ruminant diets and lead to greater integration of common vetch within agro-pastoral farming systems in this region, thereby reducing the need for protein feed imports and enhancing the income of farmers. Against this background, the objectives of this study were: (i) to evaluate the potential of common vetch for use as a protein feed source in ruminant diets, by studying grain chemical composition, ruminal degradability kinetics of CP, and intestinal protein digestibility of four common vetch varieties over two cropping years; (ii) to investigate relationships between grain protein, fiber fractions, and ash concentrations, and grain ruminal degradability parameters of CP and intestinally digestible protein of common vetch varieties grown in rainfed conditions on the Tibetan Plateau.

2. Materials and Methods

2.1. Experimental Site

Common vetch was planted in rainfed conditions during the 2015 and 2016 cropping seasons at the Xiahe experimental station of Lanzhou University, China (35°19′ N, 102°58′E; 2880 m above sea level (asl)), which is located on the eastern margin of the Tibetan Plateau. The soil type of the site is classified as chernozem and is slightly acidic (pH 6.8), low in phosphorous (0.73 g/kg) but adequate in potassium (13.5 g/kg). The preceding crop was rape (*Brassica campestris* L.) in both years. The daily meteorological data (air temperature and rainfall) were recorded via an automatic meteorological station (PC200W, Campbell Scientific) installed near the experimental field. During the cropping period (April through September) in 2015 and 2016, mean air temperature was 10.8 and 12.0 °C, respectively, and total precipitation was 314 and 417 mm, respectively. Monthly mean temperature and total precipitation in July and August were 2.1 to 3.4 °C, respectively, and 22 to 34 mm greater in 2016 than 2015 (Table 1).

Table 1. Monthly mean air temperature and total precipitation during the 2015 and 2016 cropping seasons at the Xiahe experimental station, Gansu, China.

	Mean Air Temperature (°C)		Total Precipitation (mm)	
Month	**2015**	**2016**	**2015**	**2016**
April	4.4	6.2	23	25
May	9.3	9.3	38	72
June	12.4	12.7	49	54
July	13.4	15.5	69	103
August	14.2	17.6	62	84
September	11.2	10.6	73	79
Sum			314	417

2.2. Plant Material, Experimental Design, and Sampling

Three improved varieties of common vetch (Lanjian No. 1, Lanjian No. 2, and Lanjian No. 3) and one local variety were evaluated in a field experiment. The local variety originated in the Gansu province of China and is now widely cultivated in Gansu, Qinghai, and provinces located in the middle and lower reaches of the Yangtze River in China. These varieties were selected because they are grown extensively in the region surrounding the research site [10]. Agronomic characteristics of these varieties are in Table 2 [23]. Plots were planted at 150 viable seeds m^{-2}, with four replicates per variety in a completely randomized design. Each plot was 8 × 5 m and contained 26 rows of plants, spaced 20 cm apart. Common vetch was planted on 6 May 2015 and 28 April 2016 following rhizobial inoculation (CCBAU01069, China Agricultural University, Beijing, China), which was recommended based on the symbiont performance for these varieties [10]. Irrigation and fertilizer were not applied to the experimental plots during either growing season and weeds were controlled by hand.

Table 2. Agronomic characteristics of the common vetch varieties used in this experiment.

Agronomic Characteristic	Local Variety	Lanjian No.1	Lanjian No.2	Lanjian No.3
Days to mature (d)	134	145	132	124
1000 grains weight (g)	54	79	71	76
Plant height (cm)	92	106	80	69
Altitude (m.a.s.l)	–	<3000	<3500	<4000
Year of release	1987	2014	2015	2011

Source: Ministry of Agriculture, Beijing, China [23].

At the pod maturity stage of common vetch, two representative 1 × 1 m sections from each plot were manually harvested and threshed to obtain grain samples. Grain was oven-dried at 60 °C for 48 h and then ground to pass through a 2-mm sieve to obtain samples for analyses of ruminal incubation and intestinal digestibility, and to pass through a 1-mm sieve to obtain samples for chemical analysis.

2.3. Chemical Analysis

Laboratory dry matter (DM) concentration was measured from 2.0 g of ground grain from each sample by drying in a forced-air oven at 135 °C for 2 h (method 930.15; AOAC) [24]. Subsequently, grain ash concentration was measured by incineration in a muffle furnace at 550 °C for 5 h (method 938.08; AOAC). Total N was measured using the Kjeldahl method (method 988.05) [24] and CP concentration was calculated as total N × 6.25. Ether extract (EE) concentration was measured by extraction with petroleum ether (method 920.85; AOAC) [24]. Concentrations of acid detergent fiber (ADF) and neutral detergent fiber (aNDF) were measured by sequential analysis with α-amylase and sodium sulfite, and expressed with residual ash excluded [25]. Mineral concentrations were determined by atomic absorption spectroscopy (method 985.01; AOAC) [26] and phosphorus (P) concentration was determined by colorimetry (method 965.17; AOAC) [24]. All measurements were performed in triplicate and chemical standards were included in each analytical run as appropriate.

2.4. Ruminal Degradability of CP

Four adult fistulated dorper sheep rams 30–31 months of age and 58 ± 1 kg live body weight were used to determine ruminal degradation of CP of common vetch grain using the nylon bag technique described by Nandra et al. [27]. Briefly, 5.0 g of dry weight grain from each sample (in duplicate) was placed in a nylon bag (9 × 5 cm; 50-μm pore size) and incubated for 0 (control), 2, 4, 8, 12, 24, and 48 h in the rumen of four adult fistulated dorper rams. After removal, the bags were washed thoroughly with tap water and frozen (−20 °C) until further analysis. Prior to analysis, all bags were defrosted and manually washed with tap water until the water ran clear, before being oven-dried at 65 °C for 48 h and weighed. The dried undigested residues of replicates per time within rams were pooled prior to analysis. Degradability parameters of CP were determined using the exponential model described by Ørskov and McDonald [28]. The effective degradable fraction of CP was calculated as effective CP degradability (EDCP) = $A + [(B \times C)/(C + k)]$, where A is the soluble fraction of grain CP, B is the potentially degradable fraction of grain CP, C is the rate of degradation of fraction B (h^{-1}), and k is the rumen outflow rate (0.031 h^{-1}) [7]. The animals were housed in individual stalls and daily fed 1200 g of 550 g/kg DM sheepgrass (*Leymus chinensis*, (Trin.) Tzvel), 294 g/kg DM maize (*Zea mays* L.) grain, 140 g/kg DM soybean meal, and 16 g/kg DM of a concentrate mix at maintenance energy level [7] in two equal meals at 08:30 am and 16:30 pm. The rams had free access to water and mineral/vitamin licks. The experimental protocols were approved by the Animal Ethics Committee of Lanzhou University (protocol number: AEC-LZU-2016-01).

2.5. Intestinal Digestibility of CP

A modified 3-step in vitro procedure described by Gargallo et al. [29] was used to determine intestinal digestibility of rumen undegradable protein (IDP). Briefly, dried duplicate undegradable residues from the 16 h in situ ruminal incubation were pooled and ground to pass through a 1-mm sieve. Six sub-samples of dried sample residues (500 mg each) were then weighed into Ankom F57 filter bags and heat-sealed. Twenty-four sample bags were incubated in each incubation jar of a Daisy^{II} incubator (ANKOM Technology, Fairport, NY, USA) containing 2 L of 0.1 M pre-warmed HCl solution (pH = 1.9) and 1 g L^{-1} of pepsin (P-7000, Sigma, St. Louis, MO, USA), with constant rotation at 39 °C for 1 h. After incubation and washing, sample bags were reintroduced into the same incubation jars containing 2 L of pre-warmed pancreatin (0.5 M KH_2PO_4 buffer, pH = 7.75, 50 mg/kg of thymol, and 3 g/L of pancreatin (P-7545, Sigma)) and incubated with constant rotation at 39 °C for 24 h. After removal, bags were rinsed with tap water until the water ran clear, dried at 55 °C for 48 h, and reweighed. Undegradable residues from ruminal and intestinal incubations were analyzed for nutrient concentration.

2.6. Statistical Analyses

Ruminal undegradable protein (RUP) concentration was calculated as 1000 – EDCP (g/kg CP), and intestinally absorbable digestible protein (IADP) was determined as RUP (g/kg of CP) × IDP (g/kg of RUP), as described by Lawrence and Anderson [30]. Total digestible protein (TDP) was calculated as TDP = EDCP + IADP. Statistical analyses were performed using SPSS software (Version 21.0. IBM Corporation, Armonk, NY, USA). Data were analyzed using analysis of variance to assess the significance of the main effects of common vetch variety and year, and their interaction. When the F-tests were significant, variances among means were compared using the Duncan significant difference test at $p < 0.05$. Pearson's correlation analysis was used to evaluate relationships between chemical composition variables and A, B, C, and IADP of grain samples across varieties and years ($n = 32$). Chemical composition data of grain samples of common vetch varieties across years ($n = 32$) were analyzed to predict the rumen degradability parameters of CP (A, B, and C) and IADP using stepwise multiple linear regression [21,22].

3. Results

3.1. Chemical Composition

The effects of year and variety significantly influenced all proximate composition variables of common vetch grain with the exception of grain ash concentration, which was only affected by variety (Table 3). The interaction between year and variety did not significantly affect proximate composition variables. Averaged across varieties, CP concentration was less and EE, aNDF, and ADF concentrations were greater in 2016 compared to 2015. Averaged over both years, CP concentration among varieties ranged from 353 g/kg DM for Lanjian No. 3 to 370 g/kg DM for the local variety. Crude protein concentration of Lanjian No.1 was less than that of the local variety, but greater than that of the other improved varieties. Inversely, EE ranged from 16.1 g/kg DM for the local variety to 19.0 g/kg DM for Lanjian No. 3. The local variety had significantly less EE concentration compared to Lanjian No. 2 and Lanjian No. 3 but was similar to Lanjian No. 1. Averaged across years, aNDF and ADF among varieties ranged from 207 to 229 g/kg DM and 59.5 to 68.8 g/kg DM, respectively. Neutral detergent fiber and ADF of the local variety were not significantly different than that of Lanjian No. 1 but were less than that of other improved varieties. Ash concentration ranged from 27.8 g/kg DM for the local variety to 31.0 g/kg DM for Lanjian No. 3. The local variety had significantly less ash concentration compared to the improved varieties. There were significant differences in ash concentration among improved varieties, which was greater for Lanjian No. 2 and Lanjian No. 3 than Lanjian No. 1.

Phosphorus (P), calcium (Ca), magnesium (Mg), and iron (Fe) concentrations in common vetch grain varied significantly between years and among varieties, but were not affected by the year × variety interaction (Table 3). Averaged across varieties, grain concentration of all measured macronutrients was greater in 2016 than in 2015. Means over both years, P, Mg, Ca, and Fe concentrations among varieties ranged from 1.25 to 3.42, 3.18 to 3.85, 1.39 to 1.95, and 0.431 to 0.801 g/kg DM, respectively.

Phosphorus and Fe concentrations were lower and Mg and Ca concentrations were greater in grain of the local variety compared with the improved varieties, and there were no significant differences among improved varieties. Micronutrient levels also varied among varieties, but were not significantly affected by year or the interaction between year and variety. Averaged across years, grain concentrations of zinc (Zn), manganese (Mn), and copper (Cu) among varieties ranged from 36.3 to 48.6, 12.2 to 14.3, and 6.41 to 9.59 mg/kg DM, respectively. Micronutrient concentrations of grain of the local variety were greater than those of the improved varieties, and there were no significant differences among improved varieties.

3.2. Ruminal Degradation Kinetics of CP

Soluble fraction of grain CP of common vetch was significantly influenced by year and variety, but not by their interaction (Table 4). Averaged across varieties, A was greater in 2016 than 2015 (346 and 325 g/kg, respectively). Averaged across years, A among varieties ranged from 312 g/kg for Lanjian No. 2 to 359 g/kg for the local variety, and A of the local variety did not differ significantly from that of Lanjian No. 1 and was greater than that of other improved varieties. There were significant differences in the B and C among varieties, although the effect of year and of year × variety interaction were not significant. Averaged across years, B among varieties ranged from 552 g/kg for Lanjian No. 3 to 608 g/kg for the local variety. Potentially degradable protein of the local variety was greater than that of Lanjian No. 2 and Lanjian No. 3, and did not differ significantly from that of Lanjian No.1. In contrast, C ranged from an average of 0.0770 h^{-1} for the local variety and Lanjian No. 1 to an average of 0.0966 h^{-1} for Lanjian No. 2 and Lanjian No. 3. The rate of protein degradation of the local variety was not significantly different from that of Lanjian No. 1, but was less than that of other improved varieties.

Effective CP degradability of common vetch grain differed significantly between years (772 and 755 g/kg CP in 2015 and 2016, respectively) and among varieties, but was not significantly influenced by the year × variety interaction. Averaged across years, EDCP among varieties ranged from 736 g/kg CP for Lanjian No. 3 to 792 g/kg CP for the local variety. The EDCP of the local variety was significantly greater than that of Lanjian No. 2 and Lanjian No. 3, but did not differ significantly from that of Lanjian No. 1.

3.3. Intestinal Digestibility of CP

The IDP and IADP of grain were significantly affected by year and variety of common vetch, but not by the year × variety interaction (Table 4). Averaged across varieties, IDP and IADP were less in 2015 than in 2016. Averaged across years, IDP and IADP among varieties ranged from 692 to 768 g/kg of RUP, and 144 to 203 g/kg CP, respectively. The IDP and IADP of the local variety were less than that of Lanjian No. 2 and Lanjian No. 3, but not significantly different than that of Lanjian No. 1. Total digestible protein of grain was not significantly affected by year, variety of common vetch, or the interaction between year and variety, and averaged 937 g/kg CP across years and varieties.

Table 3. Effects of cropping year, variety, and the year × variety interaction on the chemical composition of four common vetch grain varieties grown on the Tibetan Plateau.

Dependent Variable	Mean across Years					Year										p-Value		
						2015					2016							
	Local Variety	Lanjian No.1	Lanjian No.2	Lanjian No.3	SEM [1]	Local Variety	Lanjian No.1	Lanjian No.2	Lanjian No.3	SEM	Local Variety	Lanjian No.1	Lanjian No.2	Lanjian No.3	SEM	Year	Variety	Year × Variety
Proximate composition (g/kg DM)																		
CP	370^{a}	360^{b}	354^{c}	353^{c}	1.57	374^{a}	365^{b}	358^{b}	358^{b}	1.88	366^{a}	356^{b}	350^{c}	347^{c}	1.96	< 0.001	<0.001	0.827
EE	16.1^{c}	17.5^{bc}	18.5^{ab}	19.0^{a}	0.278	15.8^{b}	17.2^{ab}	17.8^{a}	18.5^{a}	0.372	17.3^{b}	17.8^{ab}	19.3^{a}	19.6^{a}	0.368	0.011	0.002	0.818
aNDF	207^{c}	217^{bc}	225^{ab}	229^{a}	2.46	201^{b}	209^{ab}	220^{a}	220^{a}	2.87	214^{b}	225^{ab}	231^{a}	237^{a}	3.19	< 0.001	<0.001	0.942
ADF	59.5^{b}	61.3^{b}	68.3^{a}	68.8^{a}	1.12	58.2^{b}	59.0^{ab}	65.8^{a}	66.1^{a}	1.38	60.8^{c}	63.5^{bc}	70.8^{ab}	71.6^{a}	1.63	0.014	<0.001	0.929
Ash	27.8^{c}	29.3^{b}	30.7^{a}	31.0^{a}	0.261	27.9^{c}	29.6^{b}	30.7^{ab}	31.0^{a}	0.353	27.6^{c}	29.1^{b}	30.8^{a}	31.0^{a}	0.395	0.548	<0.001	0.845
Macronutrient minerals (g/kg DM)																		
P	1.25^{b}	3.06^{a}	3.42^{a}	3.33^{a}	0.192	0.993^{b}	2.78^{a}	3.24^{a}	3.08^{a}	0.268	1.50^{b}	3.33^{a}	3.61^{a}	3.57^{a}	0.270	0.042	<0.001	0.993
Mg	3.85^{a}	3.20^{b}	3.18^{b}	3.23^{b}	0.0736	3.75^{a}	3.05^{b}	3.09^{b}	3.11^{b}	0.108	3.96^{a}	3.35^{b}	3.28^{b}	3.35^{b}	0.0947	0.047	<0.001	0.989
Ca	1.95^{a}	1.39^{b}	1.54^{b}	1.47^{b}	0.0655	1.90^{a}	1.20^{b}	1.36^{b}	1.29^{b}	0.105	2.00^{a}	1.58^{b}	1.72^{ab}	1.65^{b}	0.0612	0.007	0.003	0.396
Fe	0.431^{b}	0.793^{a}	0.801^{a}	0.750^{a}	0.0479	0.345	0.630	0.635	0.600	0.0467	0.517^{b}	0.955^{a}	0.967^{a}	0.900^{a}	0.0684	< 0.001	0.002	0.820
Micronutrient minerals (mg/kg DM)																		
Zn	48.6^{a}	38.4^{b}	37.7^{b}	36.3^{b}	1.30	50.0^{a}	39.2^{b}	38.7^{b}	40.2^{b}	1.76	47.2^{a}	37.7^{b}	36.7^{b}	32.5^{b}	1.88	0.097	<0.001	0.625
Mn	14.3^{a}	12.2^{b}	13.2^{b}	13.1^{b}	0.220	14.0	12.2	13.3	14.1	0.291	14.6^{a}	12.2^{b}	13.0^{b}	12.0^{b}	0.332	0.201	0.002	0.056
Cu	9.59^{a}	6.81^{b}	6.68^{b}	6.41^{b}	0.357	10.1^{a}	7.79^{ab}	6.30^{b}	6.05^{b}	0.589	9.10^{a}	5.83^{b}	7.05^{b}	6.77^{b}	0.420	0.527	0.002	0.271

[a,b] Within a row, different letters represent the significant differences at p-value < 0.05 by the Duncan significant difference test. [1] SEM, standard error of the mean; aNDF, neutral detergent fiber; ADF, acid detergent fiber; Ca, calcium; CP, crude protein; Cu, copper; DM, dry matter; EE, ether extract; Fe, iron; Mg, magnesium; Mn, manganese; P, phosphorus; Zn, zinc.

Table 4. Effects of cropping year, variety, and the year × variety interaction on the ruminal degradation kinetics of crude protein and intestinal protein digestibility of four common vetch varieties grown on the Tibetan Plateau.

Dependent Variable	Mean across Years					Year										p-Value		
						2015					2016							
	Local Variety	Lanjian No.1	Lanjian No.2	Lanjian No.3	SEM [1]	Local variety	Lanjian No.1	Lanjian No.2	Lanjian No.3	SEM	Local Variety	Lanjian No.1	Lanjian No.2	Lanjian No.3	SEM	Year	Variety	Year × Variety
A, g/kg CP	359^{a}	352^{a}	312^{b}	320^{b}	5.03	367^{a}	360^{a}	327^{b}	331^{b}	6.13	352^{a}	343^{a}	298^{b}	309^{b}	7.22	0.004	<0.001	0.873
B, g/kg CP	608^{a}	604^{a}	570^{b}	552^{b}	5.86	610^{a}	610^{a}	565^{ab}	537^{b}	10.3	606^{a}	597^{ab}	575^{bc}	567^{c}	5.88	0.508	<0.001	0.355
C, h^{-1}	0.0766^{b}	0.0774^{b}	0.0975^{a}	0.0956^{a}	0.00202	0.0727^{b}	0.0759^{b}	0.0981^{a}	0.0985^{a}	0.00321	0.0761^{b}	0.0788^{b}	0.0968^{a}	0.0928^{a}	0.00256	0.560	<0.001	0.587
EDCP, g/kg CP	792^{a}	782^{a}	744^{b}	736^{b}	5.02	801^{a}	793^{a}	755^{b}	739^{b}	7.38	782^{a}	771^{a}	732^{b}	733^{b}	6.30	<0.001	<0.001	0.514
IDP, g/kg of RUP	692^{b}	706^{b}	756^{a}	768^{a}	6.73	683^{b}	705^{b}	746^{a}	754^{a}	8.39	702^{b}	708^{b}	765^{a}	782^{a}	10.3	0.023	<0.001	0.514
IADP, g/kg CP	144^{b}	154^{b}	194^{a}	203^{a}	4.99	136^{b}	146^{b}	183^{a}	197^{a}	7.00	153^{b}	162^{b}	205^{a}	208^{a}	6.70	<0.001	<0.001	0.661
TDP, g/kg CP	936	937	937	939	1.03	937	940	938	936	1.12	935	934	937	942	1.76	0.700	0.861	0.796

[a,b] Within a row, different letters represent the significant differences at p-value < 0.05 by the Duncan significant difference test. [1] SEM, standard error of the mean; A, soluble fraction of grain CP; B, potentially degradable fraction of grain CP; C, rate of degradation of fraction B; CP, crude protein; EDCP, effective CP degradability (k = 0.031 h^{-1}); RUP, rumen undegraded protein; IDP, intestinal digestible protein after 16 h of rumen incubation and 1 h of pepsin and 24 h of pancreatin digestion; IADP, intestinally absorbable digestible protein; TDP, total digestible protein. RUP = 1000 − EDCP (g/kg of CP); IADP = RUP (g/kg of CP) × IDP (g/kg of RUP); TDP = EDCP + IADP.

3.4. Prediction of A, B, C, and IADP

Correlations between grain chemical composition and *A*, *B*, *C*, and IADP were all significant at $p < 0.05$, with absolute values of the correlation coefficient ranging from 0.350 to 0.787 (Table 5). Therefore, stepwise multiple linear regression was used to predict *A*, *B*, *C*, and IADP. Goodness of fit of the regression models was evaluated based on the coefficient of determination (R^2) and root mean squared error (RMSE). For each parameter, the selected regression model produced the highest R^2 and lowest RMSE, with R^2 values of 0.805, 0.712, 0.748, and 0.891 for prediction of *A*, *B*, *C*, and IADP, respectively (Table 6).

Table 5. Pearson's correlations between chemical composition, ruminal degradability parameters of crude protein, and intestinally absorbable digestible protein of common vetch grains.

Dependent Variable [1]	CP	EE	NDF	ADF	Ash
A	0.581 ***	−0.463 **	−0.751 ***	−0.672 ***	−0.597 ***
B	0.605 ***	−0.524 **	−0.350 *	−0.461 **	−0.619 ***
C	−0.611 ***	0.549 **	0.533 **	0.620 ***	0.710 ***
IADP	−0.787 ***	0.663 ***	0.735 ***	0.727 ***	0.781 ***

[1] *A*, soluble fraction of grain CP; *B*, potentially degradable fraction of grain CP; *C*, rate of degradation of fraction *B*; aNDF, neutral detergent fiber; ADF, acid detergent fiber; CP, crude protein; EE, ether extract; IADP, intestinally absorbable digestible protein. * $p < 0.05$; ** $p < 0.01$; *** $p < 0.001$.

Table 6. Equations to predict the ruminal degradability parameters *A*, *B*, (g/kg CP), and *C* (h^{-1}) of crude protein and intestinally absorbable digestible protein (IADP; g/kg CP) of common vetch grains.

Dependent Variable [1]	Equation	R^2	RMSE
A	= −1.53 aNDF + 672	0.751	18.8
	= −1.23aNDF − 5.73ash + 777	0.795	17.3
	= −1.38aNDF − 7.18ash + 3.27EE + 794	0.805	16.9
B	= −13.9ash + 996	0.619	26.1
	= −8.81ash + 1.33CP + 367	0.676	24.4
	= −8.99ash + 2.01CP + 0.591aNDF − 0.231	0.696	23.8
	= −6.99ash + 2.11CP + 1.22 aNDF − 1.73ADF − 125	0.712	23.3
C	= 0.00549ash − 0.0765	0.710	0.00806
	= 0.00409ash + 0.000536ADF − 0.0692	0.748	0.00759
IADP	= −2.51CP + 1076	0.787	17.4
	= −1.55CP + 9.00ash − 465	0.867	14.1
	= −0.828CP + 8.80ash + 0.635aNDF + 70.2	0.891	12.8

[1] Units: g/kg dry matter for grain chemical composition of CP, EE, aNDF, and ash; *A*, soluble fraction of grain CP; *B*, potentially degradable fraction of grain CP; *C*, rate of degradation of fraction *B* (h^{-1}); aNDF, neutral detergent fiber; CP, crude protein; EE, ether extract; R^2, coefficients of determination; RMSE, root mean squared error.

4. Discussion

4.1. Chemical Composition

In this study, CP concentration of common vetch grain was greater in 2015 than 2016, while concentrations of EE, aNDF, and ADF were less in 2015 compared to 2016. Dornbos and Mullen [31] reported that drought stress during grain filling of soybean increased grain protein concentration and decreased lipid concentration, and Al-Karaki and Ereifej [32] reported lower protein and greater EE concentration of field pea (*Pisum sativum* L.) with greater precipitation during pod filling. Ítavo et al. [33] reported that EE concentration of legume grain is positively associated with energy release. These observations help to explain differences in the grain chemical composition between years in this study, which may be attributed to greater mean air temperature and total precipitation during July and August of 2016 when common vetch was in flowering to pod-filling stages of phenological development.

In this study, all varieties produced relatively high CP, with values ranging from 353 to 370 g/kg DM. These CP values were greater than those reported from Syria (266–316 g/kg DM) [11] and Turkey (249–279 g/kg DM) [34], but less than that of varieties grown in Jordan (mean = 393 g/kg DM) [12]. Neutral detergent fiber (207–229 g/kg DM) and ADF (59.5–68.8 g/kg DM) concentrations of grain of all varieties fell within the range of values for aNDF and ADF reported by Ramos-Morales et al. [7], Seifdavati and Taghizadeh [13], and Huang et al. [5] (aNDF = 144–423 g/kg DM, ADF = 52–124 g/kg DM). Ether extract of grain of common vetch varieties in this study (16.1–19.0 g/kg DM) was greater than that reported from Spain (14.3 g/kg DM) [7], but within the range reported by Karadag and Yavus [34] in Turkey (11.6–32.3 g/kg DM). Such differences in the proximate composition of grain of common vetch among studies may be attributed to variation in varieties, site, growing environment, and crop management [4,11,34]. The relative high levels of grain protein for all common vetch varieties evaluated in this study demonstrate that they could be a valuable supplement to low-quality ruminant diets, particularly for the local variety [7].

Macronutrient concentrations of common vetch grain in this study were greater in 2016 than 2015, which may be partially due to greater precipitation and warmer air temperature during grain filling in 2016, resulting in accelerated translocation of macronutrients from vegetative tissues and roots to the developing grain. Similarly, Thavarajah et al. [35] reported that warmer air temperature during grain filling was associated with greater Fe concentration of lentil (*Lens culinaris* L.) grain. Uzun et al. [36] reported that 1000-seed weight of common vetch grain was negatively correlated with most mineral elements, such as Ca, Mg, and Mn. Previous studies of these varieties tested showed that seed weight of the local variety was less than that the improved varieties (Table 2) [23]. In this study, large varietal differences in grain concentrations for most minerals may be attributed to a starch dilution effect in lager seeds [36]. Macro and micronutrient concentrations of grain of common vetch varieties in this study are within the ranges reported from studies in Turkey [36] and China [17]. Concentrations of P, Mg, Ca, and Fe in grain of common vetch grown in Cyprus [37] were 5.7, 1.70, 1.40, and 0.14 g/kg DM, respectively, in contrast to those in this study (mean = 2.77, 3.37, 1.59, and 0.69 g/kg DM, respectively). Grain concentrations of Cu and Mn in the present study are similar to values reported for common vetch grown in Jordan [12], although the concentration of Zn was greater in the present study. Grain Zn concentration in this study was similar to that of common vetch grown in Cyprus [37], but the levels of Cu and Mn were comparatively greater in that study. Differences in grain mineral concentrations among studies may be related to differences in varieties planted [36], the stage of grain maturity at harvest [12], and other factors such as site conditions, growing environment, and agronomic practices [17].

Overall, the large varietal differences in chemical composition of common vetch grain indicate that farmers can integrate grain of these varieties into ruminant diets in terms of quality traits to meet the nutrient demands of diverse livestock classes.

4.2. *Ruminal Degradability of CP*

In this study, the values for *A* and *B* are in accordance with earlier reports on grain of common vetch [7,38], pea, and lupin (*Lupinus albus* var. *multolupa*) [7,39]. The average value for *C* ($0.0867\ h^{-1}$) of grain of the common vetch varieties in this study were in line with the previous reports on common vetch ($0.081\ h^{-1}$) as well as soybean meal ($0.082\ h^{-1}$) [38]. Mean EDCP of common vetch grain in this study (764 g/kg CP) was less than that (857 g/kg CP) reported from Greece studies by Zagorakis et al. [40], but in agreement with earlier observations from Spain (753 g/kg CP; [41]. Such differences in the EDCP of common vetch grain could be ascribed in part to methodological differences such as animal species [42], basal diet [38], rumen outflow rate [39], and protein availability among varieties.

Effective CP degradability of grain was relatively high for all varieties in this study (736–792 g/kg CP). Therefore, these varieties could be considered as a source of degradable protein for ruminal microorganisms and then for microbial protein synthesis [7]. Zagorakis et al. [40] reported an optimum rumen degradable: undegradable protein ratio of 60:40 in feeds for highly productive ruminants. As a consequence, results of the present study suggest that if the tested varieties are used as a protein source in highly productive ruminant diets, they should first be treated to reduce CP degradability, such as with heat treatment [39]. Seifdavati and Taghizadeh [13] reported that autoclaved common vetch grain had 7.07% less EDCP compared to the raw grain.

4.3. Intestinal Digestibility of CP

In this study, large variability in grain IDP among years and varieties (683–782 g/kg RUP) may be partially related to varietal differences in concentrations of ash and ADF. Fu et al. [43] reported that IDP of concentrated feedstuffs is positively associated with ADF concentration, while it was negatively associated with ash concentration. Mean IDP (731 g/kg RUP) of common vetch varieties in this study was greater than that reported for other varieties (600 g/kg RUP by Ramos-Morales et al. [7] and 668 g/kg RUP by Seifdavati and Taghizadeh [13]), but less than that reported for soybean meal (905 g/kg RUP) [30]. These differences among studies of common vetch may be ascribed to differences in common vetch varieties.

In ruminants, the nutritive value of a protein supplementation depends on protein digestion in the small intestine [7,19]. In the present study, IADP of the local variety was less than that of Lanjian No. 2 and Lanjian No. 3, but did not differ significantly from that of Lanjian No. 1. This implies that Lanjian No. 2 and Lanjian No. 3 may be more suitable sources of protein in highly productive ruminant diets than the local variety and Lanjian No. 1. Reports on IADP in grain of common vetch are scarce. However, grain IADP of common vetch varieties in this study was similar to that reported for other annual grains such as camelina (*Camelina sativa* (L.) Crantz; 191 g/kg CP) [30], but lower than for soybean meal (376 g/kg CP) [30].

4.4. Prediction of Degradability Parameters and Intestinal Digestibility of Grain

A significant positive or negative correlation was observed between chemical composition and *A*, *B*, *C*, and IADP of common vetch grain in this study. Limited information has been reported on these relationships for the grain of legumes. To our knowledge, the information available is limited, with the exception of that from studies in Ireland, where concentrations of ash, CP, EE, NDF, ADF, and acid detergent lignin were significantly correlated with degradability parameters of CP in 12 concentrated feedstuffs [21,22].

The use of the chemical composition of concentrated feedstuffs to predict the degradability parameters and intestinal digestibility was previously proposed by Woods et al. [21,22]. Woods et al. [21,22] indicated that R^2 values greater than 0.90 are desirable for equations predicting degradability parameters and intestinal digestibility of concentrated feeds. In the present study, the best fit equations for predicting *A*, *B*, and *C* based on chemical composition had R^2 values of 0.805, 0.712, and 0.748, respectively. Therefore, the equations developed from this study may not be sufficiently precise to accurately predict degradability parameters based on chemical analyses. For equations predicting IADP, the R^2 value increased from 0.787 when only CP was included in the prediction equation to 0.891 when CP, ash, and aNDF were included. Since the R^2 value of this equation is close to 0.9, it may be feasible to use these variables, derived from chemical analysis, to predict grain IADP of common vetch. This result may be incorporated into future breeding programs for improving grain quality of common vetch.

5. Conclusions

All common vetch varieties evaluated in this study produced high grain CP concentration. However, when grain of these varieties is used as a protein source in diets of highly productive ruminants, it should be pre-treated to reduce ruminal degradability. This study also shows that CP, ash, and aNDF can be used to predict IADP of common vetch grain. Additionally, large varietal differences in nutritive value in this and other studies demonstrate that it is necessary to analyze grain of common vetch varieties for nutritive value before using in ration formulations. In terms of grain EDCP and

IADP, the varieties Lanjian No. 2 and Lanjian No. 3 have great potential for supplementing ruminant diets when grown on the Tibetan Plateau.

Author Contributions: Y.H. and Z.N. conceived and designed the experiment; Y.H. and R.L. performed the experiments; Y.H. and Z.Z. analyzed the data; Y.H., J.A.C., and Z.N. wrote the manuscript.

References

1. Herrero, M.; Havlík, P.; Valin, H.; Notenbaert, A.; Rufino, M.C.; Thornton, P.K.; Blümmel, M.; Weiss, F.; Grace, D.; Obersteiner, M. Biomass use, production, feed efficiencies, and greenhouse gas emissions from global livestock systems. *PNAS* **2013**, *110*, 20888–20893. [CrossRef]
2. Wathes, C.M.; Buller, H.; Maggs, H.; Campbell, M.L. Livestock production in the UK in the 21st century: A perfect storm averted? *Animals* **2013**, *3*, 574–583. [CrossRef]
3. Shang, Z.H.; Gibb, M.J.; Leiber, F.; Ismail, M.; Ding, L.M.; Guo, X.S.; Long, R.J. The sustainable development of grassland-livestock systems on the Tibetan plateau: Problems, strategies and prospects. *Rangeland J.* **2014**, *36*, 267–296. [CrossRef]
4. Larbi, A.; Hassan, S.; Kattash, G.; El-Moneim, A.M.; Jammal, B.; Nabil, H.; Nakkul, H. Annual feed legume yield and quality in dryland environments in north-west Syria: 2. grain and straw yield and straw quality. *Anim. Feed Sci. Technol.* **2010**, *160*, 90–97. [CrossRef]
5. Huang, Y.F.; Gao, X.L.; Nan, Z.B.; Zhang, Z.X. Potential value of the common vetch (*Vicia sativa* L.) as an animal feedstuff: A review. *J. Anim. Physiol. An. N.* **2017**, *101*, 1–17. [CrossRef] [PubMed]
6. Machado Filho, L.C.P.; D'Ávila, L.M.; da Silva Kazama, D.C.; Bento, L.L.; Kuhnen, S. Productive and economic responses in grazing dairy cows to grain supplementation on family farms in the South of Brazil. *Animals* **2014**, *4*, 463–475. [CrossRef]
7. Ramos-Morales, E.; Sanz-Sampelayo, M.; Molina-Alcaide, E. Nutritive evaluation of legume seeds for ruminant feeding. *J. Anim. Physiol. An. N.* **2010**, *94*, 55–64. [CrossRef] [PubMed]
8. Zhang, J.; Guo, G.; Chen, L.; Yuan, X.J.; Yu, C.Q.; Shimojo, M.; Shao, T. Effect of applying lactic acid bacteria and propionic acid on fermentation quality and aerobic stability of oats-common vetch mixed silage on the Tibetan plateau. *Anim. Sci. J.* **2015**, *86*, 595–602. [CrossRef]
9. Nan, Z.B.; Abd El-Moneim, A.M.; Larbi, A.; Nie, B. Productivity of vetches (*Vicia* spp.) under alpine grassland conditions in China. *Trop. Grassl.* **2006**, *40*, 177–182.
10. Fu, P. Effects of inoculation with rhizobium and phosphate on growth of vetch. M.S. Thesis, Lanzhou University, Lanzhou, China, 2015; pp. 1–52.
11. Larbi, A.; Abd El-Moneim, A.M.; Nakkoul, H.; Jammal, B.; Hassan, S. Intra-species variations in yield and quality determinants in Vicia species: 3. Common vetch (*Vicia sativa* ssp. *sativa* L.). *Anim. Feed Sci. Technol.* **2011**, *164*, 241–251. [CrossRef]
12. Samarah, N.H.; Ereifej, K. Chemical composition and mineral content of common vetch seeds during maturation. *J. Plant Nutr.* **2009**, *32*, 177–186. [CrossRef]
13. Seifdavati, J.; Taghizadeh, A. Effects of moist heat treatment on ruminal nutrient degradability of and *in vitro* intestinal digestibility of crude protein from some of legume seeds. *J. Food Agric. Env.* **2012**, *10*, 390–397.
14. Koumas, A.; Economides, S. Replacement of soybean meal by broad bean or common vetch grain in lamb and kid fattening diets. *Cyprus Agric. Res. I. Tech. Bull.* **1987**, *88*, 1–7.
15. Gül, M.; Yörük, M.A.; Macit, M.; Esenbuga, N.; Karaoglu, M.; Aksakal, V.; Irfan Aksu, M. The effects of diets containing different levels of common vetch (*Vicia sativa*) seed on fattening performance, carcass and meat quality characteristics of Awassi male lambs. *J. Sci. Food Agric.* **2005**, *85*, 1439–1443. [CrossRef]
16. Musco, N.; Cutrignelli, M.I.; Calabrò, S.; Tudisco, R.; Infascelli, F.; Grazioli, R.; Lo Presti, V.; Gresta, F.; Chiofalo, B. Comparison of nutritional and antinutritional traits among different species (*Lupinus albus* L. *Lupinus luteus* L. *Lupinus angustifolius* L.) and varieties of lupin seeds. *J. Anim. Physiol. An. N.* **2017**, *101*, 1–15. [CrossRef] [PubMed]
17. Mao, Z.X. Studies on the nutritive quality of common vetch. M.S. Thesis, Lanzhou University, Lanzhou, China, 2012; pp. 1–107.

18. NRC (Nutrient Requirements of Small Ruminants). *Nutrient Requirements of Dairy Cattle*, 7th ed.; National Academy of Sciences: Washington, DC, USA, 2001.
19. McNiven, M.A.; Prestløkken, E.; Mydland, L.T.; Mitchell, A.W. Laboratory procedure to determine protein digestibility of heat-treated feedstuffs for dairy cattle. *Anim. Feed Sci. Technol.* **2002**, *96*, 1–13. [CrossRef]
20. Phipps, D.A. Metal ion physiology: Role, action and mechanism. In *Handbook of Metal–Ligand Interactions in Biological Fluids*, 1st ed.; Berthon, G., Ed.; Marcel Dekker: New York, NY, USA, 1995; Volume 1, pp. 89–106.
21. Woods, V.B.; Moloney, A.P.; Calsamiglia, S.; Mara, F.P.O. The nutritive value of concentrate feedstuffs for ruminant animals Part II: In situ ruminal degradability of crude protein. *Anim. Feed Sci. Technol.* **2003**, *110*, 131–143. [CrossRef]
22. Woods, V.B.; Moloney, A.P.; Calsamiglia, S.; Mara, F.P.O. The nutritive value of concentrate feedstuffs for ruminant animals Part III. Small intestinal digestibility as measured by in vitro or mobile bag techniques. *Anim. Feed Sci. Technol.* **2003**, *110*, 145–157. [CrossRef]
23. Gao, X.L. Responses of four *Vicia sativa* cultivars to phosphorus and potassium fertilizers. M.S. Thesis, Lanzhou University, Lanzhou, China, 2018; pp. 1–97.
24. AOAC (Association of Official Analytical Chemists). *Official Methods of Analysis*, 15th ed.; AOAC International: Arlington, VA, USA, 1990; pp. 71–90.
25. Van Soest, P.J.; Robertson, J.B.; Lewis, B.A. Methods for dietary fiber, neutral detergent fiber, and non-starch polysaccharides in relation to animal nutrition. *J. Dairy Sci.* **1991**, *74*, 3583–3597. [CrossRef]
26. AOAC (Association of Official Analytical Chemists). *Official Methods of Analysis*, 17th ed.; AOAC: Gaithersburg, MD, USA, 2000; pp. 84–90.
27. Nandra, K.S.; Dobos, R.C.; Orchard, B.A.; Neutze, S.A.; Oddy, V.H.; Cullis, B.R.; Jones, A.W. The effect of animal species on *in sacco* degradation of dry matter and protein of feeds in the rumen. *Anim. Feed Sci. Technol.* **2000**, *83*, 273–285. [CrossRef]
28. Ørskov, E.R.; McDonald, I. The estimation of protein degradability in the rumen from incubation measurements weighted according to rate of passage. *J. Agric. Sci.* **1979**, *92*, 499–503. [CrossRef]
29. Gargallo, S.; Calsamiglia, S.; Ferret, A. Technical note: A modified three-step *in vitro* procedure to determine intestinal digestion of protein. *J. Anim. Sci.* **2006**, *84*, 2163–2167. [CrossRef]
30. Lawrence, R.D.; Anderson, J.L. Ruminal degradation and intestinal digestibility of camelina meal and carinata meal compared with other protein sources. *Prof. Anim. Sci.* **2017**, *34*, 10–18. [CrossRef]
31. Dornbos, D.L., Jr.; Mullen, R.E. Soybean seed protein and oil contents and fatty acid composition adjustments by drought and temperature. *J. Am. Oil Chem. Soc.* **1992**, *69*, 228–231. [CrossRef]
32. Al-Karaki, G.N.; Ereifej, K.I. Relationships between seed yield and chemical composition of field peas grown under semi-arid mediterranean conditions. *J. Agron. Crop Sci.* **1999**, *182*, 279–284. [CrossRef]
33. Ítavo, L.C.V.; Soares, C.M.; Ítavo, C.C.B.F.; Dias, A.M.; Petit, H.V.; Leal, E.S.; Souza, A.D.V.V. Calorimetry, chemical composition and *in vitro* digestibility of oilseeds. *Food Chem.* **2015**, *185*, 219–225. [CrossRef]
34. Karadag, Y.; Yavus, M. Seed yields and biochemical compounds of common vetch (*Vicia sativa* L.) lines grown in semi-arid regions of Turkey. *Afr. J. Biotechnol.* **2010**, *9*, 8334–8338.
35. Thavarajah, D.; Thavarajah, P.; See, C.T.; Vandenberg, A. Phytic acid and Fe and Zn concentration in lentil (*Lens culinaris* L.) seeds is influenced by temperature during seed filling period. *Food Chem.* **2010**, *122*, 254–259. [CrossRef]
36. Uzun, A.; Gücer, S.; Acikgoz, E. Common vetch (*Vicia sativa* L.) germplasm: Correlations of crude protein and mineral content to seed traits. *Plant Foods Hum. Nutr.* **2011**, *66*, 254–260. [CrossRef]
37. Hadjipanayiotou, M.; Economides, S.; Koumas, A. Chemical composition, digestibility and energy content of leguminous grains and straws grown in a Mediterranean region. *Ann. Zootech.* **1985**, *34*, 23–30. [CrossRef]
38. Rotger, A.; Ferret, A.; Calsamiglia, S.; Manteca, X. *In situ* degradability of seven plant protein supplements in heifers fed high concentrate diets with different forage to concentrate ratio. *Anim. Feed Sci. Technol.* **2006**, *125*, 73–87. [CrossRef]
39. Aguilera, J.F.; Bustos, M.; Molina, E. The degradability of legume seed meals in the rumen: Effect of heat treatment. *Anim. Feed Sci. Technol.* **1992**, *36*, 101–112. [CrossRef]
40. Zagorakis, K.; Liamadis, D.; Milis, C.; Dotas, V.; Dotas, D. Nutrient digestibility and *in situ* degradability of alternatives to soybean meal protein sources for sheep. *Small Rumin. Res.* **2015**, *124*, 38–44. [CrossRef]

41. González, J.; Andrés, S. Rumen degradability of some feed legume seeds. *Anim. Res.* **2003**, *52*, 17–25. [CrossRef]
42. Molina Alcaide, E.; YáñezRuiz, D.R.; Moumen, A.; Martín García, A.I. Ruminal degradability and *in vitro* intestinal digestibility of sunflower meal and *in vitro* digestibility of olive by-products supplemented with urea or sunflower meal Comparison between goats and sheep. *Anim. Feed Sci. Technol.* **2003**, *110*, 3–15. [CrossRef]
43. Fu, L.X.; Ma, T.; Diao, Q.Y.; Cheng, S.R.; Sun, Z.L.; Li, C. Correlation analysis of ruminal degradation characteristics and *in vitro* small intestinal digestibility of rumen undegraded protein of common concentrates for mutton sheep. *Chinese J. Anim. Nutr.* **2018**, *30*, 2641–2651.

PERMISSIONS

All chapters in this book were first published by MDPI; hereby published with permission under the Creative Commons Attribution License or equivalent. Every chapter published in this book has been scrutinized by our experts. Their significance has been extensively debated. The topics covered herein carry significant findings which will fuel the growth of the discipline. They may even be implemented as practical applications or may be referred to as a beginning point for another development.

The contributors of this book come from diverse backgrounds, making this book a truly international effort. This book will bring forth new frontiers with its revolutionizing research information and detailed analysis of the nascent developments around the world.

We would like to thank all the contributing authors for lending their expertise to make the book truly unique. They have played a crucial role in the development of this book. Without their invaluable contributions this book wouldn't have been possible. They have made vital efforts to compile up to date information on the varied aspects of this subject to make this book a valuable addition to the collection of many professionals and students.

This book was conceptualized with the vision of imparting up-to-date information and advanced data in this field. To ensure the same, a matchless editorial board was set up. Every individual on the board went through rigorous rounds of assessment to prove their worth. After which they invested a large part of their time researching and compiling the most relevant data for our readers.

The editorial board has been involved in producing this book since its inception. They have spent rigorous hours researching and exploring the diverse topics which have resulted in the successful publishing of this book. They have passed on their knowledge of decades through this book. To expedite this challenging task, the publisher supported the team at every step. A small team of assistant editors was also appointed to further simplify the editing procedure and attain best results for the readers.

Apart from the editorial board, the designing team has also invested a significant amount of their time in understanding the subject and creating the most relevant covers. They scrutinized every image to scout for the most suitable representation of the subject and create an appropriate cover for the book.

The publishing team has been an ardent support to the editorial, designing and production team. Their endless efforts to recruit the best for this project, has resulted in the accomplishment of this book. They are a veteran in the field of academics and their pool of knowledge is as vast as their experience in printing. Their expertise and guidance has proved useful at every step. Their uncompromising quality standards have made this book an exceptional effort. Their encouragement from time to time has been an inspiration for everyone.

The publisher and the editorial board hope that this book will prove to be a valuable piece of knowledge for researchers, students, practitioners and scholars across the globe.

LIST OF CONTRIBUTORS

Lauren Ford, Chloe Curry and Gary Sheldrake
School of Chemistry and Chemical Engineering, Queen's University Belfast, Northern Ireland BT9 5AG, UK

Mairead Campbell and Katerina Theodoridou
Institute of Global Food Security, Queen's University Belfast, 19 Chlorine Gardens, Northern Ireland, Belfast BT9 5DL, UK
School of Biological Sciences, Queen's University Belfast, 19 Chlorine Gardens, Northern Ireland, Belfast BT9 5DL, UK

Jaimie Dick
Institute of Global Food Security, Queen's University Belfast, 19 Chlorine Gardens, Northern Ireland, Belfast BT9 5DL, UK
School of Biological Sciences, Queen's University Belfast, 19 Chlorine Gardens, Northern Ireland, Belfast BT9 5DL, UK
Queen's Marine Laboratory (QML) Queen's University Belfast, 12-13 The Strand, Northern Ireland, Portaferry BT22 1PF, UK

Lorenzo Stella
School of Chemistry and Chemical Engineering, Queen's University Belfast, Northern Ireland BT9 5AG, UK
Atomistic Simulation Centre (ASC), School of Mathematics and Physics, Queen's University Belfast, University Road, Belfast BT7 1NN, UK

Pamela J. Walsh
School of Chemistry and Chemical Engineering, Queen's University Belfast, Northern Ireland BT9 5AG, UK
Queen's Marine Laboratory (QML) Queen's University Belfast, 12-13 The Strand, Northern Ireland, Portaferry BT22 1PF, UK
School of Mechanical Engineering, Queen's University Belfast, The Asbhy Building, Stranmillis Road, Northern Ireland, Belfast BT9 5AJ, UK

Francesc González-Solé, Carmen Villodre, David Solà-Oriol and José F. Pérez
Department of Animal and Food Sciences, Animal Nutrition and Welfare Service, Autonomous University of Barcelona, 08193 Bellaterra, Spain

Lourdes Criado-Mesas
Plant and Animal Genomics, Centre for Research in Agricultural Genomics (CRAG), CSIC-IRTA-UAB-UB, 08193 Bellaterra, Spain

Josep M. Folch
Department of Animal and Food Sciences, Animal Nutrition and Welfare Service, Autonomous University of Barcelona, 08193 Bellaterra, Spain
Plant and Animal Genomics, Centre for Research in Agricultural Genomics (CRAG), CSIC-IRTA-UAB-UB, 08193 Bellaterra, Spain

Wellington C. García
Department of Animal and Food Sciences, Animal Nutrition and Welfare Service, Autonomous University of Barcelona, 08193 Bellaterra, Spain
Department of Animal Production, Agrarian University of Ecuador, Guayaquil 090108, Ecuador

Mercè Farré
Department of Mathematics, Area of Statistics and Operations Research, Autonomous University of Barcelona, 08193 Bellaterra, Spain

Elisabet Borda
R&D Animal Nutrition Department, Bioiberica S.A.U., 08389 Palafolls, Spain

Francisco J. Pérez-Cano
Physiology Section, Department of Biochemistry and Physiology, Faculty of Pharmacy and Food Science, University of Barcelona (UB), 08028 Barcelona, Spain
Nutrition and Food Safety Research Institute (INSA-UB), 08921 Santa Coloma de Gramenet, Spain

Zhenwei Zhang, Yanlu Wang, Xuemeng Si, Zhijun Cao, Shengli Li and Hongjian Yang
State Key Laboratory of Animal Nutrition, College of Animal Science and Technology, China Agricultural University, Beijing 100193, China

José Daza, Oscar Balocchi and Juan Keim
Animal Production Institute, Faculty of Agricultural Sciences, Universidad Austral de Chile, Valdivia, Chile

Daniel Benavides
Graduate School, Faculty of Agricultural Sciences, Universidad Austral de Chile, Valdivia, Chile

Rubén Pulido
Animal Science Institute, Faculty of Veterinary Sciences, Universidad Austral de Chile, Valdivia, Chile

Annick Bertrand
Soils and Crops Research and Development Centre, Agriculture and Agri-Food Canada, Québec City, QC G1V 2J3, Canada

Dervan D.S.L. Bryan and Henry L. Classen
Department of Animal and Poultry Science, University of Saskatchewan, Saskatoon, SK S7N5A8, Canada

Zahia Amanzougarene, Susana Yuste and Manuel Fondevila
Departamento de Producción Animal y Ciencia de los Alimentos, Instituto Agroalimentario de Aragón (IA2), Universidad de Zaragoza-CITA, M. Servet 177, 50013 Zaragoza, Spain

Carlos Farias-Kovac, Nuria Nicodemus, Rebeca Delgado, Cesar Ocasio-Vega, Tamia Noboa, Rosa Carabafio and Javier Garcia
Departamento de Producci6n Agraria, Escuela Tecnica Superior de Ingenieria Agron6mica, Agroalimentaria y de Biosistemas, Universidad Politecnica de Madrid, C/Senda del Rey 18, 28040 Madrid, Spain

Ramadan Allam-Sayed Abdelrasoul
Poultry Department, Faculty of Agriculture, Fayoum University, Fayoum 63514, Egypt

Zhulin Xue, Nan Liu, Yuqi Wei and Yingjun Zhang
Key Laboratory of Grasslands Management and Utilization, Ministry of Agriculture and Rural Affairs, College of Grassland Science and Technology, China Agricultural University, Beijing 100193, China

Philipe Moriel and Elizabeth Palmer
IFAS-Range Cattle Research and Education Center, University of Florida, Gainesville, FL 32611, USA

Mauro Spanghero, Carla Fabro and Chiara Sarnataro
Dipartimento di Scienze Agroalimentari, Ambientali e Animali, University of Udine, 33100 Udine, Italy

Maria Chiaravalli and Stefania Colombini
Dipartimento di Scienze Agrarie e Ambientali, University of Milan, 20122 Milano, Italy

Federico Froldi and Maurizio Moschini
Dipartimento di Scienze Animali, della Nutrizione e degli Alimenti, Università Cattolica del Sacro Cuore of Piacenza, 29122 Piacenza, Italy

Federico Mason
Department of Biodiversity Protection, Institute of Animal Reproduction and Food Research of Polish Academy of Sciences (IARFR PAS), 10-748 Olsztyn, Poland

Stefano Schiavon and Franco Tagliapietra
Dipartimento di Agronomia, Animali, Alimenti, Risorse naturali e Ambiente, University of Padova, 35122 Padova, Italy

Christopher D. Powell, Jennifer L. Ellis and James France
Department of Animal Biosciences, University of Guelph, Guelph, ON N1G 2W1, Canada

Mewa S. Dhanoa
Institute of Grassland and Environmental Research, Plas Gogerddan, Aberystwyth SY23 3EB, UK

Anna Garber and Jo-Anne M. D. Murray
College of Medical, Veterinary and Life Sciences, School of Veterinary Medicine, University of Glasgow, Glasgow G61 1QH, UK

Secundino López
Departamento de Producción Animal, Universidad de León, E-24007 León, Spain
Instituto de Ganadería de Montaña, CSIC-Universidad de León, Finca Marzanas s/n, 24346 Grulleros, Spain

Guangyu Li
Institute of Special Economic Animal and Plant Science, Chinese Academy of Agricultural Sciences, Changchun 130112, China

Keyuan Liu
Institute of Special Economic Animal and Plant Science, Chinese Academy of Agricultural Sciences, Changchun 130112, China
Department of Animal Science and Technology, Northeast Agricultural University, Harbin 150030, China

Yang Li, Hangshu Xin and Yonggen Zhang
Department of Animal Science and Technology, Northeast Agricultural University, Harbin 150030, China

Guobin Luo
Department of Animal Science and Technology, Northeast Agricultural University, Harbin 150030, China
Zhejiang NHU Company Ltd., Shaoxing 312500, China

Monika Pogány Simonova and Andrea Lauková
Institute of Animal Physiology, Centre of Biosciences of Slovak Academy of Sciences, Šoltésovej 4-6, 04001 Kosice, Slovakia

L'ubica Chrastinová
Institute for Nutrition, National Agricultural and Food Centre, Hlohovecká 2,951 41 Nitra-Lužianky, Slovakia

Pablo José Rufino-Moya, Mireia Blanco, Juan Ramón Bertolín and Margalida Joy
Centro de Investigación y Tecnología Agroalimentaria de Aragón (CITA), Instituto Agroalimentario de Aragón–IA2 (CITA-Universidad de Zaragoza), Avda, Montañana 930, 50059 Zaragoza, Spain

Yafeng Huang, Rui Li, Zhixin Zhang and Zhibiao Nan
State Key Laboratory of Grassland Agro-ecosystems, Lanzhou University, Lanzhou 730020, China
Key Laboratory of Grassland Livestock Industry Innovation of Ministry of Agriculture and Rural Affairs, Lanzhou University, Lanzhou 730020, China
College of Pastoral Agriculture Science and Technology, Lanzhou University, Lanzhou 730020, China

Jeffrey A. Coulter
Department of Agronomy and Plant Genetics, University of Minnesota, St. Paul, MN 55108, USA

Index

www.ingramcontent.com/pod-product-compliance
Lightning Source LLC
LaVergne TN
LVHW060512210926
848485LV00005B/1067